AI源码解读

卷积神经网络（CNN）深度学习案例
（Python版）

李永华◎编著

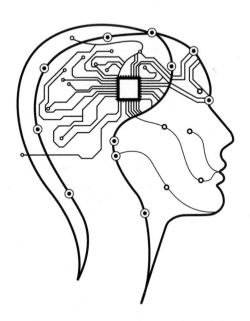

清华大学出版社
北京

内 容 简 介

本书以人工智能发展为时代背景,通过20个机器学习模型和算法案例,为读者提供较为详细的实战方案,以便进行深度学习。

在编排方式上,全书侧重对创新项目的过程进行介绍,分别从整体设计、系统流程和实现模块等角度论述数据处理、模型训练和模型应用等过程,并剖析模块的功能、使用及程序代码。为便于读者高效学习,快速掌握人工智能程序开发方法,本书配套提供项目设计工程文档、程序代码、实现过程中出现的问题及解决方法等资料,可供读者举一反三、二次开发。

本书语言简洁,深入浅出,通俗易懂,不仅适合对 Python 编程有兴趣的爱好者,而且可作为高等院校相关专业的参考教材,还可作为从事智能应用创新开发专业人员的技术参考书。

本书封面贴有清华大学出版社防伪标签,无标签者不得销售。
版权所有,侵权必究。举报: 010-62782989,beiqinquan@tup.tsinghua.edu.cn。

图书在版编目(CIP)数据

AI 源码解读: 卷积神经网络(CNN)深度学习案例: Python 版/李永华编著. —北京: 清华大学出版社,2021.9(2022.4重印)
(人工智能科学与技术丛书)
ISBN 978-7-302-57066-0

Ⅰ. ①A… Ⅱ. ①李… Ⅲ. ①人工智能-算法 Ⅳ. ①TP18

中国版本图书馆 CIP 数据核字(2020)第 252139 号

责任编辑:	盛东亮　钟志芳
封面设计:	李召霞
责任校对:	时翠兰
责任印制:	沈　露

出版发行:	清华大学出版社
网　　址:	http://www.tup.com.cn,http://www.wqbook.com
地　　址:	北京清华大学学研大厦 A 座　　　邮　编: 100084
社 总 机:	010-83470000　　　邮　购: 010-83470235
投稿与读者服务:	010-62776969,c-service@tup.tsinghua.edu.cn
质量反馈:	010-62772015,zhiliang@tup.tsinghua.edu.cn
课件下载:	http://www.tup.com.cn,010-83470236

印 装 者:	三河市龙大印装有限公司				
经　　销:	全国新华书店				
开　　本:	186mm×240mm	印　张:	24	字　数:	541 千字
版　　次:	2021 年 10 月第 1 版			印　次:	2022 年 4 月第 2 次印刷
印　　数:	1501~2500				
定　　价:	99.00 元				

产品编号: 090187-01

前 言
PREFACE

  Python 作为人工智能和大数据领域的主要开发语言，具有灵活性强、扩展性好、应用面广、可移植、可扩展、可嵌入等特点，近年来发展迅速，热度不减，相关人才需求量逐年攀升，已经成为高等院校的专业课程。

  为适应当前教学改革的要求，更好地践行人工智能模型与算法的应用，本书以实践教学与创新能力培养为目标，采取了创新方式，从不同难度、不同类型、不同算法，融合了同类教材的优点，将实际智能应用案例进行总结，希望起到抛砖引玉的作用。

  本书的主要内容和素材来自开源网站的人工智能经典模型算法、信息工程专业创新课程内容、作者所在学校近几年承担的科研项目成果及作者指导学生完成的创新项目。通过这些创新项目同学们不仅学到了知识，提高了能力，而且为本书提供了第一手素材和相关资料。

  本书内容由总述到分述、先理论后实践，采用系统整体架构、系统流程与代码实现相结合，对于从事人工智能开发、机器学习和算法实现的专业技术人员可以作为技术参考书、提高其工程创新能力；也可以作为信息通信工程及相关专业本科生的参考书，为机器学习模型分析、算法设计和实现提供帮助。

  本书的编写得到了教育部电子信息类专业教学指导委员会、信息工程专业国家第一类特色专业建设项目、信息工程专业国家第二类特色专业建设项目、教育部 CDIO 工程教育模式研究与实践项目、教育部本科教学工程项目、信息工程专业北京市特色专业建设、北京市教育教学改革项目以及北京邮电大学教育教学改革项目（2020JC03）的大力支持，在此表示感谢！

  由于作者经验与水平有限，书中疏漏之处在所难免，衷心地希望各位读者多提宝贵意见及具体整改措施，以便作者进一步修改和完善。

编 者

2021 年 7 月

目 录
CONTENTS

项目1　电影推荐小程序 ·· 1
  1.1　总体设计 ··· 1
    1.1.1　系统整体结构 ··· 1
    1.1.2　系统流程 ·· 2
  1.2　运行环境 ··· 3
    1.2.1　Python 环境 ··· 3
    1.2.2　TensorFlow 环境 ·· 3
  1.3　模块实现 ··· 4
    1.3.1　数据预处理 ·· 4
    1.3.2　模型设计 ·· 7
    1.3.3　模型训练及测试 ·· 11
    1.3.4　特征矩阵提取 ··· 14
    1.3.5　推荐电影 ·· 16
    1.3.6　客户端 ··· 19
  1.4　系统测试 ·· 24
    1.4.1　训练准确率 ·· 24
    1.4.2　运行结果 ··· 25

项目2　服装分类助手 ··· 28
  2.1　总体设计 ·· 28
    2.1.1　系统整体结构 ··· 28
    2.1.2　系统流程 ··· 29
  2.2　运行环境 ·· 31
    2.2.1　Python 环境 ··· 31
    2.2.2　PyTorch 环境 ··· 31
    2.2.3　Django 环境 ·· 31
  2.3　模块实现 ·· 32
    2.3.1　数据预处理 ·· 32
    2.3.2　模型创建与编译 ··· 33
    2.3.3　模型训练及保存 ··· 35
    2.3.4　模型生成 ··· 37

2.4 系统测试 ······ 50
   2.4.1 训练准确率 ······ 50
   2.4.2 测试效果 ······ 51
   2.4.3 模型应用 ······ 51

项目 3   检索式模型聊天机器人 ······ 59

3.1 总体设计 ······ 59
   3.1.1 系统整体结构 ······ 59
   3.1.2 系统流程 ······ 60
3.2 运行环境 ······ 61
   3.2.1 Python 环境 ······ 61
   3.2.2 TensorFlow 环境 ······ 61
3.3 模块实现 ······ 62
   3.3.1 数据预处理 ······ 63
   3.3.2 模型创建与编译 ······ 66
   3.3.3 模型训练及保存 ······ 69
   3.3.4 模型生成 ······ 71
3.4 系统测试 ······ 79
   3.4.1 训练准确率 ······ 79
   3.4.2 测试效果 ······ 80
   3.4.3 模型应用 ······ 80

项目 4   方言种类识别 ······ 82

4.1 总体设计 ······ 82
   4.1.1 系统整体结构 ······ 82
   4.1.2 系统流程 ······ 83
4.2 运行环境 ······ 83
   4.2.1 Python 环境 ······ 83
   4.2.2 TensorFlow 环境 ······ 83
   4.2.3 Jupyter Notebook 环境 ······ 84
   4.2.4 PyCharm 环境 ······ 84
4.3 模块实现 ······ 85
   4.3.1 数据预处理 ······ 85
   4.3.2 模型构建 ······ 92
   4.3.3 模型训练及保存 ······ 94
   4.3.4 模型生成 ······ 96
4.4 系统测试 ······ 98
   4.4.1 训练准确率 ······ 98
   4.4.2 测试效果 ······ 99

项目 5   行人检测与追踪计数 ······ 101

5.1 总体设计 ······ 101

		5.1.1 系统整体结构 ………………………………………………… 101
		5.1.2 系统流程 …………………………………………………… 101
	5.2	运行环境 ………………………………………………………… 102
		5.2.1 Python 环境 ……………………………………………… 102
		5.2.2 TensorFlow 环境 ………………………………………… 102
		5.2.3 安装所需的软件包 ………………………………………… 103
		5.2.4 硬件环境 …………………………………………………… 103
	5.3	模块实现 ………………………………………………………… 104
		5.3.1 准备数据 …………………………………………………… 104
		5.3.2 数据预处理 ………………………………………………… 109
		5.3.3 目标检测 …………………………………………………… 110
		5.3.4 目标追踪 …………………………………………………… 111
		5.3.5 主函数 ……………………………………………………… 112
	5.4	系统测试 ………………………………………………………… 116

项目 6 智能果实采摘指导系统 …………………………………………… 117

	6.1	总体设计 ………………………………………………………… 117
		6.1.1 系统整体结构 ……………………………………………… 117
		6.1.2 系统流程 …………………………………………………… 118
	6.2	运行环境 ………………………………………………………… 120
		6.2.1 Python 环境 ……………………………………………… 120
		6.2.2 TensorFlow 环境 ………………………………………… 120
		6.2.3 Jupyter Notebook 环境 ………………………………… 120
		6.2.4 PyCharm 环境 …………………………………………… 121
		6.2.5 微信开发者工具 …………………………………………… 121
		6.2.6 OneNET 云平台 …………………………………………… 121
	6.3	模块实现 ………………………………………………………… 123
		6.3.1 数据预处理 ………………………………………………… 123
		6.3.2 创建模型与编译 …………………………………………… 125
		6.3.3 模型训练及保存 …………………………………………… 126
		6.3.4 上传结果 …………………………………………………… 128
		6.3.5 小程序开发 ………………………………………………… 130
	6.4	系统测试 ………………………………………………………… 133
		6.4.1 训练准确率 ………………………………………………… 133
		6.4.2 测试效果 …………………………………………………… 133
		6.4.3 外部访问效果 ……………………………………………… 134

项目 7 基于 CNN 的猫种类识别 …………………………………………… 136

	7.1	总体设计 ………………………………………………………… 136
		7.1.1 系统整体结构 ……………………………………………… 136

7.1.2 系统流程 ... 137
7.2 运行环境 ... 137
7.2.1 计算型云服务器 ... 137
7.2.2 Python 环境 ... 138
7.2.3 TensorFlow 环境 ... 139
7.2.4 MySQL 环境 ... 140
7.2.5 Django 环境 ... 140
7.3 模块实现 ... 140
7.3.1 数据预处理 ... 141
7.3.2 数据增强 ... 143
7.3.3 普通 CNN 模型 ... 145
7.3.4 残差网络模型 ... 147
7.3.5 模型生成 ... 153
7.4 系统测试 ... 167
7.4.1 训练准确率 ... 167
7.4.2 测试效果 ... 169
7.4.3 模型应用 ... 169

项目 8 基于 VGG-16 的驾驶行为分析 ... 171
8.1 总体设计 ... 171
8.1.1 系统整体结构 ... 171
8.1.2 系统流程 ... 172
8.2 运行环境 ... 173
8.2.1 Python 环境 ... 173
8.2.2 TensorFlow 环境 ... 173
8.2.3 Android 环境 ... 173
8.3 模块实现 ... 174
8.3.1 数据预处理 ... 174
8.3.2 模型构建 ... 175
8.3.3 模型训练及保存 ... 177
8.3.4 模型生成 ... 178
8.4 系统测试 ... 186
8.4.1 训练准确率 ... 186
8.4.2 测试效果 ... 186
8.4.3 模型应用 ... 188

项目 9 基于 Mask R-CNN 的娱乐视频生成器 ... 189
9.1 总体设计 ... 189
9.1.1 系统整体结构 ... 189
9.1.2 系统流程 ... 189
9.2 运行环境 ... 190

		9.2.1 Python 环境	190
		9.2.2 PyTorch 环境	190
		9.2.3 Detectron2 平台	191
		9.2.4 MoviePy 的安装	192
		9.2.5 PyQt 的安装	192
	9.3	模块实现	192
		9.3.1 数据处理	192
		9.3.2 视频处理	194
		9.3.3 PyQt 界面	202
	9.4	系统测试	207
		9.4.1 训练准确率	207
		9.4.2 运行效率	208
		9.4.3 应用使用说明	209

项目10 基于 CycleGAN 的图像转换 · 213

10.1	总体设计		213
	10.1.1	系统整体结构	213
	10.1.2	系统流程	214
10.2	运行环境		214
	10.2.1	Python 环境	214
	10.2.2	TensorFlow GPU 环境	214
	10.2.3	Android 环境	216
10.3	模块实现		218
	10.3.1	数据集预处理	218
	10.3.2	模型构建	220
	10.3.3	模块分析	222
	10.3.4	模型训练及保存	225
	10.3.5	模型生成	228
10.4	系统测试		233

项目11 交通警察——车辆监控系统 · 235

11.1	总体设计		235
	11.1.1	系统整体结构	235
	11.1.2	系统流程	236
11.2	运行环境		236
	11.2.1	Python 环境	236
	11.2.2	TensorFlow 环境	236
	11.2.3	PyCharm IDE 配置	237
	11.2.4	Protoc 配置	237
11.3	模块实现		237
	11.3.1	API 下载及载入	238

	11.3.2 识别训练	239
	11.3.3 导入模型与编译	241
	11.3.4 模型生成	243
11.4	系统测试	245

项目 12 验证码的生成与识别 248

12.1	总体设计	248
	12.1.1 系统整体结构	248
	12.1.2 系统流程	249
12.2	运行环境	249
	12.2.1 Python 环境	249
	12.2.2 TensorFlow 环境	249
	12.2.3 VsCode 环境	250
12.3	模块实现	251
	12.3.1 数据预处理	251
	12.3.2 模型搭建	253
	12.3.3 模型训练及保存	255
	12.3.4 模型测试	259
12.4	系统测试	263
	12.4.1 训练准确率	263
	12.4.2 测试效果	263

项目 13 基于 CNN 的交通标志识别 266

13.1	总体设计	266
	13.1.1 系统整体结构	266
	13.1.2 系统流程	267
13.2	运行环境	267
13.3	模块实现	268
	13.3.1 数据预处理	268
	13.3.2 模型构建	270
	13.3.3 模型训练及保存	273
13.4	系统测试	274
	13.4.1 训练准确率	274
	13.4.2 测试效果	275

项目 14 图像风格转移 277

14.1	总体设计	277
	14.1.1 系统整体结构	277
	14.1.2 系统流程	278
14.2	运行环境	278
	14.2.1 Python 环境	278
	14.2.2 TensorFlow 环境	278

           14.2.3 库安装 ··· 279
           14.2.4 VGG-19 网络下载 ·· 279
      14.3 模块实现 ··· 279
           14.3.1 实时风格转移 ··· 279
           14.3.2 非实时风格转移 ·· 286
           14.3.3 交互界面设计 ··· 289
      14.4 系统测试 ··· 294
           14.4.1 非实时风格转移测试 ·· 294
           14.4.2 实时风格转移测试 ··· 294

项目 15 口罩识别系统 ·· 296
      15.1 总体设计 ··· 296
           15.1.1 系统整体结构 ··· 296
           15.1.2 系统流程 ··· 296
      15.2 运行环境 ··· 297
      15.3 模块实现 ··· 298
           15.3.1 数据预处理 ·· 298
           15.3.2 模型训练及保存 ·· 298
           15.3.3 页面显示和视频流输入 ··· 302
           15.3.4 模型生成 ··· 304
      15.4 系统测试 ··· 306
           15.4.1 训练准确率 ·· 306
           15.4.2 测试效果 ··· 307

项目 16 垃圾分类微信小程序 ··· 310
      16.1 总体设计 ··· 310
           16.1.1 系统整体结构 ··· 310
           16.1.2 系统流程 ··· 311
      16.2 运行环境 ··· 311
           16.2.1 Python 环境 ··· 311
           16.2.2 TensorFlow 环境 ·· 312
           16.2.3 微信小程序及后台服务器环境 ···································· 312
      16.3 模块实现 ··· 313
           16.3.1 数据预处理 ·· 313
           16.3.2 创建模型与编译 ·· 314
           16.3.3 模型训练及保存 ·· 315
           16.3.4 模型生成 ··· 316
      16.4 系统测试 ··· 322
           16.4.1 训练准确率 ·· 322
           16.4.2 测试效果 ··· 322
           16.4.3 模型应用 ··· 322

项目 17　基于 OpenCV 的人脸识别程序 ……… 325
17.1　总体设计 ……… 325
17.1.1　系统整体结构 ……… 325
17.1.2　系统流程 ……… 325
17.2　运行环境 ……… 326
17.2.1　Python 环境 ……… 326
17.2.2　TensorFlow 环境 ……… 326
17.3　模块实现 ……… 327
17.3.1　数据预处理 ……… 327
17.3.2　模型构建 ……… 328
17.3.3　模型训练 ……… 330
17.4　系统测试 ……… 330

项目 18　基于 CGAN 的线稿自动上色 ……… 331
18.1　总体设计 ……… 331
18.1.1　系统整体结构 ……… 331
18.1.2　系统流程 ……… 331
18.2　运行环境 ……… 332
18.2.1　Python 环境 ……… 332
18.2.2　TensorFlow 环境 ……… 332
18.3　模块实现 ……… 333
18.3.1　数据预处理 ……… 333
18.3.2　模型构建 ……… 335
18.3.3　模型训练及保存 ……… 339
18.3.4　模型应用 ……… 342
18.4　系统测试 ……… 343
18.4.1　训练效果 ……… 343
18.4.2　测试效果 ……… 343
18.4.3　模型使用说明 ……… 345

项目 19　基于 ACGAN 的动漫头像生成 ……… 346
19.1　总体设计 ……… 346
19.1.1　系统整体结构 ……… 346
19.1.2　系统流程 ……… 346
19.2　运行环境 ……… 347
19.2.1　Python 环境 ……… 348
19.2.2　TensorFlow 环境 ……… 348
19.2.3　OpenCV 环境 ……… 348
19.2.4　Illustration2Vec ……… 348
19.3　模块实现 ……… 348
19.3.1　数据获取 ……… 349

		19.3.2	数据处理	349
		19.3.3	模型构建	352
		19.3.4	模型训练及保存	355
	19.4	系统测试		358
		19.4.1	模型导入及调用	358
		19.4.2	生成指定标签	358

项目20 手势语言识别 360

- 20.1 总体设计 360
  - 20.1.1 系统整体结构 360
  - 20.1.2 系统流程 361
- 20.2 运行环境 361
  - 20.2.1 Python 环境 361
  - 20.2.2 TensorFlow 环境 362
  - 20.2.3 OpenCV-Python 环境 362
- 20.3 模块实现 362
  - 20.3.1 设置直方图 362
  - 20.3.2 载入手势图片 365
  - 20.3.3 模型训练及保存 367
- 20.4 系统测试 368
  - 20.4.1 测试准确率 368
  - 20.4.2 测试效果 368

项目 1 电影推荐小程序

PROJECT 1

本项目采用协同过滤推荐算法,基于卷积神经网络,通过 MovieLens 数据集实现在移动端完成电影推荐。

1.1 总体设计

本部分包括系统整体结构和系统流程。

1.1.1 系统整体结构

系统整体结构如图 1-1 所示。

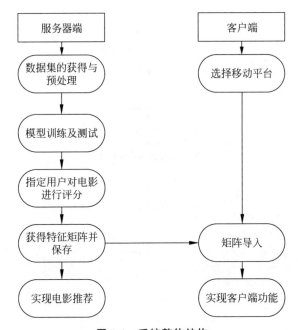

图 1-1 系统整体结构

1.1.2 系统流程

系统流程如图1-2所示,模型设计流程如图1-3所示。

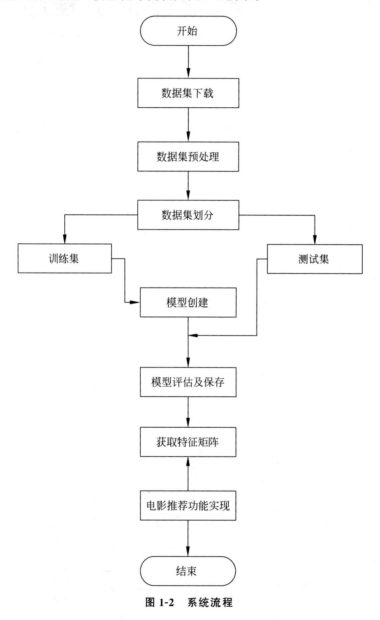

图1-2 系统流程

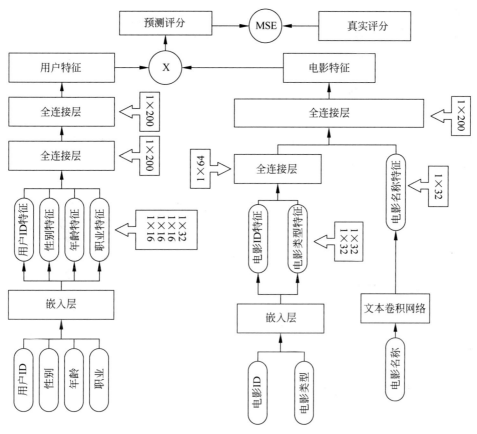

图 1-3　模型设计流程

1.2　运行环境

本部分包括 Python 环境和 TensorFlow 环境。

1.2.1　Python 环境

采用 Python 3.6 在 Windows 环境下运行,直接从 Anaconda 官网下载并完成环境的配置,下载地址为:https://www.anaconda.com/。

1.2.2　TensorFlow 环境

打开 Anaconda Prompt,输入清华仓库镜像,输入命令:

```
conda config --add channels https://mirrors.tuna.tsinghua.edu.cn/anaconda/pkgs/free/
```

```
conda config -set show_channel_urls yes
```

创建 Python 环境,名称为 TensorFlow,输入命令:

```
conda create -n tensorflow python = 3.5
```

有需要确认的地方,都输入 y。

在 Anaconda Prompt 中激活 TensorFlow 环境,输入命令:

```
activate tensorflow
```

安装 CPU 版本的 TensorFlow,输入命令:

```
pip install -upgrade --ignore-installed tensorflow
```

安装完毕。安装 TensorFlow 时注意与 Python 版本之间的对应,否则安装成功也无法调用。

1.3 模块实现

本部分包括 6 个模块:数据预处理、模型设计、模型训练及测试、特征矩阵提取、推荐电影、移动端。下面分别给出各模块的功能介绍及相关代码。

1.3.1 数据预处理

本部分包括数据集获取和数据预处理。

1. 数据集获取

MovieLens 1M 数据集包括 6000 个用户对 4000 部电影的 1 亿条评论。下载地址为: http://files.grouplens.org/datasets/movielens/ml-1m.zip。数据集分为用户数据、电影数据和评分数据。相关代码如下:

```python
# 导入本项目需要的数据包
import pandas as pd
from sklearn.model_selection import train_test_split    # 数据集划分
import numpy as np
from collections import Counter                          # 统计字符出现的次数
import tensorflow as tf
import os                                                # 处理文件和目录
import pickle                                            # 保存数据到本地
import re                                                # 正则表达式
from tensorflow.python.ops import math_ops
from urllib.request import urlretrieve                   # 将 URL 下载到本地
from os.path import isfile, isdir                        # 判断是否存在文件和文件夹
from tqdm import tqdm                                    # 提供进度条信息
```

```python
import zipfile                                          # zip 解压提取文件
import hashlib                                          # 进行哈希加密
# 定义提取函数,三个参量分别为压缩包存放路径、数据集名称和提取后的存放路径
def unzip(save_path, _, database_name, data_path):
    print('Extracting{}...'.format(database_name))
    with zipfile.ZipFile(save_path) as zf:
        zf.extractall(data_path)                        # 提取文件到该路径文件夹
# 定义下载压缩包函数
def download_extract(database_name, data_path):
    DATASET_ML1M = 'ml-1m'                              # 定义数据集名称
    if database_name == DATASET_ML1M:
        url = 'http://files.grouplens.org/datasets/movielens/ml-1m.zip'
        hash_code = 'c4d9eecfca2ab87c1945afe126590906'
        extract_path = os.path.join(data_path, 'ml-1m') # 提取到文件夹路径
        save_path = os.path.join(data_path, 'ml-1m.zip')# 从压缩文件提取路径
        extract_fn = _unzip
    if os.path.exists(extract_path):                    # 判断路径文件是否存在
        print('Found{}Data'.format(database_name))
        return
    if not os.path.exists(data_path):                   # 路径文件不存在则创建一个
        os.makedirs(data_path)
    if not os.path.exists(save_path):
        with \
        DLProgress(unit = 'B', unit_scale = True, miniters = 1, desc = 'Downloading{}'.format\
(database_name)) as pbar:                               # 设置参数
            urlretrieve(
                url,
                save_path,
                pbar.hook)                              # 将远程数据下载到本地
assert hashlib.md5(open(save_path, 'rb').read()).hexdigest() == hash_code, \
'{}file is corrupted. Remove the file and try again.'.format(save_path)  # 如果 md5 出现异常则报错
    os.makedirs(extract_path)
    try:
        extract_fn(save_path, extract_path, database_name, data_path)  # 解压
    except Exception as err:                            # 如果有错,删除所有文件夹
        shutil.rmtree(extract_path)
        raise err                                       # 抛出异常
    print('Done.')
class DLProgress(tqdm):                                 # 定义下载时处理进度条类
    last_block = 0
    def hook(self, block_num = 1, block_size = 1, total_size = None):
        self.total = total_size
        self.update((block_num - self.last_block) * block_size)
        self.last_block = block_num
# 给出下载路径和文件名,开始下载
data_dir = './'
download_extract('ml-1m', data_dir)
```

成功下载并提取数据集的提示信息如图1-4所示。

```
Downloading ml-1m: 5.92MB [01:15, 78.7kB/s]
Extracting ml-1m...
Done.
```

图1-4　成功下载并提取数据集

2. 数据预处理

本数据集由3个文件构成，不同文件内数据的格式不同，需要不同的方法对其进行预处理，如图1-5所示。

Movie	ID	Title	Genres
0	1	Toy Story (1995)	Animation|Children's|Comedy
1	2	Jumanji (1995)	Adventure|Children's|Fantasy
2	3	Grumpier Old Men (1995)	Comedy|Romance
3	4	Waiting to Exhale (1995)	Comedy|Drama
4	5	Father of the Bride Part II (1995)	Comedy

图1-5　预处理前的电影数据

电影ID(Movie ID)本身为数字不用处理，电影类别(Genres)是分类字段，需要转换为数字。首先，将Genres中的类别转换成字符串到数字的字典；其次，将每个电影的Genres字段转换成数字列表，因为有些电影是多个Genres的组合。去掉电影名称(Title)中的年份，方法与类别字段一样，创建文本到数字的字典；最后，将Title中的描述转换成数字的列表。

注意：Genres和Title字段长度应统一，以便于在神经网络中处理，不足部分用数字填充。相关代码如下：

```python
#读取Movie数据集
movies_title = ['MovieID','Title','Genres']
movies = pd.read_table('./ml-1m/movies.dat', sep = '::', header = None,
                        names = movies_title, engine = 'python')
movies_orig = movies.values
#将Title中的年份去掉
pattern = re.compile(r'^(.*)\(((\d+)\)$ ')
title_map = {val:pattern.match(val).group(1) for ii,val in
                        enumerate(set(movies['Title']))}
movies['Title'] = movies['Title'].map(title_map)        #将处理后的Title重新导入
#电影Title转数字字典
title_set = set()                                        #创建一个无序不重复元素集,返
                                                         #回一个可迭代对象
```

```python
for val in movies['Title'].str.split():          # val 是电影 Title 中的所有单词
    title_set.update(val)                         # 更新 val 中的元素
    title_set.add('<PAD>')                        # 填充表示
    title2int = {val:ii for ii, val in enumerate(title_set)}  # 创造字典
# 将电影 Title 转成等长数字列表,长度是 15
title_count = 15
title_map = {val:[title2int[row] for row in val.split()] for ii,val in
             enumerate(set(movies['Title'])))}
# 字典格式为"电影字符串:[Title 全部单词构成的数值列表]"
for key in title_map:
    for cnt in range(title_count-len(title_map[key])):
        title_map[key].insert(len(title_map[key]) + cnt,title2int['<PAD>'])
movies['Title'] = movies['Title'].map(title_map)  # Title 将名称转换为数组
# 电影类型转数字字典
genres_set = set()
for val in movies['Genres'].str.split('|'):
# 对于电影类型进行字符串转换,并用"\"分割遍历
    genres_set.update(val)                        # 更新 val 的值
genres_set.add('<PAD>')
genres2int = {val:ii for ii, val in enumerate(genres_set)}
# 格式为"电影字符串:数字"
# 将电影类型转成等长数字列表,长度是 18
genres_map = {val:[genres2int[row] for row in val.split('|')] for ii,val in enumerate(set
(movies['Genres']))}
for key in genres_map:
    for cnt in range(max(genres2int.values()) - len(genres_map[key])):
        genres_map[key].insert(len(genres_map[key]) + cnt,genres2int['<PAD>'])
movies['Genres'] = movies['Genres'].map(genres_map)
```

经过预处理后的电影数据如图 1-6 所示,全部由数字组成,便于处理。

Movie ID		Title	Genres
0	1	[2182, 4463, 1799, 1799, 1799, 1799, 1799, 179...	[13, 0, 3, 9, 9, 9, 9, 9, 9, 9, 9, 9, 9, 9,...
1	2	[534, 1799, 1799, 1799, 1799, 1799, 1799, 1799...	[18, 0, 14, 9, 9, 9, 9, 9, 9, 9, 9, 9, 9, 9,...
2	3	[2907, 4910, 3791, 1799, 1799, 1799, 1799, 179...	[3, 1, 9, 9, 9, 9, 9, 9, 9, 9, 9, 9, 9, 9,...
3	4	[2056, 2588, 3848, 1799, 1799, 1799, 1799, 179...	[3, 2, 9, 9, 9, 9, 9, 9, 9, 9, 9, 9, 9, 9,...
4	5	[3085, 2477, 5062, 4071, 813, 1259, 1799, 1799...	[3, 9, 9, 9, 9, 9, 9, 9, 9, 9, 9, 9, 9, 9,...

图 1-6 经过预处理后的电影数据

1.3.2 模型设计

获得数据集并进行预处理之后,进行神经网络的构建和设计,包括定义嵌入矩阵、构建

全连接层和损失函数。

1. 定义嵌入矩阵

神经网络的起点是嵌入层，包括用户特征和电影特征。其中，对于用户特征，只需将预处理后的数据作为嵌入矩阵的索引即可。电影特征需要分类讨论，然后多个矩阵相加求和，因为一部电影可能有多种类型特征，对电影名称的分析则需要用到文本卷积网络。

文本卷积网络的第一层是词嵌入层，由每一个单词的嵌入向量组成嵌入矩阵。第二层是使用多个不同窗口大小的卷积核在嵌入矩阵上做卷积（窗口大小是指每次卷积覆盖几个单词）运算，文本卷积要覆盖整个单词的嵌入向量。第三层网络通过最大池化得到一个长向量，最后使用 Dropout 进行正则化，得到电影名称的特征。

```python
#定义用户数据的嵌入矩阵,原始矩阵经过嵌入层后得到的输入
def get_user_embedding(uid,user_gender,user_age,user_job):    #用户四个嵌入矩阵
    with tf.name_scope("user_embedding"):
        #用户 ID
        #输出均匀分布的区间[-1,1]中的随机值,返回矩阵大小为用户数×特征数
        uid_embed_matrix = tf.Variable(tf.random_uniform([uid_max,
                          embed_dim],-1,1),name = "uid_embed_matrix")
        #选取 uid_embed_matrix 中用户 ID 对应的向量
        uid_embed_layer = tf.nn.embedding_lookup(uid_embed_matrix,uid,name =
                          "uid_embed_layer")
        #性别
        #矩阵中的特征值降为一半
        gender_embed_matrix = tf.Variable(tf.random_uniform([gender_max,
                          embed_dim//2],-1,1),name = "gender_embed_matrix")
        #选取 gender_embed_matrix 中用户性别对应的向量
        gender_embed_layer = tf.nn.embedding_lookup(gender_embed_matrix,
                          user_gender,name = "gender_embed_layer")
        #年龄
        age_embed_matrix = tf.Variable(tf.random_uniform([age_max,
                          embed_dim//2],-1,1),name = "age_embed_matrix")
        #选取 age_embed_matrix 中用户年龄对应的向量
        age_embed_layer = tf.nn.embedding_lookup(age_embed_matrix,user_age,
                          name = "age_embed_layer")
        #工作
        job_embed_matrix = tf.Variable(tf.random_uniform([job_max,
                          embed_dim//2],-1,1),name = "job_embed_matrix")
        #选取 job_embed_matrix 中用户工作对应的向量
        job_embed_layer = tf.nn.embedding_lookup(job_embed_matrix,user_job,
                          name = "job_embed_layer")
    return uid_embed_layer,gender_embed_layer,age_embed_layer,
                          job_embed_layer
#定义电影 ID 的嵌入矩阵
```

```python
def get_movie_id_embed_layer(movie_id):
    with tf.name_scope("movie_embedding"):
        movie_id_embed_matrix = tf.Variable(tf.random_uniform([movie_id_max,
            embed_dim], -1, 1), name = "movie_id_embed_matrix")
        movie_id_embed_layer = tf.nn.embedding_lookup(movie_id_embed_matrix,
            movie_id, name = "movie_id_embed_layer")
    return movie_id_embed_layer
def get_movie_categories_layers(movie_categories):
    with tf.name_scope("movie_categories_layers"):
        movie_categories_embed_matrix = tf.Variable(tf.random_uniform([movie_categories_max,
embed_dim], -1, 1), name = "movie_categories_embed_matrix")
        movie_categories_embed_layer = tf.nn.embedding_lookup(movie_categories_embed_matrix,
movie_categories, name = "movie_categories_embed_layer")
        if combiner == "sum":
            #压缩求和,但不降低维度
            movie_categories_embed_layer = tf.reduce_sum(movie_categories_embed_layer, axis = 1,
keepdims = True)
    return movie_categories_embed_layer
#电影名称的文本卷积网络
def get_movie_cnn_layer(movie_titles):
    #从嵌入矩阵中得到电影名称对应各个单词的嵌入向量
    with tf.name_scope("movie_embedding"):
        movie_title_embed_matrix = tf.Variable(tf.random_uniform([movie_title_max,
            embed_dim], -1, 1), name = "movie_title_embed_matrix")
        movie_title_embed_layer = tf.nn.embedding_lookup(movie_title_embed_m
            atrix, movie_titles, name = "movie_title_embed_layer")
        movie_title_embed_layer_expand = tf.expand_dims
            (movie_title_embed_layer, -1)    #给电影名称嵌入层加一维
#对文本嵌入层使用不同尺寸的卷积核进行卷积和最大池化
pool_layer_lst = []
for window_size in window_sizes:
    with tf.name_scope("movie_txt_conv_maxpool_{}".format
        (window_size)):    #格式化字符串
        #初始化卷积核参数
        filter_weights = tf.Variable(tf.truncated_normal([window_size,
            embed_dim, 1, filter_num], stddev = 0.1), name = "filter_weights")
        #初始化偏置
        filter_bias = tf.Variable(tf.constant(0.1, shape = [filter_num]),
            name = "filter_bias")
        #函数参量分别为卷积输入图像、卷积核和步长
        conv_layer = tf.nn.conv2d(movie_title_embed_layer_expand,
            filter_weights, [1, 1, 1, 1], padding = "VALID", name = "conv_layer")
        #最终迭代到windowsize = 5
        relu_layer = tf.nn.relu(tf.nn.bias_add(conv_layer, filter_bias),
            name = "relu_layer")
```

```python
    #池化操作,参数依次为输入、池化窗口的大小和每个维度上滑动的步长
    maxpool_layer = tf.nn.max_pool(relu_layer,[1,sentences_size-
                                    window_size + 1,1,1],[1,1,1,1],
                                    padding = "VALID",name = "maxpool_layer")
    pool_layer_lst.append(maxpool_layer)
#Dropout 层
with tf.name_scope("pool_dropout"):
    pool_layer = tf.concat(pool_layer_lst,3,name = "pool_layer")
    max_num = len(window_sizes) * filter_num
    #将 pool_layer 降到三维
    pool_layer_flat = tf.reshape(pool_layer,
                                    [ -1,1,max_num],name = "pool_layer_flat")
    #执行 Dropout 操作,目的是防止过拟合
    dropout_layer = tf.nn.dropout(pool_layer_flat,dropout_keep_prob,
                    name = "dropout_layer")
    return pool_layer_flat, dropout_layer
```

2. 构建全连接层

从嵌入层索引出特征后,将各特征传入全连接层,将得到的输出结果再次传入全连接层,最终得到用户特征和电影特征向量。

```python
#将用户数据的嵌入矩阵全部连接生成用户特征
def get_user_feature_layer(uid_embed_layer,gender_embed_layer,
                            age_embed_layer,job_embed_layer):
    with tf.name_scope("user_fc"):
        #第一层全连接
        #函数参量依次是该层输入/输出的维数、名称和使用的激活函数
        uid_fc_layer = tf.layers.dense(uid_embed_layer,embed_dim,name = "uid_fc_layer",
            activation = tf.nn.relu)
        gender_fc_layer = tf.layers.dense(gender_embed_layer,embed_dim,name = "gender_fc_
            layer",activation = tf.nn.relu)
        age_fc_layer = tf.layers.dense(age_embed_layer,embed_dim,name = "age_fc_layer",
            activation = tf.nn.relu)
        job_fc_layer = tf.layers.dense(job_embed_layer,embed_dim,name = "job_fc_layer",
            activation = tf.nn.relu)
        #第二层全连接
        #将第一层输出的四个矩阵连接,返回连接后的张量
        user_combine_layer = tf.concat([uid_fc_layer,gender_fc_layer,age_fc_layer,job_fc_
            layer],2)#(?, 1, 128)
        #将 128 维的矩阵扩展到 200 维,用于构建全连接层
        user_combine_layer = tf.contrib.layers.fully_connected(user_combine_layer,200,tf.tanh)
        #将矩阵重新构形,-1 表示任意值
        user_combine_layer_flat = tf.reshape(user_combine_layer,[ -1,200])
    return user_combine_layer,user_combine_layer_flat
#将电影数据的各个层一起做全连接
```

```
Def get_movie_feature_layer(movie_id_embed_layer,
                            movie_categories_embed_layer, dropout_layer):
    with tf.name_scope("movie_fc"):
    #第一层全连接
    movie_id_fc_layer = tf.layers.dense(movie_id_embed_layer, embed_dim,
                        name = "movie_id_fc_layer", activation = tf.nn.relu)
    movie_categories_fc_layer = tf.layers.dense(movie_categories_embed_l
    ayer, embed_dim, name = "movie_categories_fc_layer", activation = tf.nn.relu)
    #第二层全连接
    movie_combine_layer = tf.concat([movie_id_fc_layer,
                          movie_categories_fc_layer, dropout_layer], 2)
    movie_combine_layer = tf.contrib.layers.fully_connected
                                     (movie_combine_layer, 200, tf.tanh)
    movie_combine_layer_flat = tf.reshape(movie_combine_layer, [-1, 200])
    return movie_combine_layer, movie_combine_layer_flat
```

3. 损失函数

直接将用户特征和电影特征进行矩阵乘法运算得到预测评分,再用 MSE 算法将预测评分与真实评分进行回归运算得到损失函数。

```
#计算出评分
with tf.name_scope("inference"):
    #将用户特征和电影特征做矩阵乘法运算得到预测评分
    inference = tf.reduce_sum(user_combine_layer_flat * movie_combine_layer_flat, axis = 1)
    inference = tf.expand_dims(inference, axis = 1)          #与 target 统一格式
with tf.name_scope("loss"):
    #MSE 损失,将计算值回归到评分
    cost = tf.losses.mean_squared_error(targets, inference)
    loss = tf.reduce_mean(cost)                              #计算均值
global_step = tf.Variable(0, name = "global_step", trainable = False)
optimizer = tf.train.AdamOptimizer(lr)                       #进行反向传播训练
#计算梯度
gradients = optimizer.compute_gradients(loss)
#应用梯度
rain_op = optimizer.apply_gradients(gradients, global_step = global_step)
```

1.3.3 模型训练及测试

定义模型框架后使用经过预处理的数据集进行训练、测试模型,并将测试后的模型保存,以便调用。

1. 模型训练

模型训练相关代码如下:

```
#训练网络
```

```python
% matplotlib inline
% config InlineBackend.figure_format = 'retina'    # 呈现高分辨率图像
import matplotlib.pyplot as plt
import time
import datetime
losses = {'train':[],'test':[]}
with tf.Session(graph = train_graph) as sess:
    # 搜集数据给 tensorBoard 使用
    # 跟踪梯度值和稀疏度
    grad_summaries = []
    for g,v in gradients:
        if g is not None:
            # 显示直方图信息,显示训练中变量的分布情况
            grad_hist_summary = tf.summary.histogram("{}/grad/hist".format
                                (v.name.replace(':','_')),g)
            # 显示标量信息,统计 0 的占比来表现稀疏度
            sparsity_summary = tf.summary.scalar("{}/grad/sparsity".format
                    (v.name.replace(':','_')),tf.nn.zero_fraction(g))
            grad_summaries.append(grad_hist_summary)
            grad_summaries.append(sparsity_summary)
    # 将上述几种类型汇总再进行合并
    grad_summaries_merged = tf.summary.merge(grad_summaries)
    # 模型和 summaries 的输出目录
    timestamp = str(int(time.time()))
    # os.path.join 函数将多个路径组合后返回
    out_dir = os.path.abspath(os.path.join(os.path.curdir,"runs",timestamp))
    print("Writing to {}\n".format(out_dir))
    # 损失和精准度的总结,显示损失标量信息
    loss_summary = tf.summary.scalar("loss", loss)
    # 训练总结
    train_summary_op = tf.summary.merge([loss_summary,grad_summaries_merged])
    train_summary_dir = os.path.join(out_dir,"summaries","train")
    train_summary_writer = tf.summary.FileWriter(train_summary_dir,
                           sess.graph)     # 指定文件来保存图像
    # 推理总结
    inference_summary_op = tf.summary.merge([loss_summary])
    inference_summary_dir = os.path.join(out_dir,"summaries","inference")
    inference_summary_writer = tf.summary.FileWriter(inference_summary_dir,
                               sess.graph)
    sess.run(tf.global_variables_initializer())     # 参数初始化
    saver = tf.train.Saver()                # 创建 Saver 对象用于保存数据
    for epoch_i in range(num_epochs):
        # 将数据集分成训练集和测试集,随机种子不固定
        train_X,test_X,train_y,test_y = train_test_split(features,
                        targets_values,test_size = 0.2,random_state = 0)
        # 在分好的训练集中再选 batch_size 个
        train_batches = get_batches(train_X,train_y,batch_size)
```

```python
        # 在分好的测试集中再选 batch_size 个
        test_batches = get_batches(test_X,test_y,batch_size)
        # 训练的迭代
        for batch_i in range(len(train_X)//batch_size):        # 结果向下取整
            x,y = next(train_batches)                           # next()返回迭代器的下一个项目
            categories = np.zeros([batch_size,18])
    for i in range(batch_size):
        categories[i] = x.take(6,1)[i]                          # x 取每一行下标为 6 的数据
    titles = np.zeros([batch_size,sentences_size])
    for i in range(batch_size):
        titles[i] = x.take(5,1)[i]                              # x 取每一行下标为 5 的数据
    feed = {
        uid:np.reshape(x.take(0,1),[batch_size,1]),
        user_gender:np.reshape(x.take(2,1),[batch_size,1]),
        user_age:np.reshape(x.take(3,1),[batch_size,1]),
        user_job:np.reshape(x.take(4,1),[batch_size,1]),
        movie_id:np.reshape(x.take(1,1),[batch_size,1]),
        movie_categories:categories,  # x.take(6,1)
        movie_titles:titles,  # x.take(5,1)
        targets:np.reshape(y,[batch_size,1]),
        dropout_keep_prob:dropout_keep,
        lr:learning_rate}
    step,train_loss,summaries,_ = sess.run([global_step,loss,
                        train_summary_op,train_op],feed)
        # 保存训练损失
        losses['train'].append(train_loss)
        # 调用 add_summary 函数将训练过程数据保存在指定文件中
        train_summary_writer.add_summary(summaries,step)
        # 展示每一个 batch
        if(epoch_i * (len(train_X)//batch_size) + batch_i) % 
                                    show_every_n_batches == 0:
            time_str = datetime.datetime.now().isoformat()
            print('{}:Epoch{:>3}Batch{:>4}/{}train_loss = {:.3f}'.format(
                time_str,
                epoch_i,
                batch_i,
                (len(train_X)//batch_size),
                train_loss))
# 使用测试数据的迭代
for batch_i in range(len(test_X)//batch_size):
    x,y = next(test_batches)
    categories = np.zeros([batch_size,18])
    for i in range(batch_size):
        categories[i] = x.take(6,1)[i]                          # x 取每一行下标为 6 的数据
    titles = np.zeros([batch_size,sentences_size])
    for i in range(batch_size):
        titles[i] = x.take(5,1)[i]                              # x 取每一行下标为 5 的数据
```

```
        feed = {
            uid:np.reshape(x.take(0,1),[batch_size,1]),
            user_gender:np.reshape(x.take(2,1),[batch_size,1]),
            user_age:np.reshape(x.take(3,1),[batch_size,1]),
            user_job:np.reshape(x.take(4,1),[batch_size,1]),
            movie_id:np.reshape(x.take(1,1),[batch_size,1]),
            movie_categories:categories, #x.take(6,1)
            movie_titles:titles, #x.take(5,1)
            targets:np.reshape(y,[batch_size,1]),
            dropout_keep_prob:1,
            lr:learning_rate}
        step,test_loss,summaries = sess.run([global_step,loss,
                                            inference_summary_op],feed)
        #保存测试损失
        losses['test'].append(test_loss)
        #调用 add_summary 函数将测试过程数据保存在指定文件中
        inference_summary_writer.add_summary(summaries,step)
        time_str = datetime.datetime.now().isoformat()        #表示日期
        if(epoch_i * (len(test_X)//batch_size) + batch_i) %
                                            show_every_n_batches == 0:
            print('{}:Epoch{:>3}Batch{:>4}/{}test_loss = {:.3f}'.format(
                time_str,
                epoch_i,
                batch_i,
                (len(test_X)//batch_size),
                test_loss))
```

2. 模型保存

模型保存的相关代码如下:

```
#模型保存
saver.save(sess,save_dir)
print('Model Trained and Saved')
```

1.3.4 特征矩阵提取

模型训练完成并保存后,需要让所有用户对电影进行预测评分,并将生成的用户特征向量和电影特征向量组合成矩阵。

1. 指定用户对电影进行评分

指定用户对电影进行评分的相关代码如下:

```
#指定用户对电影进行评分
def rating_movie(user_id_val,movie_id_val):
    loaded_graph = tf.Graph()                                #新建图
    with tf.Session(graph = loaded_graph) as sess:           #在 Session 中导入新建图
```

```python
# 导入训练好的模型
loader = tf.train.import_meta_graph(load_dir + '.meta')  # 创建网络
loader.restore(sess, load_dir)                            # 恢复网络参数
# 从加载的模型中获取tensors
uid, user_gender, user_age, user_job, movie_id, movie_categories, \
            movie_titles, targets, lr, dropout_keep_prob, inference, \
            _, __ = get_tensors(loaded_graph)
categories = np.zeros([1, 18])
categories[0] = movies.values[movieid2idx[movie_id_val]][2]
titles = np.zeros([1, sentences_size])
titles[0] = movies.values[movieid2idx[movie_id_val]][1]
feed = {
    uid: np.reshape(users.values[user_id_val-1][0], [1, 1]),
    user_gender: np.reshape(users.values[user_id_val-1][1], [1, 1]),
    user_age: np.reshape(users.values[user_id_val-1][2], [1, 1]),
    user_job: np.reshape(users.values[user_id_val-1][3], [1, 1]),
    movie_id: np.reshape(movies.values[movieid2idx[movie_id_val]][0], [1, 1]),
    movie_categories: categories,  # x.take(6,1)
    movie_titles: titles,  # x.take(5,1)
    dropout_keep_prob: 1}
# 预测评分
inference_val = sess.run([inference], feed)
return(inference_val)
```

2. 生成电影特征矩阵

生成电影特征矩阵的相关代码如下：

```python
# 生成电影特征矩阵
loaded_graph = tf.Graph()                                 # 新建图
movie_matrics = []
with tf.Session(graph=loaded_graph) as sess:              # 在Session中引入新建图
    # 导入训练好的模型
    loader = tf.train.import_meta_graph(load_dir + '.meta')  # 创建网络
    loader.restore(sess, load_dir)                            # 恢复网络参数
    # 从加载的模型中获取tensors
    # 使用函数get_tensors_by_name从loaded_graph中获取已保存的操作、占位符和变量
    uid, user_gender, user_age, user_job, movie_id, movie_categories, movie_tit\
                les, targets, lr, dropout_keep_prob, _, movie_combine_layer_flat, \
                __ = get_tensors(loaded_graph)
    for item in movies.values:  # item为[movie_id, list[电影名], list[电影类型]]
        categories = np.zeros([1, 18])
        categories[0] = item.take(2)                      # 取电影类型list[]
        titles = np.zeros([1, sentences_size])
        titles[0] = item.take(1)                          # 取电影名称list[]
        feed = {
            movie_id: np.reshape(item.take(0), [1, 1]),
```

```
                movie_categories:categories,  # x.take(6,1)
                movie_titles:titles,  # x.take(5,1)
                dropout_keep_prob:1}
            # 完成全部的数据流动,得到输出的电影特征
            movie_combine_layer_flat_val = sess.run([movie_combine_layer_flat],feed)
            # 为每个电影生成一个特征矩阵并保存
            movie_matrices.append(movie_combine_layer_flat_val)
pickle.dump((np.array(movie_matrices).reshape(-1,200)),
        open('movie_matrices.p','wb'))
# 将所有电影特征保存到 movie_matrices.p 文件里
movie_matrices = pickle.load(open('movie_matrices.p',mode = 'rb'))
```

3. 生成用户特征矩阵

生成用户特征矩阵的相关代码如下:

```
# 生成用户特征矩阵
loaded_graph = tf.Graph()                                       # 新建图
users_matrics = []
with tf.Session(graph = loaded_graph) as sess:                  # 在 Session 中引入新建图
    # 导入训练好的模型
    loader = tf.train.import_meta_graph(load_dir + '.meta')     # 创建网络
    loader.restore(sess,load_dir)                               # 恢复网络参数
    # 从加载的模型中获取 tensors
    # 使用函数 get_tensors_by_name 从 loaded_graph 中获取已保存的操作、占位符和变量
    uid,user_gender,user_age,user_job,movie_id,movie_categories,movie_tit
                            les,targets,lr,dropout_keep_prob,_,__,
                    user_combine_layer_flat = get_tensors(loaded_graph)
    for item in users.values:
        feed = {
            uid:np.reshape(item.take(0),[1,1]),
            user_gender:np.reshape(item.take(1),[1,1]),
            user_age:np.reshape(item.take(2),[1,1]),
            user_job:np.reshape(item.take(3),[1,1]),
            dropout_keep_prob:1}
        # 完成全部的数据流动,得到输出的用户特征
        user_combine_layer_flat_val = sess.run([user_combine_layer_flat],feed)
        # 为每个用户生成一个用户特征矩阵并保存
        users_matrics.append(user_combine_layer_flat_val)
pickle.dump((np.array(users_matrics).reshape(-1,200)),
open('users_matrics.p','wb'))
# 将所有用户特征保存到 users_matrics.p 文件
users_matrics = pickle.load(open('users_matrics.p',mode = 'rb'))
```

1.3.5 推荐电影

得到电影特征矩阵和用户特征矩阵后,进行电影推荐,可推荐同类型电影、用户喜欢的

电影和看过这部电影的用户还喜欢看哪些电影。

1. 推荐同类型电影

本部分相关代码如下：

```python
#推荐同类型电影
def recommend_same_type_movie(movie_id_val, top_k = 20):
    loaded_graph = tf.Graph()
    with tf.Session(graph = loaded_graph) as sess:
        #导入训练好的模型
        loader = tf.train.import_meta_graph(load_dir + '.meta')   #创建网络
        loader.restore(sess, load_dir)                            #恢复网络参数
        #向量单位化
        norm_movie_matrics = tf.sqrt(tf.reduce_sum(tf.square(movie_matrics),
                                                   1, keep_dims = True))
        normalized_movie_matrics = movie_matrics/norm_movie_matrics
        #推荐同类型的电影
        #将输入的电影ID转换为电影特征向量
        probs_embeddings = (movie_matrics[movieid2idx[movie_id_val]]).
                                                      reshape([1, 200])
        #用得到的特征向量与整个电影矩阵相乘得到余弦相似度
        probs_similarity = tf.matmul(probs_embeddings, tf.transpose
                                     (normalized_movie_matrics))
        sim = (probs_similarity.eval())                 #转换为字符串
        print("您看的电影是:{}".
                          format(movies_orig[movieid2idx[movie_id_val]]))
        print("以下是给您的推荐:")
        p = np.squeeze(sim)                             #将向量数组转换为一维数组
        #将p中元素从小到大排列后依次索引,将20个元素后面的所有元素置为0
        p[np.argsort(p)[: - top_k]] = 0
        p = p/np.sum(p)                                 #将p标准化
        results = set()
        while len(results)!= 5:                         #推荐5个
            c = np.random.choice(3883, 1, p = p)[0]     #从3883中以概率p随机选取1个
            results.add(c)
        for val in (results):
            print(val)                                  #输出转换后的字符串
            print(movies_orig[val])
        return results
```

2. 推荐用户喜欢的电影

本部分相关代码如下：

```python
#推荐您喜欢的电影
def recommend_your_favorite_movie(user_id_val, top_k = 10):
    loaded_graph = tf.Graph()
    with tf.Session(graph = loaded_graph) as sess:
```

```python
        # 导入训练好的模型
        loader = tf.train.import_meta_graph(load_dir + '.meta')
        loader.restore(sess, load_dir)
        # 推荐您喜欢的电影
        # 将输入的用户ID转换为用户特征向量
        probs_embeddings = (users_matrics[user_id_val - 1]).reshape([1, 200])
        # 将用户特征向量与电影特征矩阵相乘得到预测评分
        probs_similarity = tf.matmul(probs_embeddings, tf.transpose
                                                        (movie_matrics))
        sim = (probs_similarity.eval())              # 转换为字符串
        print("以下是给您的推荐:")
        p = np.squeeze(sim)
        p[np.argsort(p)[:-top_k]] = 0
        p = p/np.sum(p)
        results = set()
        while len(results) != 5:                     # 推荐5个
            c = np.random.choice(3883, 1, p=p)[0]
            results.add(c)
        for val in (results):
            print(val)
            print(movies_orig[val])
        return results
```

3. 看过这部电影的人还喜欢看哪些电影

本部分相关代码如下:

```python
# 看过这部电影的人还喜欢看哪些电影
import random
def recommend_other_favorite_movie(movie_id_val, top_k = 20):
    loaded_graph = tf.Graph()
    with tf.Session(graph=loaded_graph) as sess:
        # 导入训练好的模型
        loader = tf.train.import_meta_graph(load_dir + '.meta')
        loader.restore(sess, load_dir)
        # 将输入的电影ID转换为电影特征向量
        probs_movie_embeddings = (movie_matrics[movieid2idx[movie_id_val]]).
                                                        reshape([1, 200])
        # 将电影特征向量和用户特征矩阵相乘得到每个人对当前电影的评分
        probs_user_favorite_similarity = tf.matmul(probs_movie_embeddings,
                                            tf.transpose(users_matrics))
        # 选出喜欢这部电影的top_k个用户
        favorite_user_id = np.argsort(probs_user_favorite_similarity.eval())
                                                            [0][-top_k:]
        print("您看的电影是:{}".format(movies_orig[movieid2idx
                                                        [movie_id_val]]))
        print("喜欢看这部电影的人是{}".format(users_orig[favorite_user_id-1]))
```

```
#将上面选出的top_k个用户转换为用户特征向量
probs_users_embeddings = (users_matrics
                          [favorite_user_id-1]).reshape([-1,200])
#将上述选出的用户特征向量和所有电影特征矩阵相乘得到评分
probs_similarity = tf.matmul(probs_users_embeddings,tf.transpose
                                                   (movie_matrics))
sim = (probs_similarity.eval())         #转换为字符串
p = np.argmax(sim,1)
print("喜欢看这部电影的人还喜欢看:")
results = set()
while len(results)!= 5: #推荐5个
    c = p[random.randrange(top_k)]
    results.add(c)
for val in (results):
    print(val)
    print(movies_orig[val])
return results
```

1.3.6 客户端

本部分包括各模块的相关代码。

1. App 模块

App 模块相关代码如下:

```
//app.js
App({
  onLaunch: function () {
    //显示本地存储能力
    var logs = wx.getStorageSync('logs') || []
    logs.unshift(Date.now())
    wx.setStorageSync('logs', logs)
    //登录
    wx.login({
      success: res => {
        //发送 res.code 到后台换取 openId, sessionKey, unionId
      }
    })
    //获取用户信息
    wx.getSetting({
      success: res => {
        if (res.authSetting['scope.userInfo']) {
          //已经授权,可以直接调用 getUserInfo 获取头像昵称,不会弹框
          wx.getUserInfo({
            success: res => {
              //可以将 res 发送给后台解码出 unionId
```

```
                    this.globalData.userInfo = res.userInfo
                    //由于 getUserInfo 是网络请求,可能会在 Page.onLoad 之后才返回
                    //所以此处加入 callback 以防止这种情况
                    if (this.userInfoReadyCallback) {
                      this.userInfoReadyCallback(res)
                    }
                  }
                })
              }
            }
          })
        },
        globalData: {
          userInfo: null
        }
      })
```

2. app.json 模块

app.json 模块相关代码如下:

```
{
  "pages":[
    "pages/about/about",
    "pages/logs/logs",
    "pages/weekly/weekly",
    "pages/detail/detail"                    //页面跳转
  ],
  "tabBar":{                                 //导航栏设置
    "list":[
      {
        "text":"推荐",
        "pagePath":"pages/weekly/weekly",
        "iconPath":"images/icons/tui.jpg",
        "selectedIconPath":"images/icons/tui-selected.jpeg"
      },
      {
        "text":"关于",
        "pagePath":"pages/about/about",
        "iconPath":"images/icons/guanyu.png",
        "selectedIconPath":"images/icons/guanyu-selected.jpg"
      }
    ]
  },
  "window":{                                 //窗口设置
    "backgroundTextStyle":"light",
    "navigationBarBackgroundColor": "#fff",
```

```
    "navigationBarTitleText": "电影推荐",
    "navigationBarTextStyle":"black"
  },
  "style": "v2",
  "sitemapLocation": "sitemap.json"
}
```

3. app.wxss 模块

app.wxss 模块相关代码如下:

```
/** app.wxss **/                                              //样本框设置
.container{
  background-color: #eee;
  height: 100vh;
  display: flex;
  flex-direction: column;
  justify-content: space-around;
  align-items: center;
}
.about-banner{
  width: 375rpx;
  height: 375rpx;
  border-radius: 50%;
}

//about 页面
about.json                                                    //页面样式
{
  "navigationBarBackgroundColor": "#fff",
    "navigationBarTitleText": "电影推荐",
    "navigationBarTextStyle":"black"
}
about.wxml                                                    //页面内容
<view class = "container">
<image class = "about-banner" src = "/images/about.jpg"></image>    //插入图片
<test style = "font-weight:bold;font-size:60rpx">电影推荐</test>     //插入文字
<view>
<test><navigator style = "display:inline;"url = "/pages/weekly/weekly"open-type =
"switchTab"
hover-class = "nav-hover"class = "nav-default">推荐</navigator>      //插入可跳转页面推荐
<text>好看的电影</text></test>
</view>
</view>
about.wxss                                                    //页面样式
.container{
  background-color: #eee;
  height: 100vh;
  display: flex;
  flex-direction: column;
```

```
        justify-content: space-around;
        align-items: center;
    }
    .about-banner{
        width: 375rpx;                                              //设置大小
        height: 375rpx;
        border-radius: 50%;
    }
    .nav-default{
        color: blue;                                                //设置颜色
    }
    .nav-hover{
        color: red;
    }
weekly.js
Page({
    data:{
        weeklyMovieList:[                                           //电影列表
            {
                name:"推荐同类型的电影",
                person:"",
                comment:"3385[3454 'Whatever It Takes (2000)' 'Comedy|Romance']707[716 'Switchblade
Sisters (1975)' 'Crime']2351[2420 'Karate Kid, The (1984)' 'Drama']2189[2258 'Master Ninja I
(1984)' 'Action']2191[2260 'Wisdom (1986)' 'Action|Crime']",
                id:77
            },
            {
                name:"推荐您喜欢的电影",
                person:"",
                comment:"1642[1688 'Anastasia (1997)' 'Animation|Children's|Musical']994Ani[1007
'Apple Dumpling Gang, The(1975)' 'Children's|Comedy|Western']667[673 'Space Jam (1996)' '
Adventure|AnimationChildren's|Fantasy']1812[1881 'Quest for Camelot (1998)' 'Adventure|
AnimationChildren's|Fantasy']1898[1967 'Labyrinth (1986)' 'Adventure|Children's|Fantasy']",
                id:88
            },
            {
                name:"看过这部电影的用户还看了(喜欢)哪些电影",
                person:"喜欢看这部电影的用户是:[[5782 'F' 35 0][5767 'M' 25 2][3936 'F' 35 12][3595 'M'
25 0][1696 'M' 35 7][2728 'M' 35 12][763 'M' 18 10][4404 'M' 25 1][3901 'M' 18 14][371 'M' 18 4]
[1855 'M' 18 4][2338 'M' 45 17][450 'M' 45 1][1130 'F' 26 17][3035 'M' 25 7][100 'M' 35 17][567 'M' 35
20][5861 'F' 50 1][4800 'M' 18 4][3281 'M' 25 17]]",
                comment:"1779[1848 'Borrowers, The (1997)' 'Adventure|Children's|ComedyFantasy']1244
[1264 'Diva (1981)' 'Action|Drama|Mystery|Romance|Thriller']1812[1881 'Quest for Camlot (1998)'
'Adventure|Animation|Children's|Fantasy']1742[1805 'Wild Things (1998)' 'Crime|Drama|Mystery|
Thriller']",
                id:1291841
            }
        ]
    },
    onLoad:function(options){                                       //加载数据
```

```
            this.setData({
                currentIndex:this.data.weeklyMovieList.length-1
            })
        },
        f0:function(event){
            this.setData({
                currentIndex:this.data.weeklyMovieList.length-1
            })
        },
        f1:function(event){
            var movieId = event.currentTarget.dataset.movieId
            console.log(movieId);
            wx.navigateTo({
                url: '/pages/detail/detail?id = ' + movieId,
            })
        }
    })
weekl.wxml                                                      //加载页面
<view class = "">
<swiper class = "movie-swiper"indicator-dots = "{{true}}" current = "{{currentIndex}}">
<swiper-item class = "movie" wx:for = "{{weeklyMovieList}}">
<view class = "container movie-card"
bindtap = "f1"
data-user-name = "blabla" data-movie-id = "{{item.id}}">
    <text>第{{index + 1}}类:{{item.name}}</text>
    <text>{{item.person}}</text>
    <text>以下是给您的推荐:{{item.comment}}</text>
<text catchtap = "f0" wx:if = "{{index < (weeklyMovieList.length-1)}}" class = "return-button">返回
</text>
    </view>
</swiper-item>
</swiper>
</view>
weekly.wxss                                                     //参数设置
.movie{
    display: flex;
}
.movie-details{
    display: flex;
    flex-direction: column;
    width: 550rpx;
}
.movie-image{
    width: 200px;
    height: 200px;
}
.movie-swiper{
    height: 90vh;
}
.movie-card{
```

```
        height: 100%;
        width: 100%;
        position: relative;
    }
    .return-button{
        position: absolute;
        right:0;
        top: 40px;
        font-size: 20rpx;
        font-style: italic;
        border: 1px solid blue;
        border-right:0 ;
        border-radius: 10%;
    }
```

1.4 系统测试

本部分包括训练准确率和运行结果。

1.4.1 训练准确率

训练模型并保存后,得到损失值随训练轮次变化的曲线,如图1-7所示。开始时训练损失值(training loss)迅速降低,随着训练轮次(batch)的增加,训练损失值缓慢下降,并最终趋于稳定。

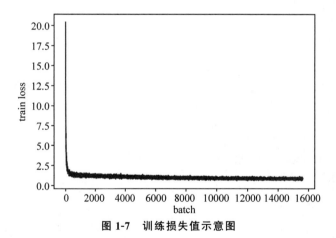

图1-7 训练损失值示意图

也可以利用类似方法得到测试损失值,如图1-8所示。测试损失值(test loss)波动比较大,随着训练轮次(batch)增加呈下降趋势。

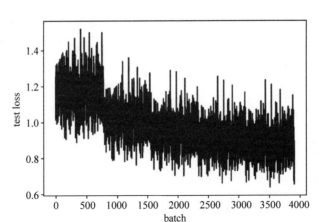

图 1-8　测试损失值示意图

1.4.2　运行结果

本部分包括三种推荐方式的运行结果。

1. 推荐同类型的电影

相关代码如下：

recommend_same_type_movie(1401, 20)

调用函数查看电影 ID 为 1401 的同类型电影，方法是计算当前电影特征向量与整个电影特征矩阵的余弦相似度，取最大的 top_k 个，最后推荐出 5 部电影，如图 1-9 所示。

```
您看的电影是：[1401 'Ghosts of Mississippi (1996)' 'Drama']
以下是给您的推荐：
1248
[1268 'Pump Up the Volume (1990)' 'Drama']
2417
[2486 '24-hour Woman (1998)' 'Drama']
3057
[3126 'End of the Affair, The (1955)' 'Drama']
2167
[2236 'Simon Birch (1998)' 'Drama']
3834
[3904 'Uninvited Guest, An (2000)' 'Drama']
```

图 1-9　推荐同类型电影结果示例

2. 推荐用户喜欢的电影

相关代码如下：

recommend_your_favorite_movie(234, 10)

调用函数查看 ID 为 234 的用户可能喜欢的电影，方法是根据用户特征向量和所有电影特征矩阵计算评分，取最大的 top_k 个，最后推荐出 5 部电影，如图 1-10 所示。

```
以下是给您的推荐：
523
[527 "Schindler's List (1993)" 'Drama|War']
3342
[3411 'Babymother (1998)' 'Drama']
1230
[1250 'Bridge on the River Kwai, The (1957)' 'Drama|War']
847
[858 'Godfather, The (1972)' 'Action|Crime|Drama']
1950
[2019
'Seven Samurai (The Magnificent Seven) (Shichinin no samurai) (1954)'
'Action|Drama']
```

图 1-10　推荐用户喜欢的电影结果示例

3. 看过这部电影的用户还喜欢看哪些电影

相关代码如下：

```
recommend_other_favorite_movie(1401, 20)
```

这种推荐方法目前最有意义，也是人们经常使用的方法。例如，很多互联网公司都采用这种协同过滤的方法，淘宝中喜欢该商品的人还喜欢哪些商品，音乐软件中喜欢这首歌的还喜欢听哪些歌等。在本例中，调用函数查看喜欢看电影 ID 为 1401 的用户还喜欢哪些电影，取最大的 top_k 个用户，如图 1-11 所示。得到特征向量之后，再计算这些用户对所有电影的评分，选取得分最高的 5 部电影，结果如图 1-12 所示。

```
您看的电影是：[1401 'Ghosts of Mississippi (1996)' 'Drama']
喜欢看这部电影的用户是：[[2617 'F' 45 15]
 [4814 'M' 18 14]
 [1216 'M' 25 2]
 [1231 'M' 25 12]
 [3833 'M' 25 1]
 [3780 'M' 1 0]
 [5335 'M' 25 14]
 [74 'M' 35 14]
 [5102 'M' 25 12]
 [1884 'M' 45 20]
 [1982 'M' 1 10]
 [2338 'M' 45 17]
 [4633 'M' 1 0]
 [2848 'M' 25 0]
 [5342 'F' 35 7]
 [3031 'M' 18 4]
 [1763 'M' 35 7]
 [3901 'M' 18 14]
 [713 'M' 35 7]
 [4085 'F' 25 6]]
```

图 1-11　喜欢电影 ID 为 1401 的用户列表

客户端首页（about 页）如图 1-13 所示，推荐同类型电影的页面如图 1-14 所示，推荐喜欢的电影页面如图 1-15 所示。

喜欢看这部电影的用户还喜欢看：
1667
[1716 'Other Voices, Other Rooms (1997)' 'Drama']
1132
[1148 'Wrong Trousers, The (1993)' 'Animation|Comedy']
847
[858 'Godfather, The (1972)' 'Action|Crime|Drama']
1950
[2019
'Seven Samurai (The Magnificent Seven) (Shichinin no samurai) (1954)'
'Action|Drama']
735
[745 'Close Shave, A (1995)' 'Animation|Comedy|Thriller']

图 1-12 推荐用户可能喜欢的其他电影示例

图 1-13 客户端首页

图 1-14 推荐同类型电影

图 1-15 推荐喜欢的电影

项目 2　服装分类助手

PROJECT 2

本项目应用机器学习模型采用卷积神经网络，部署在 Web 环境中，通过 Fashion-MNIST 数据集进行模型训练和改进，实现网页端服装类别精准识别。

2.1　总体设计

本部分包括系统整体结构和系统流程。

2.1.1　系统整体结构

系统整体结构如图 2-1 所示。

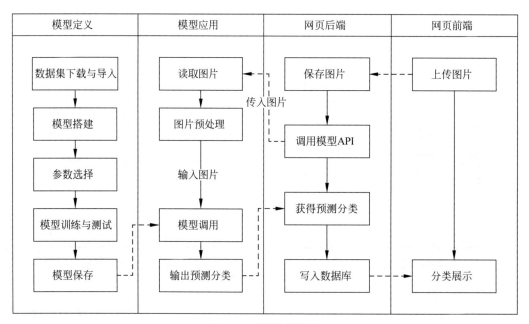

图 2-1　系统整体结构

2.1.2 系统流程

系统流程如图2-2所示，Web端部署流程如图2-3所示，VGG模型结构如图2-4所示。

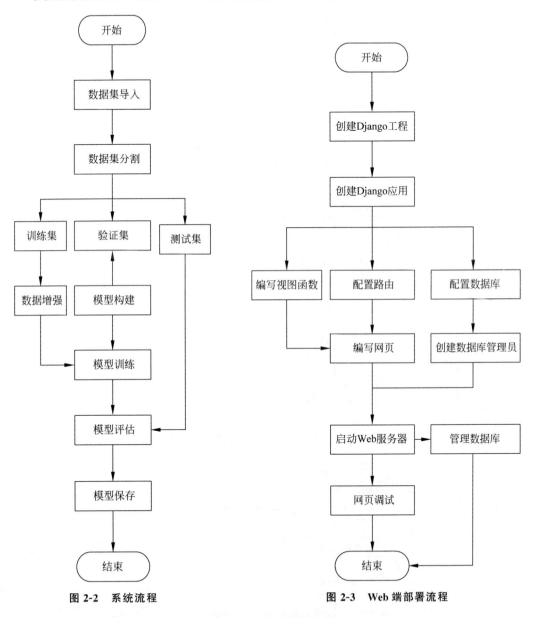

图 2-2 系统流程　　　　　　图 2-3 Web 端部署流程

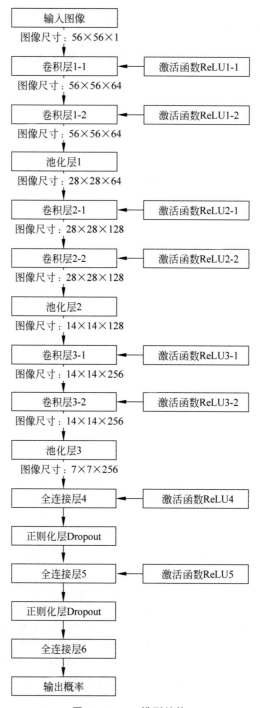

图 2-4 VGG 模型结构

2.2 运行环境

本部分包括 Python 环境、PyTorch 环境和 Django 环境。

2.2.1 Python 环境

在 Windows 系统中下载 Anaconda,下载地址为:https://www.anaconda.com/。本项目采用 Python 3.6 版本,完成 Anaconda 的安装并配置好环境变量,也可以下载虚拟机在 Linux 环境下运行代码。

2.2.2 PyTorch 环境

打开 Anaconda Prompt,添加清华仓库镜像,输入命令:

```
conda config --add channels https://mirrors.tuna.tsinghua.edu.cn/anaconda/pkgs/free/
conda config --add channels https://mirrors.tuna.tsinghua.edu.cn/anaconda/pkgs/main/
conda config -set show_channel_urls yes
```

创建新的 Python 环境,名称为 PyTorch,Python 版本号要对应,输入命令:

```
conda create -n pytorch python=3.6
```

有需要确认的地方,都输入 y。

在 Anaconda Prompt 中激活 PyTorch 环境,输入命令:

```
activate pytorch
```

在 PyTorch 官网查看可供选择 PyTorch 版本、系统版本、安装方式以及编程语言,获得下载命令,还可选择 CUDA 加速,下载地址为:https://pytorch.org/get-started/locally/。

选用 Stable 1.4 版本、Windows 系统、conda 安装包、Python 语言以及 CUDA 10.1 版本的 PyTorch,在 Anaconda Prompt 中安装,输入命令:

```
conda install pytorch torchvision cudatoolkit=10.1
```

有需要确认的地方,都输入 y,安装完成。

2.2.3 Django 环境

打开 Anaconda Prompt,激活 PyTorch 环境,输入命令:

```
activate pytorch
```

下载 Django,输入命令:

```
conda install Django
```

有需要确认的地方,都输入 y,安装完成后检查是否成功安装了 Django,输入命令:

```
conda list
```

若有显示结果,则 Django 成功安装。

2.3 模块实现

本项目包括 4 个模块:数据预处理、模型创建与编译、模型训练及保存、模型生成。下面分别给出各模块的功能介绍及相关代码。

2.3.1 数据预处理

Fashion-MNIST 是德国研究机构 Zalando Research 推出的一个全新的数据集,其中训练集包含 60000 个样例,测试集包含 10000 个样例,每个样本都为 28×28 的单通道灰度图像。该数据集的样本来自日常穿着的时尚单品,共 10 个类别,分别为[0]T-shirt/top、[1]Trouser、[2]Pullover、[3]Dress、[4]Coat、[5]Sandal、[6]Shirt、[7]Sneaker、[8]Bag、[9]Ankle boot。

Fashion-MNIST 在 PyTorch 中是内置的数据集,对 datasets 类使用 API 下载即可,对于 datasets 类的对象,可以使用 DataLoader()方法加载数据集。

```
#下载和导入训练集
trainset = torchvision.datasets.FashionMNIST(root = './data', train = True, download = True, transform = transform)
trainloader = torch.utils.data.DataLoader(trainset, batch_size = 50, shuffle = True, num_workers = 2)
#下载和导入测试集
testset = torchvision.datasets.FashionMNIST(root = './data', train = False, download = True, transform = transform1)
testloader = torch.utils.data.DataLoader(testset, batch_size = 50, shuffle = False, num_workers = 2)
```

由于在线下载速度较慢,可以选择离线导入数据集,在 GitHub 上下载 Fashion-MNIST,下载地址为:https://github.com/zalandoresearch/fashion-mnist。在当前工作目录下创建 \FashionMNIST 文件夹,并在其中创建\raw 和\processed 文件夹,将下载好的数据集放入\raw 下,再使用上述代码导入即可,如图 2-5 所示。

为丰富图像训练集,更好地提取图像特征,提高模型泛化能力,防止模型过拟合,提升模型的健壮性,需要对训练集数据进行增强。常用的方式有旋转图像、剪切图像、改变图像色差、扭曲图像特征、改变图像尺寸大小、增强图像噪音等。在 PyTorch 中,tansforms 类提供了 20 余种图像处理的方法,可用于数据增强。

```
Using downloaded and verified file: ./mnist/FashionMNIST\raw\train-images-idx3-
ubyte.gz
Extracting ./mnist/FashionMNIST\raw\train-images-idx3-ubyte.gz to ./mnist/
FashionMNIST\raw
Using downloaded and verified file: ./mnist/FashionMNIST\raw\train-labels-idx1-
ubyte.gz
Extracting ./mnist/FashionMNIST\raw\train-labels-idx1-ubyte.gz to ./mnist/
FashionMNIST\raw
Using downloaded and verified file: ./mnist/FashionMNIST\raw\t10k-images-idx3-
ubyte.gz
Extracting ./mnist/FashionMNIST\raw\t10k-images-idx3-ubyte.gz to ./mnist/
FashionMNIST\raw
Using downloaded and verified file: ./mnist/FashionMNIST\raw\t10k-labels-idx1-
ubyte.gz
Extracting ./mnist/FashionMNIST\raw\t10k-labels-idx1-ubyte.gz to ./mnist/
FashionMNIST\raw
Processing...
Done!
```

图 2-5 读取代码成功示意图

对训练集进行随机灰度变换和随机水平翻转处理,根据模型输入要求重置图像大小,并转换成 PyTorch 需要的数据格式 Tensor 张量。

```
#训练集预处理
transform = transforms.Compose([
    transforms.Resize((56,56)),
    transforms.RandomHorizontalFlip(),
    transforms.RandomGrayscale(),
    transforms.ToTensor()])
#对测试集则不需要进行数据增强,重置图像大小后转为张量即可
#测试集预处理
transform1 = transforms.Compose([
    transforms.Resize((56,56)),
    transforms.ToTensor()])
```

2.3.2 模型创建与编译

数据加载进模型之后,需要定义模型结构并优化损失函数。

1. 定义模型结构

原 VGG-16 模型要求输入 $224 \times 224 \times 3$ 的图片,限于 GPU 的计算能力,选择将 $28 \times 28 \times 1$ 的数据集图片大小重置为 $56 \times 56 \times 1$,由此计算出进入第一个全连接层的图像尺寸为 $7 \times 7 \times 256$;最后一个全连接层输出值设为类别数量 10。按设计好的参数定义模型结构,代码如下:

```
class Net(nn.Module):
    #定义模型结构
    def __init__(self):
        super(Net,self).__init__()
```

```python
        self.conv11 = nn.Conv2d(1,64,3,padding = 1)
        self.conv12 = nn.Conv2d(64,64,3,padding = 1)
        self.pool1 = nn.MaxPool2d(2, 2)
        self.conv21 = nn.Conv2d(64,128,3,padding = 1)
        self.conv22 = nn.Conv2d(128,128,3,padding = 1)
        self.pool2 = nn.MaxPool2d(2, 2)
        self.conv31 = nn.Conv2d(128,256,3,padding = 1)
        self.conv32 = nn.Conv2d(256, 256,3,padding = 1)
        self.pool3 = nn.MaxPool2d(2, 2)
        self.fc6 = nn.Linear(256 * 7 * 7,1024)
        self.drop1 = nn.Dropout2d()
        self.fc7 = nn.Linear(1024,1024)
        self.drop2 = nn.Dropout2d()
        self.fc8 = nn.Linear(1024,10)
    # 前向传播和损失函数
    def forward(self,x):
        x = F.relu(self.conv11(x))
        x = F.relu(self.conv12(x))
        x = self.pool1(x)
        x = F.relu(self.conv21(x))
        x = F.relu(self.conv22(x))
        x = self.pool2(x)
        x = F.relu(self.conv31(x))
        x = F.relu(self.conv32(x))
        x = self.pool3(x)
        x = x.view( -1,256 * 7 * 7)
        x = F.relu(self.fc6(x))
        x = self.drop1(x)
        x = F.relu(self.fc7(x))
        x = self.drop2(x)
        x = self.fc8(x)
        return x
```

2. 优化损失函数

为了评估实际情况和预测情况的差距,引入相对熵来描述这一差距。本模型是多类别的分类问题,因此选用经典的交叉熵作为损失函数,代码如下:

```python
# 损失函数
loss = nn.CrossEntropyLoss()
```

为了进行参数的调整和更新,以达到损失函数的最小化或最大化,使模型产生更好更快的效果,需要选择训练策略。本模型选择 Adam 优化算法,它具有能计算每个参数的自适应学习率的特点。因此,可以设置学习率为默认值 0.001。

```python
# 优化算法
optimizer = optim.Adam(net.parameters(), lr = 0.001)
```

2.3.3 模型训练及保存

定义模型结构及损失函数后,需要对模型进行训练,使其具有服装图像分类的能力。经过训练的模型需要测试,以评估模型的训练效果。根据训练和测试的效果进行参数的调整后,保存模型。

1. 模型训练

训练前加载参数及相关代码如下:

```
#初始化训练参数
initepoch = 0
loss_list = []
acc_list = []
#优化算法
optimizer = optim.Adam(net.parameters(), lr = 0.001)
if os.path.exists(path) is not True:
    #损失函数
    loss = nn.CrossEntropyLoss()
else:
    #加载训练参数
    checkpoint = torch.load(path)
    net.load_state_dict(checkpoint['model_state_dict'])
    optimizer.load_state_dict(checkpoint['optimizer_state_dict'])
    initepoch = checkpoint['epoch']
    loss = checkpoint['loss']
    loss_list = checkpoint['loss_list']
acc_list = checkpoint['acc_list']
```

设定训练集 batch_size=50,则每训练 50 张图片后进行一次迭代,根据损失函数前向传播,完成参数更新。

```
#按规定的周期数进行训练
for epoch in range(initepoch, epochs):
    #重置参数
    timestart = time.time()
    running_loss = 0.0
    correct = 0
    for data in trainloader:
        #输入数据集数据
        inputs, labels = data
        inputs, labels = inputs.to(device), labels.to(device)
        #梯度置零
        optimizer.zero_grad()
        # 向前传播
        outputs = net(inputs)
```

```python
#计算损失
l = loss(outputs, labels)
running_loss += l.item()
#计算正确率
_, predicted = torch.max(outputs.data, 1)
correct += (predicted == labels).sum().item()
#向后传播
l.backward()
#梯度更新
optimizer.step()
```

训练过程中,输出当前周期下的训练准确度、损失值,可以观察训练过程的效果,如图 2-6 所示。

```
epoch 22 cost 132.478006 sec
[epoch 23] loss: 0.002125
Accuracy of the network on the 60000 train images: 96.250 %
epoch 23 cost 140.675475 sec
[epoch 24] loss: 0.002055
Accuracy of the network on the 60000 train images: 96.343 %
epoch 24 cost 142.117363 sec
[epoch 25] loss: 0.001997
Accuracy of the network on the 60000 train images: 96.497 %
epoch 25 cost 127.151308 sec
[epoch 26] loss: 0.002103
Accuracy of the network on the 60000 train images: 96.240 %
epoch 26 cost 126.909165 sec
[epoch 27] loss: 0.001929
Accuracy of the network on the 60000 train images: 96.625 %
epoch 27 cost 127.073644 sec
[epoch 28] loss: 0.002013
Accuracy of the network on the 60000 train images: 96.540 %
epoch 28 cost 127.063076 sec
[epoch 29] loss: 0.001847
Accuracy of the network on the 60000 train images: 96.757 %
epoch 29 cost 128.047564 sec
Finished Training
```

图 2-6　训练效果

为使训练效果更加直观,可以借助画图工具对训练过程可视化,模型将在每一个训练周期结束后保存损失值和准确度。

```python
import matplotlib.pyplot as plt
#显示中文字体
plt.rcParams['font.sans-serif'] = ['SimHei']
plt.rcParams['axes.unicode_minus'] = False
x1 = range(0, epochs)
x2 = range(0, epochs)
y1 = acc_list
y2 = loss_list
plt.subplot(2, 1, 1)
plt.plot(x1, y1)
plt.title('训练准确率')
plt.xlabel('训练周期数')
plt.ylabel('训练准确率(%)')
```

```
plt.subplot(2, 1, 2)
plt.plot(x2, y2)
plt.title('训练损失')
plt.xlabel('训练周期数')
plt.ylabel('训练损失')
plt.show()
```

2. 模型保存

在模型训练过程中,同时保存模型当前已训练周期数、权重、损失函数、优化算法,便于在训练终止后从当前进度恢复训练,并同时保存每个周期的损失值和正确率,以方便数据可视化。

```
#保存训练参数的格式
path = 'weights.tar'
#保存当前周期的损失值和精确度
if epoch <= len(loss_list) - 1:
    loss_list[epoch] = running_loss / len(trainset)
else:
    loss_list.append(running_loss / len(trainset))
if epoch <= len(acc_list) - 1:
    acc_list[epoch] = 100.0 * correct / len(trainset)
else:
    acc_list.append(100.0 * correct / len(trainset))
#保存训练参数
torch.save({'epoch':epoch,
            'model_state_dict':net.state_dict(),
            'optimizer_state_dict':optimizer.state_dict(),
            'loss':loss,
            'loss_list':loss_list,
            'acc_list':acc_list
            },path)
```

2.3.4 模型生成

该应用分两部分:一是网页端交互功能,用户可以上传需要分类的图片并查看分类结果;二是图片预处理,将图片转换为 PyTorch 能够处理的格式并输入模型中,获取图片分类结果。

1. 上传图片

获得用户上传的图片并输入至模型获得对应的分类,将标签和图片写入数据库中。为了提高数据库的稳定性,采用 bulk_create() 方法批量写入数据,避免每写入一条数据就需要调用和关闭数据库的烦琐操作。

```
#上传图片
def upload_accept(request):
```

```python
        upload_list = []
        photos = request.FILES.getlist('image')
        for photo in photos:
            #获取类别
            lable = imageclassify(photo)
            picture = Image(photo = photo,lable = lable)
            upload_list.append(picture)
    #批量写入数据库
    Image.objects.bulk_create(upload_list)
    del upload_list
    return render(request, 'form_file_upload.html')
```

2. 分类展示

在每一个类别的页面下,由 filter()方法从数据库中选取对应分类下的所有图片,并将选出的图片传递到前端页面进行显示。

```python
#首页,默认显示 T-shirt 类
def index(request):
    #根据标签值筛选分类
    Is_Tshirt = Image.objects.filter(lable = "T-shirt")
    #传回图片对象到前端页面并跳转
    return render(request,'index.html',{'img':Is_Tshirt})
```

3. 图像预处理

使用 PIL 库作为打开图片的方式,类型为 Image,并将图片转为灰度图像。

```python
#加载图片
from PIL import Image as Imggg
item = Imggg.open(item).convert('L')
```

为使图片符合输入模型的数据格式,对图像进行预处理。模型输入图像大小要求为 56×56,因此,修改用户输入的图像分辨率为 56×56;为了对图像进行数据化处理,将 PIL Image 对象转换为 numpy 类型;原始数据集图片为黑底色白图案,将输入图片进行黑白色反转;PyTorch 要求的数据输入格式为 $[b,c,h,w]$,需要扩展 numpy 的维数,再转换成 Tensor 张量。

```python
#图像预处理
def process_image(img):
    #重置分辨率
    img1 = img.resize((56,56))
    # image to numpy
    img1 = np.array(img1).astype(np.float32)
    #黑白反转
    img1 = (255-img1)/255
    #扩展维数,输出的 img 格式为[b,c,h,w],为[1,1,56,56]
```

```
            img1 = np.expand_dims(img1,0)
            img1 = np.expand_dims(img1,0)
            #numpy to tensor
            img2 = torch.from_numpy(img1)
    return img2
```

4. 模型调用与导入

首先,进行实例化;其次,使用 load()方法加载模型的权重;最后,使用 load_state_dict()方法将参数加载到网络上。

```
#加载模型
model = Net()
checkpoint = torch.load('./imageupload/weights.tar'
map_location = 'cpu')
model.load_state_dict(checkpoint['model_state_dict'])
将经过预处理的图像数据输入模型,取概率最大的一类作为分类结果.
#输入至模型
score = model(process_image(item))
#找到最大概率对应的索引号,该图片即索引号对应的类别
_,lable = torch.max(score,1)
#将模型预测保存为 predict_batch.py 作为 API,在视图函数调用该 API
from imageupload import predict_batch
#对上传的图片进行分类识别
def imageclassify(photo):
    lable = predict_batch.get_results(photo)
    return lable
```

5. 相关代码

本部分包括布局文件、模型预测文件、视图函数、路由配置、静态文件、数据库的相关代码。

1) 布局文件

布局文件相关代码如下:

```html
//分类展示页面(以 T-shirt 类展示为例,其他类与此类似)
<!DOCTYPE html>
<html lang = "en">
<head>
    <meta charset = "utf-8">
    <meta content = "IE = edge" http-equiv = "X-UA-Compatible">
    <meta content = "width = device-width, initial-scale = 1" name = "viewport">
    <meta content = "" name = "description">
    <meta content = "" name = "author">
    <title>fashion-classification</title>
    <link href = "../static/assets/bootstrap/css/bootstrap.min.css" rel = "stylesheet">
    <!--[if lt IE 9]>
```

```html
<script src="https://oss.maxcdn.com/html5shiv/3.7.2/html5shiv.min.js"></script>
<script src="https://oss.maxcdn.com/respond/1.4.2/respond.min.js"></script>
<![endif]-->
<link href='https://fonts.googleapis.com/css?family=Hind:400,500,700|Anonymous+Pro:400,400italic,700,700italic&subset=latin,latin-ext' rel='stylesheet' type='text/css'>
<link href="../static/assets/font-awesome/css/font-awesome.min.css" rel="stylesheet">
<link href="../static/assets/pe-icons/css/pe-icon-7-stroke.css" rel="stylesheet">
<link href="../static/assets/css/animate.css" rel="stylesheet">
<link href="../static/assets/css/plugins.css" rel="stylesheet">
<link href="../static/style.css" rel="stylesheet">
</head>
<body class="top-navigation pushy-right-side borderless">
<!-- Site Overlay -->
<div class="nav-wrapper smoothie">
    <div class="nav-inner smoothie">
        <div class="container-fluid special-max-width">
            <div class="row">
                <div class="col-sm-3 col-xs-6">
                    <a class="logo-dark smoothie" href="{% url 'index' %}"><img alt="" class="logo img-responsive" src="../static/assets/images/logo.png"></a>
                </div>
                <div class="col-sm-9 col-xs-6">
                    <div class="collapse navbar-collapse" id="navbar-collapse-1">
                        <ul class="nav navbar-nav navbar-right">
                            <li>
                                <a class="button" href="{% url 'upload_accept' %}"><span class="nav-label">上传商品</span></a>
                            </li>
                        </ul>
                    </div>
                </div>
            </div>
        </div>
    </div>
</div>
<div id="master-wrapper">
    <div class="nav-wrapper smoothie">
        <div class="nav-inner smoothie">
            <div class="container-fluid special-max-width"><div class="row"></div>
            </div>
        </div>
    </div>
    <section id="our-courses" class="white-wrapper">
        <div class="section-inner">
            <div class="container">
                <div class="row mb60 text-left">
                    <div class="col-sm-12">
```

```html
                    <h3 class="section-title">fashion-classification</h3>
                </div>
            </div>
        </div>
        <div class="container">
            <div class="row">
                <div class="masonry-portfolio">
                    <div class="col-sm-12 mb60">
                        <ul class="menu list-unstyled clearfix list-inline wow fadeIn text-left" data-wow-delay="0.2s" style="display:block">
                            <li><a class="btn btn-primary btn-transparent active" href="{% url 'index' %}">T-shirt</a></li>
                            <li><a class="btn btn-primary btn-transparent" href="{% url 'trouser' %}">trouser</a></li>
                            <li><a class="btn btn-primary btn-transparent" href="{% url 'pullover' %}">pullover</a></li>
                            <li><a class="btn btn-primary btn-transparent" href="{% url 'dress' %}">dress</a></li>
                            <li><a class="btn btn-primary btn-transparent" href="{% url 'coat' %}">coat</a></li>
                            <li><a class="btn btn-primary btn-transparent" href="{% url 'sandal' %}">sandal</a></li>
                            <li><a class="btn btn-primary btn-transparent" href="{% url 'shirt' %}">shirt</a></li>
                            <li><a class="btn btn-primary btn-transparent" href="{% url 'sneaker' %}">sneaker</a></li>
                            <li><a class="btn btn-primary btn-transparent" href="{% url 'bag' %}">bag</a></li>
                            <li><a class="btn btn-primary btn-transparent" href="{% url 'ankleboot' %}">ankleboot</a></li?>
                        </ul>
                    </div>
                    <div class="col-sm-12">
                        <div class="masonry-portfolio-items row">
                            <div class="row">
                                {% for i in img %}
                                <div class="col-sm-4 masonry-portfolio-item mb30 apps">
                                    <div class="hover-effect smoothie">
                                        <a href="#" class="smoothie">
                                            <img src="{{ MEDIA_URL }}{{ i.photo }}" alt="Image" class="img-responsive smoothie"></a>
                                    </div>
                                </div>
                                {% endfor %}
                                <div class="col-sm-4 masonry-portfolio-item mb30 design"></div>
                            </div>
```

```html
                                    </div>
                                </div>
                            </div>
                        </div>
                    </div>
                </div>
            </section>
            <a href="#" id="back-to-top"><i class="fa fa-angle-up"></i></a>
                <div class="top-frame"></div>
                    <div class="bottom-frame"></div>
        </div>
        <script src="../static/assets/js/jquery.min.js"></script>
        <script src="../static/assets/bootstrap/js/bootstrap.min.js"></script>
        <script src="../static/assets/js/owl-carousel.js"></script>
        <script src="../static/assets/js/plugins.js"></script>
        <script src="../static/assets/js/init.js"></script>
</body>
</html>
```

//上传图片页面
```html
{% load static %}
<!DOCTYPE html>
<html>
<head>
    <meta charset="utf-8">
    <meta name="viewport" content="width=device-width, initial-scale=1.0">
    <title>上传商品-fashion-classification</title>
    <meta name="keywords" content="H">
    <meta name="description" content="H">
    <link href="../static/css/bootstrap.min.css?v=3.3.6" rel="stylesheet">
    <link href="../static/css/font-awesome.css?v=4.4.0" rel="stylesheet">
    <link href="../static/css/animate.css" rel="stylesheet">
    <link href="../static/css/plugins/dropzone/basic.css" rel="stylesheet">
    <link href="../static/css/plugins/dropzone/dropzone.css" rel="stylesheet">
    <link href="../static/css/style.css?v=4.1.0" rel="stylesheet">
</head>
<body class="gray-bg">
    <div class="wrapper wrapper-content animated fadeIn">
        <div class="row">
            <div class="col-sm-12">
                <div class="ibox float-e-margins">
                    <div class="ibox-title">
                        <h5>上传商品</h5>
                    </div>
                    <div class="ibox-content">
                        <form id="filedropzone" method="post" action="/upload_accept/" class="dropzone dz-clickable">
                            {% csrf_token %}
```

```html
                    <div class="dz-default dz-message">
                    </div>
                </form>
            </div>
        </div>
    </div>
</div>
<!-- 全局 js -->
<script src="../static/js/jquery.min.js?v=2.1.4"></script>
<script src="../static/js/bootstrap.min.js?v=3.3.6"></script>
<!-- 自定义 js -->
<script src="../static/js/content.js?v=1.0.0"></script>
<!-- DROPZONE -->
<script src="../static/js/plugins/dropzone/dropzone.js"></script>
```
```javascript
<script>
$(document).ready(function () {
    $("#filedropzone").dropzone({
        maxFiles: 10,
        maxFilesize: 4096,
        acceptedFiles: ".jpg,.jpeg,.png,.bmp",
        paramName: "image",
        dictFileTooBig:"文件过大,上传文件最大支持.",
        dictResponseError: '文件上传失败!',
        dictDefaultMessage: "拖入需要上传的文件",
        init: function () {
            var myDropzone = this, submitButton = document.querySelector("#qr"),
                cancelButton = document.querySelector("#cancel");
            myDropzone.on('addedfile', function (file) {
                //添加上传文件的过程,可再次弹出弹框,添加上传文件的信息
            });
            myDropzone.on('sending', function (data, xhr, formData) {
                //向后台发送该文件的参数
                formData.append('watermark', jQuery('#info').val());
            });
            myDropzone.on('success', function (files, response) {
                //文件上传成功之后的操作
            });
            myDropzone.on('error', function (files, response) {
                //文件上传失败后的操作
            });
            myDropzone.on('totaluploadprogress', function (progress, byte, bytes) {
                //progress 为进度百分比
                $("#pro").text("上传进度:" + parseInt(progress) + "%");
                //计算上传速度和剩余时间
                var mm = 0;
                var byte = 0;
```

```
                    var tt = setInterval(function () {
                        mm++;
                        var byte2 = bytes;
                        var remain;
                        var speed;
                        var byteKb = byte / 1024;
                        var bytesKb = bytes / 1024;
                        var byteMb = byte / 1024 / 1024;
                        var bytesMb = bytes / 1024 / 1024;
                        if (byteKb <= 1024) {
                            speed = (parseFloat(byte2 - byte) / (1024) / mm).toFixed(2) + " KB/s";
                            remain = (byteKb - bytesKb) / parseFloat(speed);
                        } else {
                            speed = (parseFloat(byte2 - byte) / (1024 * 1024) / mm).toFixed(2) + " MB/s";
                            remain = (byteMb - bytesMb) / parseFloat(speed);
                        }
                        $("#dropz #speed").text("上传速率:" + speed);
                        $("#dropz #time").text("剩余时间" + arrive_timer_format(parseInt(remain)));
                        if (bytes >= byte) {
                            clearInterval(tt);
                            if (byteKb <= 1024) {
                                $("#dropz #speed").text("上传速率:0 KB/s");
                            } else {
                                $("#dropz #speed").text("上传速率:0 MB/s");
                            }
                            $("#dropz #time").text("剩余时间:0:00:00");
                        }
                    }, 1000);
                });
                submitButton.addEventListener('click', function () {
                    //单击上传文件
                    myDropzone.processQueue();
                });
                cancelButton.addEventListener('click', function () {
                    //取消上传
                    myDropzone.removeAllFiles();
                });
            }
        });
        //剩余时间格式转换
        function arrive_timer_format(s) {
            var t;
            if (s > -1) {
                var hour = Math.floor(s / 3600);
                var min = Math.floor(s / 60) % 60;
```

```javascript
            var sec = s % 60;
            var day = parseInt(hour / 24);
            if (day > 0) {
                hour = hour - 24 * day;
                t = day + "day " + hour + ":";
            }
            else t = hour + ":";
            if (min < 10) {
                t += "0";
            }
            t += min + ":";
            if (sec < 10) {
                t += "0";
            }
            t += sec;
        }
        return t;
    }
})
</script>
</body>
</html>
```

2）模型预测文件

模型预测文件相关代码如下：

```python
import torch                                    # 导入模块
import torch.nn as nn
from imageupload import model
from imageupload.model import Net
import numpy as np
from PIL import Image as Imggg
classes = ('T-shirt', 'Trouser', 'Pullover', 'Dress',
           'Coat', 'Sandal', 'Shirt', 'Sneaker', 'Bag', 'Ankle boot')
# 预处理图片
def process_image(img):
    # 重置分辨率
    img1 = img.resize((56,56))
    # image to numpy
    img1 = np.array(img1).astype(np.float32)
    # 黑白反转
    img1 = (255 - img1)/255
    # 扩展维数,输出的img格式为[b,c,h,w],为[1,1,56,56]
    img1 = np.expand_dims(img1,0)
    img1 = np.expand_dims(img1,0)
    # numpy to tensor
    img2 = torch.from_numpy(img1)
```

```python
        return img2
def get_results(item):
    #加载模型
    model = Net()
    checkpoint = torch.load('./imageupload/weights.tar',
                            map_location = 'cpu')
    model.load_state_dict(checkpoint['model_state_dict'])
    #加载图片
    item = Imggg.open(item).convert('L')
    #输入至模型
    score = model(process_image(item))
    #找到最大概率对应的索引号,该图片即索引号对应的类别
    _,lable = torch.max(score,1)
    return classes[lable]
```

3）视图函数

视图函数相关代码如下：

```python
from django.shortcuts import render
from .models import Image
from imageupload import predict_batch
#默认显示 T-shirt 类
def index(request):
    #根据标签值筛选分类
    Is_Tshirt = Image.objects.filter(lable = "T-shirt")
    #传回图片对象到前端页面并跳转
    return render(request,'index.html',{'img':Is_Tshirt})
def trouser(request):                      #裤子
    Is_Trouser = Image.objects.filter(lable = "Trouser")
    return render(request,'trouser.html',{'img':Is_Trouser})
def pullover(request):                     #套衫
    Is_Pullover = Image.objects.filter(lable = "Pullover")
    return render(request,'pullover.html',{'img':Is_Pullover})
def dress(request):                        #连衣裙
    Is_Dress = Image.objects.filter(lable = "Dress")
    return render(request,'dress.html',{'img':Is_Dress})
def coat(request):                         #大衣
    Is_Coat = Image.objects.filter(lable = "Coat")
    return render(request,'coat.html',{'img':Is_Coat})
def sandal(request):                       #凉鞋
    Is_Sandal = Image.objects.filter(lable = "Sandal")
    return render(request,'sandal.html',{'img':Is_Sandal})
def shirt(request):                        #衬衫
    Is_Shirt = Image.objects.filter(lable = "Shirt")
    return render(request,'shirt.html',{'img':Is_Shirt})
def sneaker(request):                      #运动鞋
    Is_Sneaker = Image.objects.filter(lable = "Sneaker")
```

```python
        return render(request,'sneaker.html',{'img':Is_Sneaker})
def bag(request):                                    #箱包
    Is_Bag = Image.objects.filter(lable = "Bag")
    return render(request,'bag.html',{'img':Is_Bag})
def ankleboot(request):                              #短靴
    Is_Ankleboot = Image.objects.filter(lable = "Ankle boot")
    return render(request,'ankleboot.html',{'img':Is_Ankleboot})
#上传图片
def upload_accept(request):
    upload_list = []
    photos = request.FILES.getlist('image')
    for photo in photos:
        #获取类别
        lable = imageclassify(photo)
        picture = Image(photo = photo,lable = lable)
        upload_list.append(picture)
    #批量写入数据库
    Image.objects.bulk_create(upload_list)
    del upload_list
    return render(request, 'form_file_upload.html')
#对上传的图片进行分类识别
def imageclassify(photo):
    lable = predict_batch.get_results(photo)
    return lable
```

4）路由配置

路由配置相关代码如下：

```python
from django.conf.urls import url                     #导入模块
from django.contrib import admin
from django.conf import settings
from django.conf.urls.static import static
from imageupload import views
urlpatterns = [                                      #链接模式
    url(r'^admin/', admin.site.urls),
    url(r'^index/', views.index, name = 'index'),
    url(r'^trouser/', views.trouser, name = 'trouser'),
    url(r'^pullover/', views.pullover, name = 'pullover'),
    url(r'^dress/', views.dress, name = 'dress'),
    url(r'^coat/', views.coat, name = 'coat'),
    url(r'^sandal/', views.sandal, name = 'sandal'),
    url(r'^shirt/', views.shirt, name = 'shirt'),
    url(r'^sneaker/', views.sneaker, name = 'sneaker'),
    url(r'^bag/', views.bag, name = 'bag'),
    url(r'^ankleboot/', views.ankleboot, name = 'ankleboot'),
    url(r'^upload_accept/',views.upload_accept,name = 'upload_accept'),
] + static(settings.MEDIA_URL, document_root = settings.MEDIA_ROOT)
```

5)静态文件

静态文件相关代码如下:

```python
import os
#在项目内部构建路径,如:os.path.join(BASE_DIR, ...)
BASE_DIR = os.path.dirname(os.path.dirname(os.path.abspath(__file__)))
#快速开发设置
#参见 https://docs.djangoproject.com/en/1.11/howto/deployment/checklist/
#安全秘钥
SECRET_KEY = 'fj_^gi4+v2g^38q*t0#u5c*l6lj#^6b2-zm_kk&nr5po*dus%('
DEBUG = True
ALLOWED_HOSTS = ['*',]
#定义应用
INSTALLED_APPS = [
    'django.contrib.admin',
    'django.contrib.auth',
    'django.contrib.contenttypes',
    'django.contrib.sessions',
    'django.contrib.messages',
    'django.contrib.staticfiles',
    #添加我的应用
    'imageupload'
]
MIDDLEWARE = [                                    #中间件
    'django.middleware.security.SecurityMiddleware',
    'django.contrib.sessions.middleware.SessionMiddleware',
    'django.middleware.common.CommonMiddleware',
    'django.middleware.csrf.CsrfViewMiddleware',
    'django.contrib.auth.middleware.AuthenticationMiddleware',
    'django.contrib.messages.middleware.MessageMiddleware',
    'django.middleware.clickjacking.XFrameOptionsMiddleware',
]
ROOT_URLCONF = 'imageclassify.urls'               #配置
TEMPLATES = [                                     #模板
    {
        'BACKEND': 'django.template.backends.django.DjangoTemplates',
        'DIRS': [os.path.join(BASE_DIR, 'imageupload/templates')],
        'APP_DIRS': True,
        'OPTIONS': {
            'context_processors': [
                'django.template.context_processors.debug',
                'django.template.context_processors.request',
                'django.contrib.auth.context_processors.auth',
                'django.contrib.messages.context_processors.messages',
                #添加 media_url
                'django.template.context_processors.media',
            ],
```

```
        },
    },
]
WSGI_APPLICATION = 'imageclassify.wsgi.application'
# 数据库
# 参见 https://docs.djangoproject.com/en/1.11/ref/settings/#databases
DATABASES = {
    'default': {
        'ENGINE': 'django.db.backends.sqlite3',
        'NAME': os.path.join(BASE_DIR, 'db.sqlite3'),
    }
}
# 密码验证,参见# https://docs.djangoproject.com/en/1.11/ref/settings/#auth-password-validators
AUTH_PASSWORD_VALIDATORS = [
    {
        'NAME': 'django.contrib.auth.password_validation.UserAttributeSimilarityValidator',
    },
    {
        'NAME': 'django.contrib.auth.password_validation.MinimumLengthValidator',
    },
    {
        'NAME': 'django.contrib.auth.password_validation.CommonPasswordValidator',
    },
    {
        'NAME': 'django.contrib.auth.password_validation.NumericPasswordValidator',
    },
]
# 国际化,参见 https://docs.djangoproject.com/en/1.11/topics/i18n/
LANGUAGE_CODE = 'en-us'
TIME_ZONE = 'UTC'
USE_I18N = True
USE_L10N = True
USE_TZ = True
# 静态文件(CSS, JavaScript, Images)
# 参见 https://docs.djangoproject.com/en/1.11/howto/static-files/
# 添加静态资源地址
STATIC_URL = '/static/'
STATICFILES_DIRS = [os.path.join(BASE_DIR, 'static')]
# 添加图片保存地址
MEDIA_URL = '/media/'
MEDIA_ROOT = os.path.join(BASE_DIR, 'media')
```

6)数据库

数据库相关代码如下:

```
from django.db import models
class Image(models.Model):
```

```
photo = models.ImageField(null = True, blank = True)
lable = models.CharField(default = '', null = True, max_length = 100)
def __str__(self):
    return self.photo.name
```

2.4 系统测试

本部分包括训练准确率、测试效果与模型应用。

2.4.1 训练准确率

假设模型训练周期数 epochs＝30，并将每个训练周期后的损失和准确率保存，绘制曲线图观察，如图 2-7 和图 2-8 所示。最终，模型训练准确率达到 96.757％。

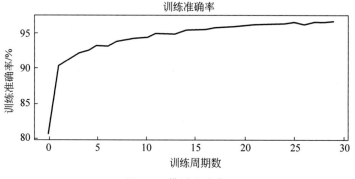

图 2-7　模型准确率

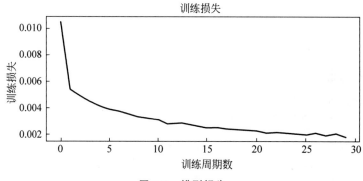

图 2-8　模型损失

2.4.2 测试效果

经过 30 个训练周期后,将测试集数据输入模型,准确率达到 92.280%,如图 2-9 所示。

```
Accuracy of the network on the 10000 test images: 92.280 %
Finished Testing
```

图 2-9 模型测试效果

2.4.3 模型应用

本部分包括启动 Web 服务器、应用使用说明和测试结果示例。

1. 启动 Web 服务器

在 Anaconda Prompt 中,启用 PyTorch 环境,输入命令:

```
activate pytorch
```

转到 Django 工程目录下,运行代码,输入命令:

```
python manage.py runserver 8000
```

服务器启动成功,如图 2-10 所示。

```
(pytorch) E:\lsj\fashion\fashion-classify>python manage.py runserver 8000
Watching for file changes with StatReloader
Performing system checks...

System check identified no issues (0 silenced).
April 19, 2020 - 15:53:24
Django version 3.0.3, using settings 'imageclassify.settings'
Starting development server at http://127.0.0.1:8000/
Quit the server with CTRL-BREAK.
```

图 2-10 服务器启动成功示意图

在浏览器中访问 http://127.0.0.1:8000+页面 URL 即可使用本应用。

2. 应用使用说明

在浏览器中访问 http://127.0.0.1:8000/index,进入首页。左上角为网页图标,单击可刷新页面;右上方有"上传商品"按钮,单击进入上传页面。首页默认显示 T-shirt 类标签下上传并分类过的图片,另有 9 个标签可供选择,单击可查看不同标签的商品图片,如图 2-11 所示。

当鼠标在页面移动时,停留处的商品图片会放大显示,如图 2-12 所示。

下拉页面时,右下角出现向上箭头的按钮,单击可回到顶部,如图 2-13 所示。

单击"上传商品"按钮进入上传图片页面,如图 2-14 所示。

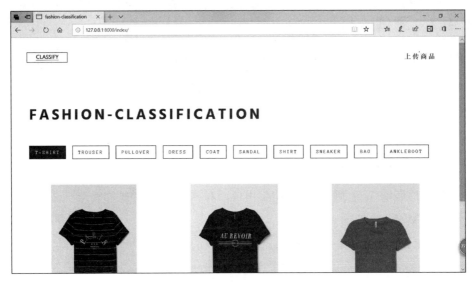

图 2-11 应用首页

图 2-12 放大对比图(鼠标停留在中间图片上方)

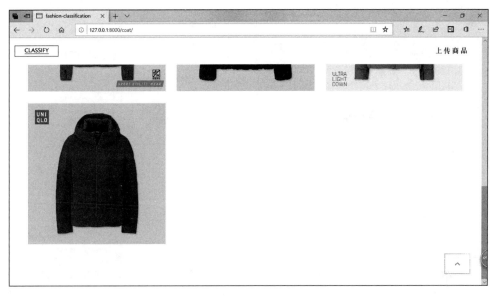

图 2-13　回到顶部按钮

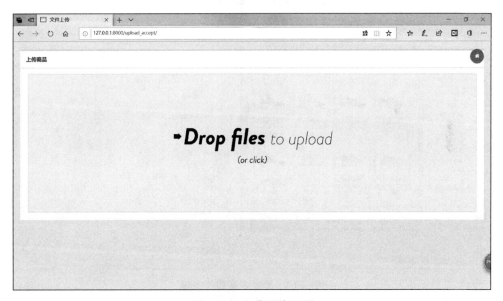

图 2-14　上传图片页面

页面支持单击和拖拽两种方式上传图片，支持格式为 jpg、jpeg、png、bmp 的图片，上传完成后在本页面显示缩略图片、图片大小以及上传成功标志（显示为绿色对钩），鼠标停留在图片上时显示图片名称，如图 2-15 和图 2-16 所示。

可以同时上传单张或者多张图片，如图 2-17 所示。

图 2-15 鼠标停留在 bag1.jpg 上方

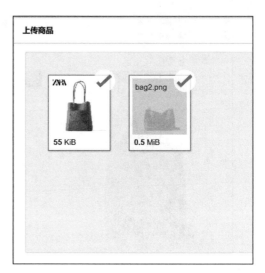
图 2-16 鼠标停留在 bag2.png 上方

图 2-17 选择多张图片上传

每次最多可上传 10 张图片，超出 10 张则会在第 11 张图片上显示"You can not upload any more files"，如图 2-18 所示。

上传页面右上方图标在鼠标停留时显示"返回商品页"，单击即可回到首页，如图 2-19 所示。

图 2-18　每次最多可上传 10 张

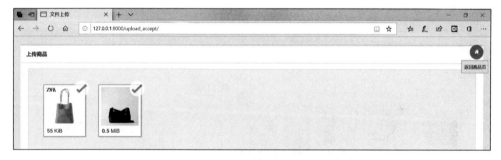

图 2-19　返回首页按钮

如果上传的图片格式不符合要求,则缩略图片位置显示为灰色,显示文件大小、文件名称以及上传失败标志(显示为红色叉号),鼠标停留在对应图片上时提示"You can't upload files of this type",如图 2-20 所示。

3．测试结果示例

测试时选择上传 3 张不同种类的商品图片,分别为包、凉鞋和连衣裙,如图 2-21 所示。

返回首页,单击进入对应的 COAT、SANDAL 和 DRESS 标签页,可以看到上传的 3 张图片已成功分类至相应的标签页面下,如图 2-22～图 2-24 所示。

图 2-20　上传图片失败

图 2-21　上传结果示例

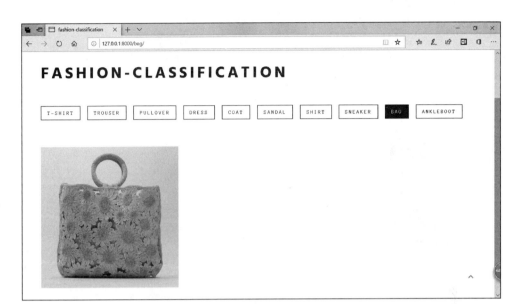

图 2-22 分类结果(BAG)示例

图 2-23 分类结果(SANDAL)示例

图 2-24　分类结果（DRESS）示例

项目 3 检索式模型聊天机器人

PROJECT 3

本项目基于 TF-IDF(Term Frequency-Inverse Document Frequency)检索的召回模型和 CNN 的精排模型构建聊天机器人,实现日常对话、情感抚慰等功能。

3.1 总体设计

本部分包括系统整体结构和系统流程。

3.1.1 系统整体结构

系统整体结构如图 3-1 所示。

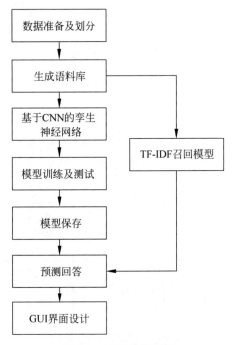

图 3-1 系统整体结构

3.1.2 系统流程

系统流程如图 3-2 所示，孪生神经网络结构如图 3-3 所示。

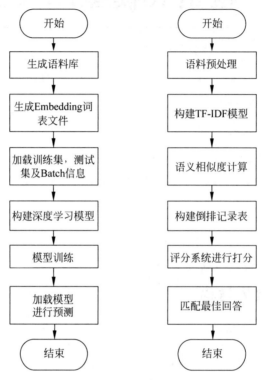

图 3-2 系统流程

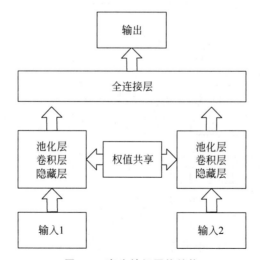

图 3-3 孪生神经网络结构

3.2 运行环境

本部分包括 Python 环境、TensorFlow 环境和 Python 包依赖关系。

3.2.1 Python 环境

需要 Python 3.6 及以上配置,在 Windows 环境下推荐下载 Anaconda 完成 Python 所需配置。

Anaconda 是开源的 Python 发行版本,包含 conda、Python 等 180 多个科学包及其依赖项。下载文件比较大,如果只需要某些包,或者需要节省带宽、存储空间,也可以使用 Miniconda 发行版(仅包含 conda 和 Python)。下载地址为:https://www.anaconda.com/。也可以下载虚拟机在 Linux 环境下运行代码。添加环境变量:单击鼠标右键,依次选择属性、高级系统设置、环境变量、新建系统变量,添加安装 Anaconda 的路径即可。

用 CMD 命令行测试,输入:

```
conda list
```

显示 Anaconda 所存的路径以及文件。Anaconda 自带 Anaconda Prompt,也可以用于其他类似安装包。

3.2.2 TensorFlow 环境

在 Anaconda 中配置 TensorFlow 环境的步骤(针对 Windows 系统)如下:
打开 Anaconda Prompt,使用语句查询 Python 版本,输入清华仓库镜像,命令为

```
conda config --add channels https://mirrors.tuna.tsinghua.edu.cn/anaconda/pkgs/free/
conda config -set show_channel_urls yes
```

创建 Python 3.x 环境,名称为 TensorFlow,此时 Python 版本和后面 TensorFlow 的版本有匹配问题,此步选择 Python 3.x,输入命令:

```
conda create -n tensorflow python=3.x
```

有需要确认的地方,都输入 y。
在 Anaconda Prompt 中激活 TensorFlow 环境,输入命令:

```
activate tensorflow
```

安装 CPU 版本的 TensorFlow,输入命令:

```
pip install -upgrade --ignore-installed tensorflow
```

安装完毕。

在 Anaconda Prompt 中激活 TensorFlow 环境，输入命令：

```
activate tensorflow
```

NumPy 是存储和处理大型矩阵的科学计算包，比 Python 自身的嵌套列表结构高效。安装命令：

```
pip install numpy
```

Matplotlib 是 Python 最著名的绘图表，提供了一整套和 MATLAB 相似的命令 API，适合交互式制图。安装命令：

```
pip install matplotlib
```

jieba 是优秀的第三方中文分词库，中文文本需要通过分词获得单个词语。安装命令：

```
pip install jieba
```

Pandas 是基于 NumPy 的一种工具，该工具是为了解决数据分析任务而创建的。Pandas 纳入了大量库和一些标准的数据模型，提供了高效操作大型数据集所需的工具。安装命令：

```
pip install pandas
```

tqdm 是快速、可扩展的 Python 进度条，在 Python 长循环中添加一个进度提示信息，用户只需封装任意的迭代器 tqdm(iterator)。安装命令：

```
pip install tqdm
```

nltk 模块中包含大量的语料库，方便完成自然语言处理的任务，包括分词、词性标注、命名实体识别及句法分析。安装命令：

```
pip install nltk
```

gensim 是开源的第三方 Python 工具包，用于从原始、非结构化的文本中无监督学习到文本隐层的主题向量表达。安装命令：

```
pip install gensim
```

PyQt 是创建 GUI 应用程序的工具包，它是 Python 编程语言和 Qt 库的成功融合。安装命令：

```
pip install pyqt5
```

3.3 模块实现

本项目包括 4 个模块：数据预处理、模型创建与编译、模型训练及保存、模型生成。下面分别给出各模块的功能介绍及相关代码。

3.3.1 数据预处理

本部分包括基础数据、数据增强和数据预处理。

1. 基础数据

数据来源于 GitHub 开源语料集，下载地址为：https://github.com/codemayq。

该库对目前市面上已有的开源中文聊天语料进行搜集和系统化整理，包括 chatterbot、豆瓣多轮、PTT 八卦、青云、电视剧对白、贴吧论坛回帖、小黄鸡、微博语料共 8 个公开闲聊常用语料和短信，并对其进行统一化规整和处理，以便于使用。

将解压后的 raw_chat_corpus 文件夹放到当前目录下，执行 python main.py，每个来源的语料分别生成一个独立的 .tsv 文件，放在新生成的 clean_chat_corpus 文件夹下。

2. 数据增强

数据增强一方面可以增加训练数据，提升模型的泛化能力；另一方面可增加噪声数据，增强模型的健壮性。本项目使用同义词替换、随机插入、随机交换、随机删除等数据增强操作。

```python
def synonym_replacement(words, n):                    #同义词替换
    new_words = words.copy()
    random_word_list = list(set([word for word in words if word not in stop_words]))
    random.shuffle(random_word_list)
    num_replaced = 0
    for random_word in random_word_list:
        synonyms = get_synonyms(random_word)          #从词林中选择同义词进行替换
        if len(synonyms) >= 1:
            synonym = random.choice(list(synonyms))
            new_words = [synonym if word == random_word else word for word in new_words]
            num_replaced += 1
        if num_replaced >= n:                         #最多替换n个词
            break
    sentence = ' '.join(new_words)
    new_words = sentence.split(' ')
    return new_words
def random_insertion(words, n):                       #随机插入
    new_words = words.copy()
    for _ in range(n):
        add_word(new_words)
    return new_words
def random_swap(words, n):                            #随机交换
    new_words = words.copy()
    for _ in range(n):
        new_words = swap_word(new_words)
    return new_words
def swap_word(new_words):                             #随机把句子里的两个单词交换n次
```

```python
            random_idx_1 = random.randint(0, len(new_words)-1)
            random_idx_2 = random_idx_1
            counter = 0
            while random_idx_2 == random_idx_1:
                random_idx_2 = random.randint(0, len(new_words)-1)
                counter += 1
                if counter > 3:
                    return new_words
            new_words[random_idx_1], new_words[random_idx_2] = new_words[random_idx_2], new_words[random_idx_1]
        return new_words
    def random_deletion(words, p):                    #随机删除
        #如果只有一个词,不用删除
        if len(words) == 1:
            return words
        #以概率 p 删除词
        new_words = []
        for word in words:
            r = random.uniform(0, 1)
            if r > p:
                new_words.append(word)
        #如果全部删除,返回随机词
        if len(new_words) == 0:
            rand_int = random.randint(0, len(words)-1)
            return [words[rand_int]]
        return new_words
```

3. 数据预处理

将文档中原始字符文本转换成 Gensim 模型所能理解的稀疏向量,进行分词处理,采用 Python 中文分词最常用的 jieba 工具,支持多种切分模式。

```python
    #采用非全模式
    def cut(self, sentence, stopword = True, cut_all = False):
        seg_list = jieba.cut(sentence, cut_all)          #对原始语料进行分词处理
        results = []
        for seg in seg_list:
            if stopword and seg in self.stopwords:       #去除停用词
                continue
            results.append(seg)
        return results
    #另一种粒度较细的 jieba 分词模式
    def cut_for_search(self, sentence, stopword = True):
        seg_list = jieba.cut_for_search(sentence)        #对原始语料进行分词处理
        results = []
        for seg in seg_list:
            if stopword and seg in self.stopwords:       #去除停用词
```

```
            continue
        results.append(seg)
    return results
```

对于一些特定的语境和特殊的词语,需要载入自定义词典,从而提高分词的准确率。

```
#载入自定义词典
def load_userdict(self, file_name):
    jieba.load_userdict(file_name)
```

停用词是指在信息检索中,为节省存储空间和提高搜索效率,在处理自然语言数据之前或之后自动过滤的某些字或词。它们没有明确意义,但出现的频率却很高,进行过滤和省略能够节省存储空间、提高搜索效率。

```
#读入停用词表
def read_in_stopword(self):
    file_obj = codecs.open(self.stopword_filepath, 'r', 'utf-8')
    while True:
        line = file_obj.readline()
        line = line.strip('\r\n')                    #去掉换行符
            if not line:
            break
        self.stopwords.add(line)
    file_obj.close()
```

使用 doc2bow()方法对每个不同单词的词频进行统计,将单词转换为编号,以稀疏向量的形式返回结果。

```
    #对句子分词
    def cut(self, seg):
        return seg.cut_for_search(self.origin_sentence)
    #获取切词后的词列表
    def get_cuted_sentence(self):
        return self.cuted_sentence
    #获取原句子
    def get_origin_sentence(self):
        return self.origin_sentence
    #设置该句子得分
    def set_score(self, score):
        self.score = score
#词袋表示方法
def sentence2vec(self, sentence):
        sentence = Sentence(sentence, self.seg)
        vec_bow = self.dictionary.doc2bow(sentence.get_cuted_sentence())
        return self.model[vec_bow]                    #返回稀疏向量形式
self.corpus_simple = [self.dictionary.doc2bow(text) for text in self.texts]   #生成语料
```

3.3.2 模型创建与编译

数据加载进模型之后,需要定义模型结构并优化损失函数。

1. 定义模型结构

在 TF-IDF 模型中定义的架构为:计算 TF-IDF 向量,通过倒排表的方式找到与当前输入类似的问题描述,针对候选问题进行余弦相似度计算。

```python
#初始化模型,将整个语料库转为 TF-IDF 表示方法,创建余弦相似度索引
#构建其他复杂模型前需要的简单模型
    def simple_model(self, min_frequency = 0):
        self.texts = self.get_cuted_sentences()
        #删除低频词
        frequency = defaultdict(int)                            #创建频率对象
        for text in self.texts:
            for token in text:
                frequency[token] += 1
        self.texts = [[token for token in text if frequency[token] > min_frequency] for text in self.texts]
        self.dictionary = corpora.Dictionary(self.texts)        #创建字典
        self.corpus_simple = [self.dictionary.doc2bow(text) for text in self.texts]
                                                                #生成语料

    #TF-IDF 模型
    def TfidfModel(self):
        self.simple_model()
        #转换模型
        self.model = models.TfidfModel(self.corpus_simple)
        self.corpus = self.model[self.corpus_simple]
        #创建相似度矩阵
        self.index = similarities.MatrixSimilarity(self.corpus)
        #对新输入的句子(作为对比的句子)进行预处理
    def sentence2vec(self, sentence):
        sentence = Sentence(sentence, self.seg)
        vec_bow = self.dictionary.doc2bow(sentence.get_cuted_sentence())
        return self.model[vec_bow]
    def bow2vec(self):
        vec = []
        length = max(self.dictionary) + 1
        for content in self.corpus:
            sentence_vectors = np.zeros(length)
            for co in content:
                sentence_vectors[co[0]] = co[1]
                #将句子出现的单词 TF-IDF 表示放入矩阵中
            vec.append(sentence_vectors)
        return vec
```

```python
# 计算最相似的句子
def similarity(self, sentence):
    sentence_vec = self.sentence2vec(sentence)
    sims = self.index[sentence_vec]
    sim = max(enumerate(sims), key = lambda item: item[1])
    index = sim[0]
    score = sim[1]
    sentence = self.sentences[index]
    sentence.set_score(score)
    return sentence                                                    # 返回一个类
```

在孪生神经网络中,每个 CNN 都有一个卷积层,卷积层连接一个池化层,进行数据的降维。在每个卷积层上都会使用多个滤波器来提取不同类型的特征。最大池化和全连接层之后,引入 Dropout 进行正则化,用以消除模型的过拟合问题。

```python
def fc_layer(self, bottom, n_weight, name):                            # 全连接层
    assert len(bottom.get_shape()) == 2
    n_prev_weight = bottom.get_shape()[1]
    initer = tf.contrib.layers.xavier_initializer()
    W = tf.get_variable(name + 'W', dtype = tf.float32, shape = [n_prev_weight,
n_weight], initializer = initer, regularizer = tf.contrib.layers.l2_regularizer(scale =
0.0000001))                                                            # y = Wx + b 线性模型
    b = tf.get_variable(name + 'b', dtype = tf.float32, initializer = tf.constant(0.01,
shape = [n_weight], dtype = tf.float32), regularizer = tf.contrib.layers.l2_regularizer
(scale = 0.0000001))
    fc = tf.nn.bias_add(tf.matmul(bottom, W), b)
    return fc
def _cnn_layer(self, input):                                           # 卷积层和池化层
    all = []
    max_len = input.get_shape()[1]
    for i, filter_size in enumerate(self.window_sizes):
        with tf.variable_scope('filter{}'.format(filter_size)):
            cnn_out = tf.layers.conv1d(input, self.n_filters, filter_size, padding = 'valid',
activation = tf.nn.relu, name = 'q_conv_' + str(i))                    # 卷积
            pool_out = tf.reduce_max(cnn_out, axis = 1, keepdims = True)  # 池化
            tanh_out = tf.nn.tanh(pool_out)                            # tanh 激活函数
            all.append(tanh_out)
    cnn_outs = tf.concat(all, axis = -1)
    dim = cnn_outs.get_shape()[-1]
    cnn_outs = tf.reshape(cnn_outs, [-1, dim])
    return cnn_outs
```

隐藏层的意义是把输入数据的特征抽象到另一个维度空间,展现其更抽象的特征,更好地进行线性划分。

```python
def _HL_layer(self, bottom, n_weight, name):                           # 隐藏层
    assert len(bottom.get_shape()) == 3
```

```python
            n_prev_weight = bottom.get_shape()[-1]
            max_len = bottom.get_shape()[1]
            initer = tf.contrib.layers.xavier_initializer()                    #初始化
            W = tf.get_variable(name + 'W', dtype=tf.float32, shape=[n_prev_weight, n_weight], initializer=initer, regularizer=tf.contrib.layers.l2_regularizer(scale=0.0000001))
            b = tf.get_variable(name + 'b', dtype=tf.float32, initializer=tf.constant(0.1, shape=[n_weight], dtype=tf.float32), regularizer=tf.contrib.layers.l2_regularizer(scale=0.0000001))                                                #y=Wx+b 线性模型
            bottom_2 = tf.reshape(bottom, [-1, n_prev_weight])
            hl = tf.nn.bias_add(tf.matmul(bottom_2, W), b)                     #y=Wx+b 单个神经元
            hl_tanh = tf.nn.tanh(hl)                                           #激活函数
            HL = tf.reshape(hl_tanh, [-1, max_len, n_weight])
            return HL
    def _build(self, embeddings):                                              #构建层
        if embeddings is not None:
            self.Embedding = tf.Variable(tf.to_float(embeddings), trainable=False, name='Embedding')
        else:                                                                  #嵌入构建
            self.Embedding = tf.get_variable('Embedding', shape=[self.vocab_size, self.embedding_size], initializer=tf.uniform_unit_scaling_initializer())
        self.q_embed = tf.nn.dropout(tf.nn.embedding_lookup(self.Embedding, self._ques), keep_prob=self.dropout_keep_prob)
        self.a_embed = tf.nn.dropout(tf.nn.embedding_lookup(self.Embedding, self._ans), keep_prob=self.dropout_keep_prob)
        with tf.variable_scope('siamese') as scope:
            #计算隐藏层和卷积层
            #hl_q = self._HL_layer(self.q_embed, self.hidden_size, 'HL_layer')
            conv1_q = self._cnn_layer(self.q_embed)
            scope.reuse_variables()                                            #权值共享
            #hl_a = self._HL_layer(self.a_embed, self.hidden_size, 'HL_layer')
            conv1_a = self._cnn_layer(self.a_embed)
        with tf.variable_scope('fc') as scope:
            con = tf.concat([conv1_q, conv1_a], axis=-1)
            logits = self.fc_layer(con, 1, 'fc_layer')
            res = tf.nn.sigmoid(logits)
        return logits, res
    def _add_loss_op(self, logits):
        #损失节点
        loss = tf.nn.sigmoid_cross_entropy_with_logits(logits=logits, labels=tf.cast(tf.reshape(self._y, [-1, 1]), dtype=tf.float32))
        reg_losses = tf.get_collection(tf.GraphKeys.REGULARIZATION_LOSSES)
        l2_loss = sum(reg_losses)
        pointwise_loss = tf.reduce_mean(loss) + l2_loss
        tf.summary.scalar('loss', pointwise_loss)
        return pointwise_loss
    def _add_acc_op(self):
```

```
            #精确度节点
            predictions = tf.to_int32(tf.round(self.res))
            correct_prediction = tf.equal(predictions, self._y)
            accuracy = tf.reduce_mean(tf.cast(correct_prediction, tf.float32))
            tf.summary.scalar('accuracy', accuracy)
            return accuracy
        def _add_train_op(self, loss):
            #训练节点
            with tf.name_scope('train_op'):
                #记录训练步骤
                self.global_step = tf.Variable(0, name = 'global_step', trainable = False)
                optimizer = tf.train.AdamOptimizer(self.learning_rate)
                #计算梯度,得到梯度和变量
                gradsAndVars = optimizer.compute_gradients(loss)
                #将梯度应用到变量下,生成训练器
                train_op = optimizer.apply_gradients(gradsAndVars, global_step = self.global_step)
                #用 summary 绘制 tensorBoard
                for g, v in gradsAndVars:
                    if g is not None:
                        tf.summary.histogram("{}/grad/hist".format(v.name), g)
                        tf.summary.scalar("{}/grad/sparsity".format(v.name), tf.nn.zero_fraction(g))
                self.summary_op = tf.summary.merge_all()
                return train_op
```

2. 优化损失函数

确定模型架构后进行编译,这是二分类问题,使用交叉熵作为损失函数。由于所有标签都带有相似的权重,通常使用精确度作为性能指标。Adam 是常用的梯度下降方法,使用它来优化模型参数。

```
#定义损失函数和优化器
loss = 
tf.nn.sigmoid_cross_entropy_with_logits(logits = logits, labels = tf.cast(tf.reshape(self._y, [-1, 1]), dtype = tf.float32))optimizer = tf.train.AdamOptimizer(self.learning_rate)
    predictions = tf.to_int32(tf.round(self.res))
    correct_prediction = tf.equal(predictions, self._y)
    accuracy = tf.reduce_mean(tf.cast(correct_prediction, tf.float32))
```

3.3.3 模型训练及保存

在定义模型架构和编译之后,使用训练集训练模型,使模型可以对语义相似的问句正确分类,用训练集来拟合并保存模型。

1. 模型训练

```python
def devStep(corpus):
    iterator = Iterator(corpus)
    dev_Loss = []                                           # 损失
    dev_Acc = []                                            # 准确率
    dev_Prec = []
    dev_Recall = []
    dev_F_beta = []
    for batch_x in iterator.next(config.batch_size, shuffle = False):
        batch_q, batch_a, batch_qmask, batch_amask, label = zip(* batch_x)
        batch_q = np.asarray(batch_q)                       # 批次
        batch_a = np.asarray(batch_a)
        loss, summary, step, predictions = sess.run(
            [model.total_loss, model.summary_op, model.global_step, model.res],
            feed_dict = {model._ques: batch_q,              # 传入训练集
                         model._ans: batch_a,
                         model._y: label,
                         model.dropout_keep_prob: 1.0})
        predictions = [1 if i >= 0.5 else 0 for i in predictions]
        acc, recall, prec, f_beta = get_binary_metrics(pred_y = predictions,
true_y = label)                                             # 预测值
        dev_Loss.append(loss)
        dev_Acc.append(acc)
        dev_Prec.append(prec)
        dev_Recall.append(recall)
        dev_F_beta.append(f_beta)
        evalSummaryWriter.add_summary(summary, step)
    return mean(dev_Loss), mean(dev_Acc), mean(dev_Recall), mean(dev_Prec),
mean(dev_F_beta)                                            # 返回参数
best_acc = 0.0
for epoch in range(config.num_epochs):                      # 轮次
    train_time1 = time.time()
    print("----- Epoch {}/{} -----".format(epoch + 1, config.num_epochs))
    train_Loss = []                                         # 损失
    train_Acc = []                                          # 准确率
    train_Prec = []
    train_Recall = []
    train_F_beta = []
    for batch_x in iterator.next(config.batch_size, shuffle = True):
        batch_q, batch_a, batch_qmask, batch_amask, label = zip(* batch_x)
        batch_q = np.asarray(batch_q)                       # 批次
        batch_a = np.asarray(batch_a)
        train_loss, train_acc, train_prec, train_recall, train_f_beta = trainStep
(batch_q, batch_a, label)                                   # 输出训练结果
        train_Loss.append(train_loss)
```

```
                train_Acc.append(train_acc)
                train_Prec.append(train_prec)
                train_Recall.append(train_recall)
                train_F_beta.append(train_f_beta)
                print(" --- epoch %d -- train loss %.3f -- train acc %.3f -- train recall
%.3f -- train precision %.3f"
                      "- train f_beta %.3f" % (
                       epoch + 1, np.mean(train_Loss), np.mean(train_Acc), np.mean
(train_Recall),np.mean(train_Prec), np.mean(train_F_beta)))     #打印准确率
                test_loss, test_acc, test_recall, test_prec, test_f_beta = devStep(test_corpus)
                print(" --- epoch %d -- test loss %.3f -- test acc %.3f -- test recall %
.3f -- test precision %.3f"
                      " -- test f_beta %.3f" % (
                       epoch + 1, test_loss, test_acc, test_recall, test_prec, test_f_beta))
```

2. 模型保存

为方便训练时读取,将模型保存为 ckpt 格式的文件,利用 TensorFlow 中的 tf.train.Saver 进行保存。

```
saver = tf.train.Saver(tf.global_variables(), max_to_keep = 10)
#定义保存的对象
best_saver = tf.train.Saver(tf.global_variables(), max_to_keep = 5)
ckpt = tf.train.get_checkpoint_state(save_path)
checkpoint_path = os.path.join(save_path, 'acc{:.3f}_{}.ckpt'.format(test_acc, epoch + 1))
bestcheck_path = os.path.join(best_path, 'acc{:.3f}_{}.ckpt'.format(test_acc, epoch + 1))
saver.save(sess, checkpoint_path, global_step = epoch)           #保存模型
```

3.3.4 模型生成

一是通过中控模块调用召回和精排模型;二是通过训练好的召回和精排模型进行语义分类,并且获取输出。

1. GUI 模块

GUI 模块是本项目的前端,提供了 2 个文本框,一个显示用户输入,另一个显示对话内容。提供 1 个"发送"按钮,调用 control.py 中的接口,返回选取的回答内容。

```
    def setupUi(self, Dialog):                           #设置界面
        Dialog.setObjectName("智能聊天机器人")
        Dialog.resize(582, 434)
        #palette = QPalette()
        #palette.setBrush(QPalette.Background, QBrush(QPixmap("./background.jpg")))
        #Dialog.setPalette(palette)
        palette = QPalette()
        pix = QPixmap("./background.jpg")
        pix = pix.scaled(Dialog.width(), Dialog.height())
```

```
                palette.setBrush(QPalette.Background, QBrush(pix))
                Dialog.setPalette(palette)
                self.label = QtWidgets.QLabel(Dialog)
                self.label.setGeometry(QtCore.QRect(40, 30, 361, 51))
                self.label.setStyleSheet("color: rgb(205, 85, 85);\n"
"font: 16pt \"黑体\";\n"
"text-decoration: underline;")
                self.label.setObjectName("dialog")
                self.plainTextEdit = QtWidgets.QPlainTextEdit(Dialog)
                self.plainTextEdit.setGeometry(QtCore.QRect(40, 80, 501, 181))
                self.plainTextEdit.setObjectName("plainTextEdit")
                self.plainTextEdit.setFocusPolicy(QtCore.Qt.NoFocus)
                self.plainTextEdit_2 = QtWidgets.QPlainTextEdit(Dialog)
                self.plainTextEdit_2.setGeometry(QtCore.QRect(40, 310, 401, 41))
                self.plainTextEdit_2.setObjectName("plainTextEdit_2")
                self.plainTextEdit.setStyleSheet("font: 14pt \"黑体\";\n")
                self.pushButton = QtWidgets.QPushButton(Dialog)
                self.pushButton.setGeometry(QtCore.QRect(480, 320, 75, 23))
                self.pushButton.setStyleSheet("font: 14pt \"黑体\";\n"
"background-color: rgb(255, 192, 203);")
                self.pushButton.setObjectName("pushButton")
                self.label_2 = QtWidgets.QLabel(Dialog)
                self.label_2.setGeometry(QtCore.QRect(50, 280, 54, 12))
                self.label_2.setText("")
                self.label_2.setObjectName("label_2")
                self.label_3 = QtWidgets.QLabel(Dialog)
                self.label_3.setGeometry(QtCore.QRect(50, 280, 71, 16))
                self.label_3.setStyleSheet("font: 75 12pt \"Aharoni\";")
                self.label_3.setObjectName("label_3")
                self.retranslateUi(Dialog)
                QtCore.QMetaObject.connectSlotsByName(Dialog)
```

2. 中控模块

中控模块设定 2 个阈值 max_sim 和 min_sim，用于缩减响应时间。如果 recall_score < min_sim，说明问答库数量少或者问句噪声大，需要复查分析；如果 min_sim < recall_score < max_sim，进行召回（recall）和精排（rerank）；如果 recall_score > max_sim，只进行召回，直接得出答案。

```
import time                                          # 导入模块
from Rerank.data_helper import *
from Recall import recall_model
from Rerank import rerank_model
class SmartQA:                                       # 定义类
    def __init__(self):                              # 初始化
        self.top_k = 5
        self.min_sim = 0.10
```

```python
        self.max_sim = 0.90
        self.embeding_size = 200
        self.vocab_file = './data/word_vocab.txt'
        self.embed_file = './word2vec/5000-small.txt'
        self.embedding = load_embedding(self.embed_file, self.embeding_size, self.vocab_file)
    # 分为 recall 和 rerank 两部分
    def search_main(self, question):
        # 粗排
        candi_questions, questionList, answerList = recall_model.main(question, self.top_k)
        answer_dict = {}
        corpus = []
        indxs = []
        matchmodel_simscore = []
        sim_questions = []
        for indx, candi in zip(*candi_questions):
            # 如果在粗排阶段就已经找到了非常相似的问题,则马上返回答案,终止循环
            if candi > self.max_sim:
                indxs.append(indx)
                break
            else:
                # 如果召回的数据噪声大,生成一个文件,复查分析
                matchmodel_simscore.append(candi)
                corpus.append((question, questionList[indx]))
                indxs.append(indx)
                sim_questions.append(questionList[indx])
        if candi_questions[1][0] < self.min_sim:
            final_answer = '我还没找到相似的答案,请说得再清楚一点'
            return final_answer, sim_questions
        if len(indxs) == 1:
            # 找到非常相似的答案
            sim = [questionList[indx] for indx, candi in zip(*candi_questions)]
            return answerList[indxs[0]], sim
        else:
            if len(indxs) != 0 :
                deepmodel_simscore = rerank_model.main(corpus, self.embedding)
                                                        # 使用精排模型
                final = list(zip(indxs, matchmodel_simscore, deepmodel_simscore))
                                                        # 输出结果
                for id, score1, score2 in final:
                    final_score = (score1 + score2) / 2
                    answer_dict[id] = final_score
                if answer_dict:                         # 如果识别成功
                    answer_dict = sorted(answer_dict.items(), key=lambda asd: asd[1], reverse=True)
                    final_answer = answerList[answer_dict[0][0]]
                else:
                    final_answer = '请说得再清楚一点.'
                return final_answer, sim_questions
```

```python
def answer(question):                                    # 定义回答的问题
    handler = SmartQA()
    final_answer, sim_questions = handler.search_main(question)
    return final_answer
if __name__ == "__main__":                               # 主函数
    handler = SmartQA()
    while (1):
        question = input('用户说:\n')
        if question == 'end':
            print('byebye~')
            break
        s1 = time.time()
        final_answer, sim_questions = handler.search_main(question)
        s2 = time.time()
        print('机器人:', final_answer)
```

3. 相关代码

本部分包括召回(Recall)模型和精排(Rerank)模型。

1) 召回模型

召回模型相关代码如下：

```python
import pandas as pd                                      # 导入模块
import matplotlib as mpl
import numpy as np
from nltk.probability import FreqDist
from .jiebaSegment import *
from .sentenceSimilarity import SentenceSimilarity
mpl.rcParams['font.sans-serif'] = ['Microsoft YaHei']  # enable chinese
# 设置外部词
seg = Seg()
seg.load_userdict('./data/userdict.txt')
def read_corpus():
    qList = []
    # 问题的关键词列表
    qList_kw = []
    aList = []
    data = pd.read_csv('./data/qa_.csv', header = None)
    data_ls = np.array(data).tolist()
    for t in data_ls:
        qList.append(t[0])
        qList_kw.append(seg.cut(t[0]))
        aList.append(t[1])
    return qList_kw, qList, aList
def invert_idxTable(qList_kw):                           # 制定一个简单的倒排表
    invertTable = {}
    for idx, tmpLst in enumerate(qList_kw):
```

```python
            for kw in tmpLst:
                if kw in invertTable.keys():
                    invertTable[kw].append(idx)
                else:
                    invertTable[kw] = [idx]
    return invertTable
def filter_questionByInvertTab(inputQuestionKW, questionList, answerList, invertTable):
                                                                    # 过滤问题
    idxLst = []
    questions = []
    answers = []
    for kw in inputQuestionKW:
        if kw in invertTable.keys():
            idxLst.extend(invertTable[kw])
    idxSet = set(idxLst)
    for idx in idxSet:
        questions.append(questionList[idx])
        answers.append(answerList[idx])
    return questions, answers
def main(question, top_k):                                          # topk 控制选出的回答个数
    qList_kw, questionList, answerList = read_corpus()
    questionList_s = questionList
    answerList_s = answerList
    # 初始化模型
    ss = SentenceSimilarity(seg)
    ss.set_sentences(questionList_s)
    ss.TfidfModel()                                                 # TF-IDF 模型
    question_k = ss.similarity_k(question, top_k)
    return question_k, questionList_s, answerList_s
if __name__ == '__main__':
    # 设置外部词
    seg = Seg()
    seg.load_userdict('./userdict/userdict.txt')
    # 读取数据
    List_kw, questionList, answerList = read_corpus()
    # 初始化模型
    ss = SentenceSimilarity(seg)
    ss.set_sentences(questionList)
    ss.TfidfModel()
    while True:
        question = input("请输入问题(q 退出): ")
        if question == 'q':
            break
        question_k = ss.similarity_k(question, 5)
        print("机器人: {}".format(answerList[question_k[0][0]]))
        for idx, score in zip(*question_k):
            print("same questions: {}, score: {}".format(questionList[idx], score))
```

2)精排模型

精排模型相关代码如下:

```python
import time                                              # 导入各种模块
import logging
import warnings
warnings.filterwarnings("ignore")
import numpy as np
import tensorflow as tf
import os
import tqdm
import sys
from copy import deepcopy
stdout = sys.stdout
from Rerank.data_helper import *
from Rerank.data_preprocess import *
from Rerank.model import SiameseQACNN
from Rerank.model_utils import *
from Rerank.metrics import *
from sklearn.metrics import accuracy_score
class NNConfig(object):                                   # 定义类
    def __init__(self, embeddings):                       # 初始化
        self.ans_length = 15
        self.num_epochs = 10
        self.ques_length = 15
        self.batch_size = 32
        self.window_sizes = [1, 1, 2]
        self.hidden_size = 128
        self.output_size = 128
        self.keep_prob = 0.5
        self.n_filters = 128
        self.embeddings = np.array(embeddings).astype(np.float32)
        self.vocab_size = 3258
        self.embedding_size = 300
        self.learning_rate = 0.0001
        self.optimizer = 'adam'
        self.clip_value = 5
        self.l2_lambda = 0.00001
        self.eval_batch = 100
def train(train_corpus, test_corpus, config):             # 定义训练
    iterator = Iterator(train_corpus)
    if not os.path.exists(save_path):
        os.makedirs(save_path)
    if not os.path.exists(best_path):
        os.makedirs(best_path)
    # 定义计算图
    with tf.Graph().as_default():
```

```python
        session_conf = tf.ConfigProto(allow_soft_placement = True, log_device_placement = False)
        with tf.Session(config = session_conf) as sess:
            # 训练
            print('Start training and evaluating ...')
            outDir = os.path.abspath(os.path.join(os.path.curdir, "summarys"))
            print("Writing to {}\n".format(outDir))
            trainSummaryDir = os.path.join(outDir, "train")
            trainSummaryWriter = tf.summary.FileWriter(trainSummaryDir, sess.graph)
            evalSummaryDir = os.path.join(outDir, "eval")
            evalSummaryWriter = tf.summary.FileWriter(evalSummaryDir, sess.graph)
            model = SiameseQACNN(config)
            # 初始化所有变量
            saver = tf.train.Saver(tf.global_variables(), max_to_keep = 10)
            best_saver = tf.train.Saver(tf.global_variables(), max_to_keep = 5)
            ckpt = tf.train.get_checkpoint_state(save_path)
            print('Configuring TensorBoard and Saver ...')
            if ckpt and tf.train.checkpoint_exists(ckpt.model_checkpoint_path):
                print('Reloading model parameters ...')
                saver.restore(sess, ckpt.model_checkpoint_path)
            else:
                print('Created new model parameters ...')
                sess.run(tf.global_variables_initializer())
            # 计算训练参数
            total_parameters = count_parameters()
            print('Total trainable parameters : {}'.format(total_parameters))
            def trainStep(batch_q, batch_a, batchY):
                _, loss, summary, step, predictions = sess.run(
                    [model.train_op, model.total_loss, model.summary_op, model.global_step, model.res],
                    feed_dict = {model._ques: batch_q,
                                model._ans: batch_a,
                                model._y: label,
                                model.dropout_keep_prob: config.keep_prob})
                predictions = [1 if i >= 0.5 else 0 for i in predictions]
                acc, recall, prec, f_beta = get_binary_metrics(pred_y = predictions, true_y = batchY)
                trainSummaryWriter.add_summary(summary, step)
                return loss, acc, prec, recall, f_beta
            def devStep(corpus):
                iterator = Iterator(corpus)                    # 定义各种参数
                dev_Loss = []
                dev_Acc = []
                dev_Prec = []
                dev_Recall = []
                dev_F_beta = []
                for batch_x in iterator.next(config.batch_size, shuffle = False):
                    batch_q, batch_a, batch_qmask, batch_amask, label = zip(*batch_x)
```

```python
                    batch_q = np.asarray(batch_q)              # 获取批次
                    batch_a = np.asarray(batch_a)
                    loss, summary, step, predictions = sess.run(    # 输出结果
                        [model.total_loss, model.summary_op, model.global_step, model.res],
                        feed_dict = {model._ques: batch_q,
                                     model._ans: batch_a,
                                     model._y: label,
                                     model.dropout_keep_prob: 1.0})
                    predictions = [1 if i >= 0.5 else 0 for i in predictions]
                    acc, recall, prec, f_beta = get_binary_metrics(pred_y = predictions,
            true_y = label)                                    # 得到参数值
                    dev_Loss.append(loss)
                    dev_Acc.append(acc)
                    dev_Prec.append(prec)
                    dev_Recall.append(recall)
                    dev_F_beta.append(f_beta)
                    evalSummaryWriter.add_summary(summary, step)
                return mean(dev_Loss), mean(dev_Acc), mean(dev_Recall), mean(dev_Prec),
    mean(dev_F_beta)
            best_acc = 0.0
            for epoch in range(config.num_epochs):             # 轮次
                train_time1 = time.time()
                print("----- Epoch {}/{} -----".format(epoch + 1, config.num_epochs))
                                                               # 输出训练参数
                train_Loss = []
                train_Acc = []
                train_Prec = []
                train_Recall = []
                train_F_beta = []
                for batch_x in iterator.next(config.batch_size, shuffle = True):
                    batch_q, batch_a, batch_qmask, batch_amask, label = zip( * batch_x)
                    batch_q = np.asarray(batch_q)              # 批次数据
                    batch_a = np.asarray(batch_a)
                    train_loss, train_acc, train_prec, train_recall, train_f_beta = trainStep
    (batch_q, batch_a, label)                                  # 获取训练参数
                    train_Loss.append(train_loss)
                    train_Acc.append(train_acc)
                    train_Prec.append(train_prec)
                    train_Recall.append(train_recall)
                    train_F_beta.append(train_f_beta)
                print(" --- epoch %d -- train loss %.3f -- train acc %.3f -- train recall
    %.3f -- train precision %.3f"
                      " -- train f_beta %.3f" % (
                          epoch + 1, np.mean(train_Loss), np.mean(train_Acc), np.mean
    (train_Recall), np.mean(train_Prec), np.mean(train_F_beta)))   # 打印训练参数值
                test_loss, test_acc, test_recall, test_prec, test_f_beta = devStep(test_corpus)
                print(" --- epoch %d -- test loss %.3f -- test acc %.3f -- test recall %.
```

```
                3f -- test precision %.3f"
                                "-- test f_beta %.3f" % (
                                epoch + 1, test_loss, test_acc, test_recall, test_prec, test_f_beta))
                                                                        #打印测试参数值
                    checkpoint_path = os.path.join(save_path, 'acc{:.3f}_{}.ckpt'.format(test
_acc, epoch + 1))                                                       #检查点路径
                    bestcheck_path = os.path.join(best_path, 'acc{:.3f}_{}.ckpt'.format(test_
acc, epoch + 1))                                                        #最佳检查路径
                    saver.save(sess, checkpoint_path, global_step = epoch)
                    if test_acc > best_acc:
                        best_acc = test_acc
                        best_saver.save(sess, bestcheck_path, global_step = epoch)
def main():                                                             #主函数
    embedding = load_embedding(embeding, embeding_size, vocab_file)
    preprocess_data1 = preprocess(train_file)                           #预处理
    preprocess_data2 = preprocess(test_file)
    train_data = read_train(preprocess_data1, stopword_file, vocab_file)    #训练数据
    test_data = read_train(preprocess_data2, stopword_file, vocab_file)     #测试数据
    train_corpus = load_train_data(train_data, max_q_length, max_a_length)
    test_corpus = load_train_data(test_data, max_q_length, max_a_length)
    config = NNConfig(embedding)                                        #配置参数
    config.ques_length = max_q_length
    config.ans_length = max_a_length
    #config.embeddings = embedding
    train(deepcopy(train_corpus), test_corpus, config)
if __name__ == '__main__':                                              #主函数
    save_path = "./model/checkpoint"
    best_path = "./model/bestval"
    train_file = '../data/train.csv'
    test_file = '../data/test.csv'
    stopword_file = '../stopwordList/stopword.txt'
    embeding = '../word2vec/5000-small.txt'
    vocab_file = '../data/word_vocab.txt'
    max_q_length = 15
    max_a_length = 15
    embeding_size = 200
    main()
```

3.4 系统测试

本部分包括训练准确率、测试效果及模型应用。

3.4.1 训练准确率

测试准确率在90%左右,损失随训练次数增多而下降,并趋于稳定,如图3-4所示。

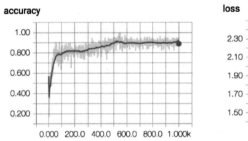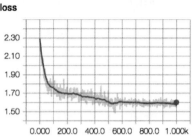

图 3-4　训练准确率和损失

3.4.2　测试效果

将文本输入模型进行测试,如图 3-5 所示。

图 3-5　模型训练效果

3.4.3　模型应用

本部分包括程序下载运行、应用使用说明和测试结果。

1. 程序下载运行

下载程序压缩包后,在 Python 环境下执行 gui.py 命令即可。

2. 应用使用说明

解压程序压缩包后,文件目录如下:

```
Recall
- jiebaSegment.py                    # jieba 分词
- sentence.py                        # 句子类
- sentenceSimilarity.py              # 相似度计算
```

```
     - recall_model.py              ＃检索模型
Rerank
     - data_preprocess.py           ＃词嵌入、划分训练和测试集
     - data_helper.py               ＃划分批次
     - model.py CNN                 ＃模型构建
  - model_utils.py                  ＃模型的依赖文件
     - qacnn.py                     ＃模型训练
     - rerank_model.py              ＃预测
control.py
GUI.py
```

其中，qacnn.py 是模型的训练文件，可以单独运行；control.py 控制 Recall 和 Rerank 模型的选择，可以单独运行；GUI.py 是本项目的图形化界面，调用 control.py 的接口。

3. 测试结果

图形化界面测试结果如图 3-6 所示。

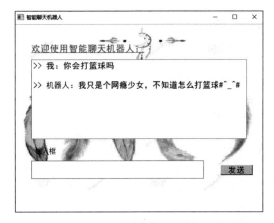

图 3-6 测试结果示例

项目 4　方言种类识别

PROJECT 4

本项目基于科大讯飞提供的数据集进行特征筛选和提取，选择 WaveNet 模型进行训练，通过语音的 MFCC 特征完成方言和类别之间的映射关系，实现方言分类问题。

4.1　总体设计

本部分包括系统整体结构和系统流程。

4.1.1　系统整体结构

系统整体结构如图 4-1 所示。

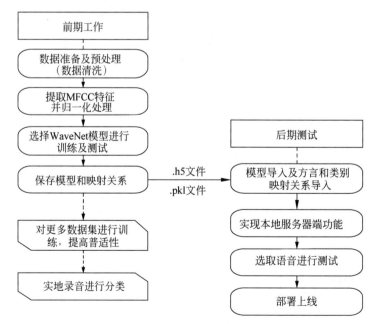

图 4-1　系统整体结构

4.1.2 系统流程

系统流程如图 4-2 所示。

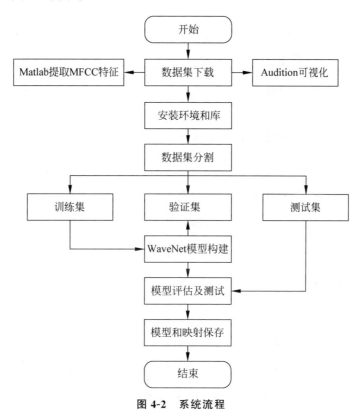

图 4-2 系统流程

4.2 运行环境

本部分包括 Python 环境、TensorFlow 环境、Jupyter Notebook 环境和 PyCharm 环境。

4.2.1 Python 环境

需要 Python 3.6 及以上配置,在 Windows 环境下通过 Anaconda 完成 Python 所需配置,下载地址为:https://www.anaconda.com/。

4.2.2 TensorFlow 环境

Anaconda 完成 TensorFlow(CPU 版本)所需环境的配置,Jupyter Notebook 编译

运行。

打开 Anaconda Prompt,输入清华仓库镜像,输入命令:

```
conda config --add channels https://mirrors.tuna.tsinghua.edu.cn/anaconda/pkgs/free/
conda config -set show_channel_urls yes
```

创建 Python 3.6 的环境,名称为 TensorFlow,此时 Python 版本和后面 TensorFlow 的版本有匹配问题,此步选择 Python 3.x,输入命令:

```
conda create -n tensorflow python = 3.6
```

有需要确认的地方,都输入 y。

在 Anaconda Prompt 中激活 TensorFlow 环境,输入命令:

```
activate tensorflow
```

安装 CPU 版本的 TensorFlow,输入命令:

```
pip install -upgrade --ignore-installed tensorflow
```

安装完毕。

4.2.3 Jupyter Notebook 环境

安装 Jupyter Notebook,前提是已安装 Python 2.7 或 Python 3.3 以上版本。一种方法是使用 Anaconda 安装,在终端输入命令:

```
conda install jupyter notebook
```

另一种方法是使用 pip 命令安装,将 pip 升级到最新版本,输入命令:

```
pip install - upgrade pip
```

再安装 Jupyter Notebook,输入命令:

```
pip install jupyter
```

安装完毕。

4.2.4 PyCharm 环境

保存模型、方言和类别之间的映射关系后,导入 PyCharm 中进行语音测试。安装 PyCharm 并激活,版本号如下:

```
PyCharm 2019.1.1 (Professional Edition)
Build #PY-191.6605.12, built on April 3, 2019
Licensed to pig6
Subscription is active until July 8, 2089
```

```
JRE: 11.0.2 + 9-b159.34 amd64
JVM: OpenJDK 64-Bit Server VM by JetBrains s.r.o
Windows 10 10.0
```

4.3 模块实现

本项目包括 4 个模块：数据预处理、模型构建、模型训练及保存、模型生成。下面分别给出各模块的功能介绍及相关代码。

4.3.1 数据预处理

本部分包括数据介绍、数据测试和数据处理。

1. 数据介绍

数据集网址为：challenge.xfyun.cn，向用户免费提供了 3 种方言（长沙话、南昌话、上海话），每种方言包括 30 人，每人 200 条数据，共计 18000 条训练数据，以及 10 人、每人 50 条，共计 1500 条验证数据。数据以 pcm 格式提供，可以理解为 .wav 文件去掉多余信息之后仅保留语音数据的格式。

2. 数据测试

使用 Audition 进行语音测试，导出 mp3 格式进行检测，如图 4-3 所示。

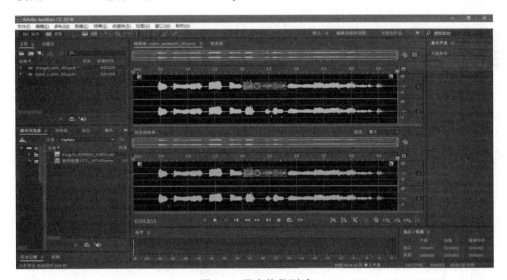

图 4-3 语音片段测试

使用 MATLAB 得到该语音片段的波形和 MFCC 特征，相关代码如下：

```
[x, fs] = audioread('test.mp3');
```

```matlab
bank = melbankm(24,256,fs,0,0.4,'m');
% melbankm 滤波器的阶数为 24
% fft 变换的长度为 256,采样频率为 8000Hz
% 归一化 melbankm 滤波器组系数
bank = full(bank);
bank = bank/max(bank(:));
for k = 1:12                          % 归一化 melbankm 滤波器组系数
n = 0:23;
dctcoef(k,:) = cos((2 * n + 1) * k * pi/(2 * 24));
end
w = 1 + 6 * sin(pi * [1:12]./12);     % 归一化倒谱提升窗口
w = w/max(w); % 预加重滤波器
xx = double(x);
xx = filter([1-0.9375],1,xx);         % 语音信号分帧
xx = enframe(xx,256,80);              % 每 256 点分为一帧
% 计算每帧的 MFCC 参数
for i = 1:size(xx,1)
y = xx(i,:);
s = y'. * hamming(256);
t = abs(fft(s));                      % fft 快速傅里叶变换
t = t.^2;
c1 = dctcoef * log(bank * t(1:129));
c2 = c1. * w';
m(i,:) = c2';
end
% 求取差分系数
dtm = zeros(size(m));
for i = 3:size(m,1)-2
dtm(i,:) = -2 * m(i-2,:)-m(i-1,:) + m(i+1,:) + 2 * m(i+2,:);
end
dtm = dtm/3;
% 合并 MFCC 参数和一阶差分 MFCC 参数
ccc = [m dtm];
% 去除首尾两帧,因为这两帧的一阶差分参数为 0
ccc = ccc(3:size(m,1)-2,:);
subplot(2,1,1)
ccc_1 = ccc(:,1);
plot(ccc_1);title('MFCC');ylabel('幅值');
[h,w] = size(ccc);
A = size(ccc);
subplot(212) ;
plot([1,w],A);
xlabel('维数');
ylabel('幅值');
title('维数与幅值的关系')
```

运行代码,成功得到一段语音的波形片段和 MFCC 特征,如图 4-4 和图 4-5 所示。

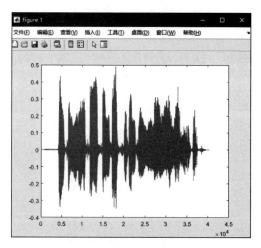

图 4-4 语音片段波形图

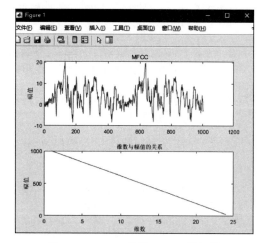

图 4-5 Matlab 提取 MFCC 特征图

3. 数据处理

本部分包括数据处理过程中的相关代码。

（1）加载库。在实验过程中，通过查找发现 Matplotlib 包内没有中文字体，加载库之后添加参数设定或者更改 Matplotlib 默认字体可以解决。相关代码如下：

```
# - * - coding:utf-8 - * -
import numpy as np
import os
from matplotlib import pyplot as plt
# 在实验过程中使用 pyplot 参数设置，使得图像可打印中文，且中文内容格式应该为 u'内容'
plt.rcParams['font.sans-serif'] = ['SimHei']    # 正常显示中文标签
plt.rcParams['axes.unicode_minus'] = False
from mpl_toolkits.axes_grid1 import make_axes_locatable
%matplotlib inline
from sklearn.utils import shuffle            # 导入各种模块
import glob
import pickle
from tqdm import tqdm
from keras.models import Model
from keras.preprocessing.sequence import pad_sequences
from keras.layers import Input, Activation, Conv1D, Add, Multiply, BatchNormalization, GlobalMaxPooling1D, Dropout
from keras.optimizers import Adam
from keras.callbacks import ModelCheckpoint, ReduceLROnPlateau
from python_speech_features import mfcc
import librosa
from IPython.display import Audio
import wave
```

(2) 加载 pcm 文件,共 18000 条训练数据及 1500 条验证数据,相关代码如下:

```python
train_files = glob.glob('data/*/train/*/*.pcm')
dev_files = glob.glob('data/*/dev/*/*.pcm')
#glob.glob()用于查找符合特定规则的文件路径名,并返回所有匹配的文件路径列表
#其中"*"为匹配符,匹配多个字符
print(len(train_files), len(dev_files), train_files[0])
#读取代码成功,打印结果
```

(3) 整理每条语音数据对应的类别标签,相关代码如下:

```python
labels = {'train': [], 'dev': []}
#使用 dict 与 list 类型嵌套存储类别标签
#tqdm 打印进度条,用于观察读取情况
for i in tqdm(range(len(train_files))):
    path = train_files[i]
    label = path.split('/')[1]
        #使用 split 将 path 以"/"分隔的字符串进行切片,并选取第 1 个分片
    labels['train'].append(label)
for i in tqdm(range(len(dev_files))):
    path = dev_files[i]
    label = path.split('/')[1]
    labels['dev'].append(label)
print(len(labels['train']), len(labels['dev']))
#读取代码成功,打印结果
```

(4) 定义处理语音、pcm 转 wav、可视化语音的三个函数,由于语音片段长短不一,所以去除少于 1s 的短片段,对长片段则切分为不超过 3s 的片段。相关代码如下:

```python
mfcc_dim = 13
sr = 16000
min_length = 1 * sr
slice_length = 3 * sr
#语音处理函数
def load_and_trim(path, sr = 16000):
    audio = np.memmap(path, dtype = 'h', mode = 'r')
        #使用 NumPy 内的 memmap()函数读写大文件,返回对象可使用 ndarray 算法操作
    audio = audio[2000: -2000]
    audio = audio.astype(np.float32)
        #astype()实现 dataframe 类型转换
    energy = librosa.feature.rmse(audio)
        #librosa 为音频处理包,使用 feature.rmse 求均方根误差
    frames = np.nonzero(energy >= np.max(energy) / 5)
        #nonzero()函数将布尔数组转为整数数组,用于进行下标运算
        #nonzero()返回参数数组中值不为 0 的元素下标,返回类型是元组
        #元组的每个元素均为一个整数数组,值为非零元素下标在对应帧的值
    indices = librosa.core.frames_to_samples(frames)[1]
        #将样本索引采样转换为帧
```

```python
        audio = audio[indices[0]:indices[-1]] if indices.size else audio[0:0]
        slices = []
        for i in range(0, audio.shape[0], slice_length):
            s = audio[i: i + slice_length]
            if s.shape[0] >= min_length:
                slices.append(s)
            # 长度>3s进行切片,<1s过滤
        return audio, slices
# 文件格式处理,pcm 转为 wav 函数
def pcm2wav(pcm_path, wav_path, channels = 1, bits = 16, sample_rate = sr):
    data = open(pcm_path, 'rb').read()
    fw = wave.open(wav_path, 'wb')
        # 设置转换为 wav 格式时相同的参数(通道数、采样率等)
    fw.setnchannels(channels)                    # 通道
    fw.setsampwidth(bits // 8)
    fw.setframerate(sample_rate)                 # 采样率
    fw.writeframes(data)                         # 写入
    fw.close()
# 语音可视化
def visualize(index, source = 'train'):
# 可视化语音,默认可视化训练集内的语音,也可以选择参数"dev"可视化测试集
    if source == 'train':
        path = train_files[index]
    else:
        path = dev_files[index]
    print(path)
    audio, slices = load_and_trim(path)
        # 读取语音信号
    print('Duration: %.2f s' % (audio.shape[0] / sr))
        # 使用 Matplotlob 库内的 pyplot 进行绘图
    plt.figure(figsize = (12, 3))
    plt.plot(np.arange(len(audio)), audio)
    plt.title(u'未处理信号')
    plt.xlabel(u'时间')
    plt.ylabel(u'信号幅度值')
    plt.show() # 展示未处理的音频信号
    feature = mfcc(audio, sr, numcep = mfcc_dim)
        # 提取 MFCC 特征
    print('Shape of MFCC:', feature.shape)
        # 画图
    fig = plt.figure(figsize = (12, 5))
        # figsize 指定 figure 的宽和高,单位为英寸
    ax = fig.add_subplot(111) # subplot()为图像分区,参数 111 表示只有一张图
    im = ax.imshow(feature, cmap = plt.cm.jet, aspect = 'auto')
        # imshow 热图绘制,通过色差、亮度展示数据的差异
    plt.title(u'归一化 MFCC')
    plt.ylabel('时间')
```

```
    plt.xlabel('MFCC 参数')
    plt.colorbar(im,cax = make_axes_locatable(ax).append_axes('right', size = '5 %', pad = 0.05))
        # 为图配渐变色时,在图旁边把色标(colorbar)标注出来
        # 其中 cax 参数表示将要绘制颜色条的轴
    ax.set_xticks(np.arange(0,13,2), minor = False);         # 设置横坐标刻度
    plt.show()
    wav_path = 'example.wav'
    pcm2wav(path, wav_path)
    return wav_path
Audio(visualize(2))                                          # 调用函数选取其中一条语音信号进行可视化
```

读取代码成功可以得到一句长沙话对应的波形和 MFCC 特征,如图 4-6 所示。

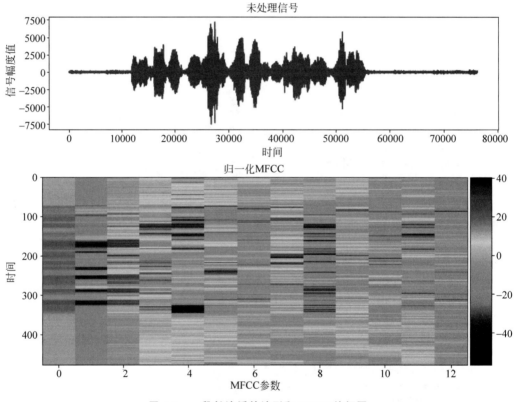

图 4-6 一段长沙话的波形和 MFCC 特征图

(5) 整理数据,查看语音片段的长度分布,最后得到 18890 个训练片段和 1632 个验证片段。相关代码如下:

```
X_train = []                                    # 参数设置
X_dev = []
Y_train = []
Y_dev = []
```

```
lengths = []
for i in tqdm(range(len(train_files))):
    path = train_files[i]
    audio, slices = load_and_trim(path)      #取语音信号
    lengths.append(audio.shape[0] / sr)       #除以 sr(=16000)换算成时间长度
    for s in slices:                           #按照片段进行训练
        X_train.append(mfcc(s, sr, numcep = mfcc_dim))
        Y_train.append(labels['train'][i])
for i in tqdm(range(len(dev_files))):          #输出参数
    path = dev_files[i]
    audio, slices = load_and_trim(path)
    lengths.append(audio.shape[0] / sr)
    for s in slices:
        X_dev.append(mfcc(s, sr, numcep = mfcc_dim))
        Y_dev.append(labels['dev'][i])
print(len(X_train), len(X_dev))
plt.hist(lengths, bins = 100)
    #Matplotlib 中的 hist()函数,用于可视化生成直方图
plt.show()
```

读取代码成功,如图 4-7 所示。

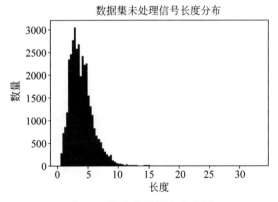

图 4-7　语音片段长度分布图

(6) 将 MFCC 特征进行归一化处理,相关代码如下:

```
#矩阵堆叠,沿竖直方向矩阵堆叠
samples = np.vstack(X_train)
#计算每一列的平均值(axis = 0)
mfcc_mean = np.mean(samples, axis = 0)
#计算每一列的标准差(axis = 0)
mfcc_std = np.std(samples, axis = 0)
print(mfcc_mean)
print(mfcc_std)
#归一化处理
```

```python
X_train = [(x - mfcc_mean) / (mfcc_std + 1e-14) for x in X_train]
X_dev = [(x - mfcc_mean) / (mfcc_std + 1e-14) for x in X_dev]
maxlen = np.max([x.shape[0] for x in X_train + X_dev])
X_train = pad_sequences(X_train, maxlen, 'float32', padding = 'post', value = 0.0)
#pad_sequences()对序列进行预处理
#maxlen 参数为序列的最大长度,返回类型 dtype 为 float32,'post'表示需要在末尾时补 0
X_dev = pad_sequences(X_dev, maxlen, 'float32', padding = 'post', value = 0.0)
print(X_train.shape, X_dev.shape)
```

(7) 对类别标签进行处理,相关代码如下:

```python
from sklearn.preprocessing import LabelEncoder
from keras.utils import to_categorical
#将标签标准化
le = LabelEncoder()
#非监督学习
Y_train = le.fit_transform(Y_train) #fit_transform()
Y_dev = le.transform(Y_dev)
print(le.classes_)
class2id = {c: i for i, c in enumerate(le.classes_)}
#enumerate()将可遍历数据对象组合为索引序列
id2class = {i: c for i, c in enumerate(le.classes_)}
num_class = len(le.classes_)
Y_train = to_categorical(Y_train, num_class)
#keras 内的 to_categorical()将类别向量转换为二进制矩阵类型表示
Y_dev = to_categorical(Y_dev, num_class)
print(Y_train.shape, Y_dev.shape)
#运行代码,获得类别标签的处理图
```

(8) 定义产生批数据的迭代器,相关代码如下:

```python
batch_size = 16
def batch_generator(x, y, batch_size = batch_size):         #批次处理
    offset = 0
    while True:
        offset += batch_size
        if offset == batch_size or offset >= len(x):
            #将序列的所有元素随机排序
            x, y = shuffle(x, y)
            offset = batch_size
        X_batch = x[offset - batch_size: offset]
        Y_batch = y[offset - batch_size: offset]
        #yield 返回生成器
        yield (X_batch, Y_batch)
```

4.3.2 模型构建

数据加载进模型之后,需要定义模型结构并优化损失函数。

1. 定义模型结构

卷积层使用带洞因果卷积，卷积后的感知范围与卷积层数呈现指数级增长关系。WaveNet 模型是一种序列生成器，用于语音建模，在语音合成的声学建模中，可以直接学习采样值序列的映射，通过先前的信号序列预测下一个时刻点值的深度神经网络模型，具有自回归的特点。相关代码如下：

```
epochs = 10  # 迭代次数
num_blocks = 3
filters = 128
# 层叠
drop_rate = 0.25
# 防止过拟合
X = Input(shape = (None, mfcc_dim,), dtype = 'float32')
# 一维卷积
def conv1d(inputs, filters, kernel_size, dilation_rate):
    return Conv1D(filters = filters, kernel_size = kernel_size, strides = 1, padding = 'causal', activation = None, dilation_rate = dilation_rate)(inputs)
# 步长 strides 为 1
# 参数 padding = 'causal' 即采用因果卷积
def batchnorm(inputs):  # 批规范化函数
    return BatchNormalization()(inputs)  # BN 算法，每一层后增加了归一化层
def activation(inputs, activation):
# 定义激活函数，实现神经元输入/输出之间的非线性化
    return Activation(activation)(inputs)
def res_block(inputs, filters, kernel_size, dilation_rate):
# 残差块
    hf = activation(batchnorm(conv1d(inputs, filters, kernel_size, dilation_rate)), 'tanh')
    hg = activation(batchnorm(conv1d(inputs, filters, kernel_size, dilation_rate)), 'sigmoid')
    h0 = Multiply()([hf, hg])
    ha = activation(batchnorm(conv1d(h0, filters, 1, 1)), 'tanh')
    hs = activation(batchnorm(conv1d(h0, filters, 1, 1)), 'tanh')
    return Add()([ha, inputs]), hs
```

2. 优化损失函数

通过 Adam() 方法进行梯度下降，动态调整每个参数的学习率，进行模型参数优化。

```
# 定义损失函数和优化器
optimizer = Adam(lr = 0.01, clipnorm = 5)
# Adam 利用梯度的一阶矩估计和二阶矩估计动态调整每个参数的学习率
model = Model(inputs = X, outputs = Y)
model.compile(loss = 'categorical_crossentropy', optimizer = optimizer, metrics = ['accuracy'])
# 模块编译，采用交叉熵损失函数
lr_decay = ReduceLROnPlateau(monitor = 'loss', factor = 0.2, patience = 1, min_lr = 0.000)
# ReduceLROnPlateau 基于训练过程中的某些测量值对学习率进行动态下降
```

```python
history = model.fit_generator(    # 使用 fit_generator()函数进行训练
    generator = batch_generator(X_train, Y_train),
    steps_per_epoch = len(X_train) // batch_size,
    epochs = epochs,
    validation_data = batch_generator(X_dev, Y_dev),
    validation_steps = len(X_dev) // batch_size,
callbacks = [checkpointer, lr_decay])
```

4.3.3 模型训练及保存

本部分包括模型训练、模型保存和映射保存。

1. 模型训练

模型训练相关代码如下:

```python
epochs = 10                                               # 参数设置
num_blocks = 3
filters = 128
drop_rate = 0.25
X = Input(shape = (None, mfcc_dim,), dtype = 'float32')   # 输入数据
def conv1d(inputs, filters, kernel_size, dilation_rate):  # 卷积
    return Conv1D(filters = filters, kernel_size = kernel_size, strides = 1, padding = 'causal',
activation = None, dilation_rate = dilation_rate)(inputs)
def batchnorm(inputs):                                    # 批标准化
    return BatchNormalization()(inputs)
def activation(inputs, activation):                       # 激活定义
    return Activation(activation)(inputs)
def res_block(inputs, filters, kernel_size, dilation_rate):  # 残差层
    hf = activation(batchnorm(conv1d(inputs, filters, kernel_size, dilation_rate)), 'tanh')
    hg = activation(batchnorm(conv1d(inputs, filters, kernel_size, dilation_rate)),
'sigmoid')
    h0 = Multiply()([hf, hg])
    ha = activation(batchnorm(conv1d(h0, filters, 1, 1)), 'tanh')
    hs = activation(batchnorm(conv1d(h0, filters, 1, 1)), 'tanh')
    return Add()([ha, inputs]), hs
# 模型训练
h0 = activation(batchnorm(conv1d(X, filters, 1, 1)), 'tanh')
shortcut = []
for i in range(num_blocks):
    for r in [1, 2, 4, 8, 16]:
        h0, s = res_block(h0, filters, 7, r)
        shortcut.append(s)                                # 直连
h1 = activation(Add()(shortcut), 'relu')
h1 = activation(batchnorm(conv1d(h1, filters, 1, 1)), 'relu')
# 参数 batch_size, seq_len, filters
h1 = batchnorm(conv1d(h1, num_class, 1, 1))
```

```
# 参数 batch_size, seq_len, num_class
# 池化
h1 = GlobalMaxPooling1D()(h1)                          # 参数 batch_size, num_class
Y = activation(h1, 'softmax')
h1 = activation(Add()(shortcut), 'relu')
h1 = activation(batchnorm(conv1d(h1, filters, 1, 1)), 'relu')
# 参数 batch_size, seq_len, filters
h1 = batchnorm(conv1d(h1, num_class, 1, 1))
# 参数 batch_size, seq_len, num_class
h1 = GlobalMaxPooling1D()(h1)                          # 参数 batch_size, num_class
Y = activation(h1, 'softmax')
optimizer = Adam(lr = 0.01, clipnorm = 5)
model = Model(inputs = X, outputs = Y)                 # 模型
model.compile(loss = 'categorical_crossentropy', optimizer = optimizer, metrics = ['accuracy'])
checkpointer = ModelCheckpoint(filepath = 'fangyan.h5', verbose = 0)
lr_decay = ReduceLROnPlateau(monitor = 'loss', factor = 0.2, patience = 1, min_lr = 0.000)
history = model.fit_generator(                         # 训练
    generator = batch_generator(X_train, Y_train),
    steps_per_epoch = len(X_train) // batch_size,
    epochs = epochs,
    validation_data = batch_generator(X_dev, Y_dev),
    validation_steps = len(X_dev) // batch_size,
    callbacks = [checkpointer, lr_decay])
```

训练输出结果如图4-8所示。

```
Epoch 1/10
1180/1180 [==============================] - 289s 245ms/step - loss: 0.5402 - acc: 0.7789 - val_loss: 0.6593 - val_acc: 0.7745
Epoch 2/10
1180/1180 [==============================] - 250s 212ms/step - loss: 0.2693 - acc: 0.9030 - val_loss: 0.4539 - val_acc: 0.8474
Epoch 3/10
1180/1180 [==============================] - 250s 212ms/step - loss: 0.1861 - acc: 0.9337 - val_loss: 0.8918 - val_acc: 0.7733
Epoch 4/10
1180/1180 [==============================] - 248s 210ms/step - loss: 0.1347 - acc: 0.9537 - val_loss: 0.4141 - val_acc: 0.8658
Epoch 5/10
1180/1180 [==============================] - 249s 211ms/step - loss: 0.1165 - acc: 0.9579 - val_loss: 0.4204 - val_acc: 0.8707
Epoch 6/10
1180/1180 [==============================] - 257s 217ms/step - loss: 0.0863 - acc: 0.9709 - val_loss: 0.3657 - val_acc: 0.9001
Epoch 7/10
1180/1180 [==============================] - 254s 215ms/step - loss: 0.0862 - acc: 0.9694 - val_loss: 0.6842 - val_acc: 0.8646
Epoch 8/10
1180/1180 [==============================] - 235s 199ms/step - loss: 0.0363 - acc: 0.9885 - val_loss: 0.4591 - val_acc: 0.8799
Epoch 9/10
1180/1180 [==============================] - 244s 206ms/step - loss: 0.0216 - acc: 0.9934 - val_loss: 0.3965 - val_acc: 0.8964
Epoch 10/10
1180/1180 [==============================] - 255s 216ms/step - loss: 0.0179 - acc: 0.9941 - val_loss: 0.4961 - val_acc: 0.8934
```

图 4-8　训练输出结果

通过观察训练集和测试集的损失函数、准确率大小来评估模型的训练程度,进行模型训练的进一步决策。训练集和测试集的损失函数(或准确率)不变且基本相等为模型训练的最佳状态。

可以将训练过程中保存的准确率和损失函数以图的形式表现出来,方便观察。

```python
import matplotlib.pyplot as plt
#解决中文显示问题
plt.rcParams['font.sans-serif'] = ['KaiTi']        #指定默认字体
plt.rcParams['axes.unicode_minus'] = False
#解决保存图像中负号"-"显示为方块的问题
```

2. 模型保存

为了能够在本地服务器调用模型,将模型保存为.h5格式的文件,Keras使用HDF5文件系统来保存模型,在使用过程中,需要Keras提供好的模型导入功能,即可加载模型..h5文件是层次结构。在数据集中还有元数据,即metadata对于每一个dataset而言,除了数据本身之外,这个数据集还有很多的属性信息。HDF5同时支持存储数据集对应的属性信息,所有属性信息的集合叫metadata。相关代码如下:

```python
model = Model(inputs = X, outputs = Y)                        #模型
model.compile(loss = 'categorical_crossentropy', optimizer = optimizer, metrics = ['accuracy'])
                                                              #参数输出
checkpointer = ModelCheckpoint(filepath = 'fangyan.h5', verbose = 0)
#保存模型,保存路径是filepath
```

3. 映射保存

保存方言与类别之间的映射关系,将映射文件保存为.pkl格式,以便调用,pkl是Python保存文件的一种格式,该存储方式可以将Python项目过程中用到的一些临时变量或者需要提取、暂存的字符串、列表、字典等数据保存,使用pickle模块可将任意一个Python对象转换成系统字节。相关代码如下:

```python
with open('resources.pkl', 'wb') as fw:
    pickle.dump([class2id, id2class, mfcc_mean, mfcc_std], fw)
```

4.3.4 模型生成

将训练好的.h5模型文件放入总目录下:信息系统设计/方言种类识别/fangyan.h5。相关代码如下:

```python
#打开映射
with open('resources.pkl', 'rb') as fr:
    [class2id, id2class, mfcc_mean, mfcc_std] = pickle.load(fr)
model = load_model('fangyan.h5')
#glob()提取路径参数
paths = glob.glob('data/*/dev/*/*.pcm')
```

将保存的方言和类别之间映射关系pkl文件放到总文件目录下:信息系统设计/方言种类识别/resources.pkl。相关代码如下:

```python
#打开保存的方言和类别之间的映射文件
with open('resources.pkl', 'rb') as fr:
    [class2id, id2class, mfcc_mean, mfcc_std] = pickle.load(fr)
```

在单机上加载训练好的模型,随机选择一条语音进行分类。新建测试主运行文件 main.py,加载库之后,调用生成的模型文件获得预测结果。相关代码如下:

```python
#glob()提取路径参数
paths = glob.glob('data/*/dev/*/*.pcm')
#通过random模块随机提取一条语音数据
path = np.random.choice(paths, 1)[0]
label = path.split('/')[1]
print(label, path)
#本部分相关代码
# -*- coding:utf-8 -*-
import numpy as np
from keras.models import load_model
from keras.preprocessing.sequence import pad_sequences
import librosa
from python_speech_features import mfcc
import pickle
import wave
import glob
#打开映射
with open('resources.pkl', 'rb') as fr:
    [class2id, id2class, mfcc_mean, mfcc_std] = pickle.load(fr)
model = load_model('fangyan.h5')
#glob()提取路径参数
paths = glob.glob('data/*/dev/*/*.pcm')
#通过random模块随机提取一条语音数据
path = np.random.choice(paths, 1)[0]
label = path.split('/')[1]
print(label, path)
#语音分片处理
mfcc_dim = 13
sr = 16000
min_length = 1 * sr
slice_length = 3 * sr
#提取语音信号的参数
def load_and_trim(path, sr=16000):
    audio = np.memmap(path, dtype='h', mode='r')
    audio = audio[2000:-2000]
    audio = audio.astype(np.float32)
    energy = librosa.feature.rmse(audio)
    frames = np.nonzero(energy >= np.max(energy) / 5)
    indices = librosa.core.frames_to_samples(frames)[1]
    audio = audio[indices[0]:indices[-1]] if indices.size else audio[0:0]
```

```python
        slices = []
        for i in range(0, audio.shape[0], slice_length):
            s = audio[i: i + slice_length]
            slices.append(s)
        return audio, slices
# 提取 MFCC 特征进行测试
audio, slices = load_and_trim(path)
X_data = [mfcc(s, sr, numcep = mfcc_dim) for s in slices]
X_data = [(x - mfcc_mean) / (mfcc_std + 1e-14) for x in X_data]
maxlen = np.max([x.shape[0] for x in X_data])
X_data = pad_sequences(X_data, maxlen, 'float32', padding = 'post', value = 0.0)
print(X_data.shape)
# 预测方言种类并输出
prob = model.predict(X_data)
prob = np.mean(prob, axis = 0)
pred = np.argmax(prob)
prob = prob[pred]
pred = id2class[pred]
print('True:', label)
print('Pred:', pred, 'Confidence:', prob)
```

4.4 系统测试

本部分包括训练准确率及测试效果。

4.4.1 训练准确率

绘制损失函数曲线和准确率曲线,经过 10 轮训练后,准确率将近 100％,验证集准确率在 89％左右。相关代码如下:

```python
train_loss = history.history['loss']                    # 训练损失
valid_loss = history.history['val_loss']                # 验证损失
plt.plot(train_loss, label = '训练集')
plt.plot(valid_loss, label = '验证集')
plt.legend(loc = 'upper right')
plt.xlabel('迭代次数')
plt.ylabel('损失')
plt.show()                                              # 绘图
train_acc = history.history['acc']
valid_acc = history.history['val_acc']
plt.plot(train_acc, label = '训练集')
plt.plot(valid_acc, label = '验证集')
plt.legend(loc = 'upper right')
plt.xlabel('迭代次数')
```

```
plt.ylabel('准确率')
plt.show()
```

随着训练次数的增多,模型在训练数据、测试数据上的损失和准确率逐渐收敛,最终趋于稳定,如图 4-9 和图 4-10 所示。

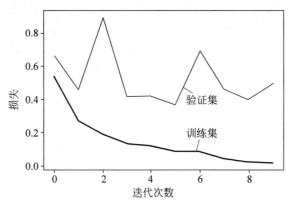

图 4-9 损失函数曲线

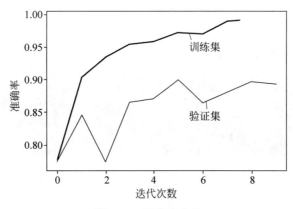

图 4-10 准确率曲线

4.4.2 测试效果

在本地服务器端进行测试,使用 PyCharm 调用保存的模型和映射。设置 PyCharm 运行环境,找到本地 Python 环境并导入,如图 4-11 所示。

从本地随机抽取一段语音进行测试,相关代码如下:

```
#glob()提取路径参数
paths = glob.glob('data/*/dev/*/*.pcm')
#通过 random 模块随机提取一条语音数据
path = np.random.choice(paths, 1)[0]
```

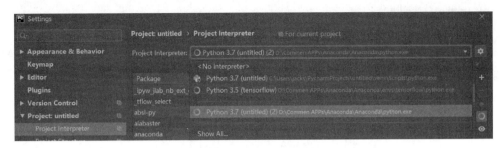

图 4-11 导入 Python 环境

```
label = path.split('/')[1]
print(label, path)
paths = glob.glob('D:/课堂导读/信息系统设计/方言种类分类/data/*/dev/*/*.pcm')
# 预测方言种类并输出
prob = model.predict(X_data)
prob = np.mean(prob, axis = 0)
pred = np.argmax(prob)
prob = prob[pred]
pred = id2class[pred]
print('True:', label)
print('Pred:', pred, 'Confidence:', prob)
```

在 PyCharm 上编辑运行,得到的分类结果与语音片段一致,如图 4-12 所示。

图 4-12 本地测试结果输出

项目 5 行人检测与追踪计数

PROJECT 5

本项目以行人的视频为研究对象,通过 YOLO V3 目标检测算法和 Deep SORT 跟踪算法对人数进行统计。

5.1 总体设计

本部分包括系统整体结构和系统流程。

5.1.1 系统整体结构

系统整体结构如图 5-1 所示。

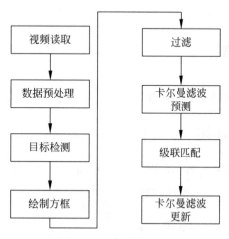

图 5-1 系统整体结构

5.1.2 系统流程

系统流程如图 5-2 所示。

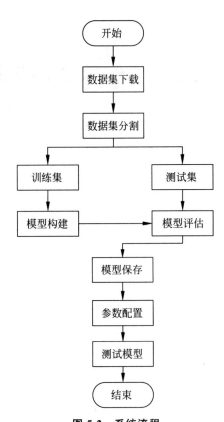

图 5-2 系统流程

5.2 运行环境

本部分包括 Python 环境、TensorFlow 环境、安装所需的软件包和硬件环境。

5.2.1 Python 环境

需要 Python 3.7 及以上配置，在 Windows 环境下推荐下载 Anaconda 完成 Python 所需配置，下载地址为：https://www.anaconda.com/。

5.2.2 TensorFlow 环境

在 https://developer.nvidia.com/cuda-downloads 下载 CUDA 并进行安装。在 https://developer.nvidia.com/rdp/cudnn-archive 下载 cuDNN 并进行安装（要与 CUDA 版本匹配），下载完成后解压到安装好的 CUDA 文件夹中。

打开 CMD，输入命令：

```
nvcc-V
```

检查安装版本,查看是否已经安装成功。

打开 Anaconda Prompt,输入清华仓库镜像,输入命令:

```
conda config -- add channels https://mirrors.tuna.tsinghua.edu.cn/anaconda/pkgs/free/
conda config -set show_channel_urls yes
```

创建 Python 3.7 的环境,名称为 TensorFlow,此时 Python 版本和后面 TensorFlow 的版本有匹配问题,此步选择 Python 3.x,输入命令:

```
conda create -n tensorflow python = 3.7
```

有需要确认的地方,都输入 y。

在 Anaconda Prompt 中激活 TensorFlow 环境,输入命令:

```
activate tensorflow
```

安装完毕。

5.2.3 安装所需的软件包

系统运行版本如下:

```
Keras == 2.3.1
tensorflow-gpu == 2.0.0
opencv-python == 4.2.0.32
scikit-learn == 0.19.2
scipy == 1.3.1
Pillow == 7.0.0
```

将上述语句写入一个.txt 文件内,命名为 Requirements,在 TensorFlow 环境中安装上述软件包,输入命令:

```
pip install -r Requirements.txt
```

安装完毕。

5.2.4 硬件环境

系统运行的硬件环境如下:

(1) 处理器:Intel®Core™ i7-6700HQ @2.60gHz(8CPUs),~2.6GHz。

(2) 显卡:NVIDIA Geforce GTX 960M。

(3) 显示内存:4055MB。

(4) 内存:16384MB RAM。

5.3 模块实现

本项目包括 5 个模块：准备数据、数据预处理、目标检测、目标追踪、主函数。下面分别给出各模块的功能介绍及相关代码。

5.3.1 准备数据

本部分包括数据准备工作及相关代码。

将 YOLO 官网的数据集通过 Keras 转换为.h5 文件，需要下载.weights 文件和 yolov3_keras 版本的源代码。下载地址分别为 https://pjreddie.com/media/files/yolov3.weights 和 https://github.com/qqwweee/keras-yolo3。

在 TensorFlow 环境中运行命令：$ python convert.py model_data/yolov3.cfg model_data/yolov3.weights model_data/yolo.h5，将需要的数据集文件放入 model_data 文件夹中，相关代码如下：

```python
import argparse                            # argparse 的作用是为.py 文件封装可以选择的参数
import configparser                        # configparser 模块读取配置文件
import io
import os
from collections import defaultdict
import numpy as np
from keras import backend as K
from keras.layers import (Conv2D, Input, ZeroPadding2D, Add,
                          UpSampling2D, MaxPooling2D, Concatenate)
from keras.layers.advanced_activations import LeakyReLU
from keras.layers.normalization import BatchNormalization
from keras.models import Model
from keras.regularizers import l2
from keras.utils.vis_utils import plot_model as plot
parser = argparse.ArgumentParser(description = 'Darknet To Keras Converter.')
                                              # 创建一个解析对象
parser.add_argument('config_path', help = 'Path to Darknet cfg file.')
# 向该对象中添加需要关注的命令行参数和选项，每个方法对应一个参数或选项
parser.add_argument('weights_path', help = 'Path to Darknet weights file.')
parser.add_argument('output_path', help = 'Path to output Keras model file.')
parser.add_argument(
    '-p',
    '-plot_model',
    help = 'Plot generated Keras model and save as image.',
    action = 'store_true')
parser.add_argument(
```

```python
        '-w',
        '-weights_only',
        help = 'Save as Keras weights file instead of model file.',
        action = 'store_true')
def unique_config_sections(config_file):
    # 转换所有config部分,使其具有唯一名称,为config解析器的兼容性向配置中添加唯一后缀
    section_counters = defaultdict(int)
    # defaultdict 的作用是:当字典里key不存在但被查找时,返回一个默认值
    output_stream = io.StringIO()           # 表示在内存中以I/O流的方式读写字符串
    with open(config_file) as fin:
        for line in fin:
            if line.startswith('['):
                section = line.strip().strip('[]')
                _section = section + '_' + str(section_counters[section])
                section_counters[section] += 1
                line = line.replace(section, _section)
            output_stream.write(line)
    output_stream.seek(0)                   # 把文件指针移动到文件开始处
    return output_stream
def _main(args):
    config_path = os.path.expanduser(args.config_path)
# 返回参数,以"~"起始的参数被替换成用户主目录;扩展失败或者参数不以"~"开始,则直接返回
# 参数(path)
    weights_path = os.path.expanduser(args.weights_path)
    weights_path = os.path.expanduser(args.weights_path)
    assert config_path.endswith('.cfg'), '{} is not a .cfg file'.format(
        config_path)                        # "断言"是声明其布尔值必须为真的判定,如果发生
                                            # 异常说明表达式为假
    assert weights_path.endswith(
        '.weights'), '{} is not a .weights file'.format(weights_path)
    output_path = os.path.expanduser(args.output_path)
    assert output_path.endswith(
        '.h5'), 'output path {} is not a .h5 file'.format(output_path)
    output_root = os.path.splitext(output_path)[0]
    # 加载权重和配置
    print('Loading weights.')
    weights_file = open(weights_path, 'rb')
    major, minor, revision = np.ndarray(
        shape = (3, ), dtype = 'int32', buffer = weights_file.read(12))
    # ndarray是多维数组对象,它的一个特点是同构,即其中所有元素的类型必须相同
    if (major * 10 + minor) >= 2 and major < 1000 and minor < 1000:
        seen = np.ndarray(shape = (1,), dtype = 'int64', buffer = weights_file.read(8))
    else:
        seen = np.ndarray(shape = (1,), dtype = 'int32', buffer = weights_file.read(4))
    print('Weights Header: ', major, minor, revision, seen)         # 打印参数
    print('Parsing Darknet config.')
    unique_config_file = unique_config_sections(config_path)        # 配置文件
```

```python
cfg_parser = configparser.ConfigParser()
cfg_parser.read_file(unique_config_file)
print('Creating Keras model.')
input_layer = Input(shape=(None, None, 3))
prev_layer = input_layer
all_layers = []
weight_decay = float(cfg_parser['net_0']['decay']          # 衰减加权
    ) if 'net_0' in cfg_parser.sections() else 5e-4
count = 0
out_index = []
for section in cfg_parser.sections():                       # 解析
    print('Parsing section {}'.format(section))
    if section.startswith('convolutional'):
        filters = int(cfg_parser[section]['filters'])
        size = int(cfg_parser[section]['size'])
        stride = int(cfg_parser[section]['stride'])
        pad = int(cfg_parser[section]['pad'])
        activation = cfg_parser[section]['activation']
        batch_normalize = 'batch_normalize' in cfg_parser[section]
        padding = 'same' if pad == 1 and stride == 1 else 'valid'
        # 设置权重
        # Darknet 将卷积权重序列化
        # 格式为[bias / beta, [gamma, mean, variance], conv_weights]
        prev_layer_shape = K.int_shape(prev_layer)
        weights_shape = (size, size, prev_layer_shape[-1], filters)
        darknet_w_shape = (filters, weights_shape[2], size, size)
        weights_size = np.product(weights_shape)             # 加权维度
        print('conv2d', 'bn'
            if batch_normalize else '  ', activation, weights_shape)
        conv_bias = np.ndarray(                              # 卷积偏置
            shape=(filters, ),
            dtype='float32',
            buffer=weights_file.read(filters * 4))
        count += filters
        if batch_normalize:                                  # 批标准化
            bn_weights = np.ndarray(                         # 加权数据
                shape=(3, filters),
                dtype='float32',
                buffer=weights_file.read(filters * 12))
            count += 3 * filters
            bn_weight_list = [                               # 加权列表
                bn_weights[0],
                conv_bias,
                bn_weights[1],
                bn_weights[2]
            ]
        conv_weights = np.ndarray(                           # 卷积加权
```

```python
            shape = darknet_w_shape,
            dtype = 'float32',
            buffer = weights_file.read(weights_size * 4))
        count += weights_size
        #Darknet conv_weights 是 Caffe 的序列化:(out_dim,in_dim,高度,宽度)
        #将它们设置为 TensorFlow 顺序:(高度,宽度,in_dim,out_dim)
        conv_weights = np.transpose(conv_weights, [2, 3, 1, 0])
        conv_weights = [conv_weights] if batch_normalize else [
            conv_weights, conv_bias
        ]
        #激活处理
        act_fn = None
        if activation == 'leaky':
            pass                                       #以后使用
        elif activation != 'linear':
            raise ValueError(
                'Unknown activation function `{}` in section {}'.format(
                    activation, section))
        #创建二维卷积层
        if stride>1:
            #Darknet 使用的左侧和顶部填充不是相同模式
            prev_layer = ZeroPadding2D(((1,0),(1,0)))(prev_layer)
        conv_layer = (Conv2D(                          #卷积层
            filters, (size, size),
            strides = (stride, stride),
            kernel_regularizer = l2(weight_decay),
            use_bias = not batch_normalize,
            weights = conv_weights,
            activation = act_fn,
            padding = padding))(prev_layer)
        if batch_normalize:                            #批标准化
            conv_layer = (BatchNormalization(
                weights = bn_weight_list))(conv_layer)
        prev_layer = conv_layer
        if activation == 'linear':  #线性激活
            all_layers.append(prev_layer)
        elif activation == 'leaky':  #使用 Leaky ReLU
            act_layer = LeakyReLU(alpha = 0.1)(prev_layer)
            prev_layer = act_layer
            all_layers.append(act_layer)
    elif section.startswith('route'):  #始于路由
        ids = [int(i) for i in cfg_parser[section]['layers'].split(',')]
        layers = [all_layers[i] for i in ids]
        if len(layers) > 1:
            print('Concatenating route layers:', layers)
            concatenate_layer = Concatenate()(layers)
            all_layers.append(concatenate_layer)
```

```python
                prev_layer = concatenate_layer
            else:
                skip_layer = layers[0]                              # 只需一层路由
                all_layers.append(skip_layer)
                prev_layer = skip_layer
        elif section.startswith('maxpool'):                         # 始于最大池化
            size = int(cfg_parser[section]['size'])
            stride = int(cfg_parser[section]['stride'])
            all_layers.append(
                MaxPooling2D(                                       # 二维最大池化
                    pool_size=(size, size),
                    strides=(stride, stride),
                    padding='same')(prev_layer))
            prev_layer = all_layers[-1]
        elif section.startswith('shortcut'):                        # 始于直连
            index = int(cfg_parser[section]['from'])
            activation = cfg_parser[section]['activation']
            assert activation == 'linear', 'Only linear activation supported.'
            all_layers.append(Add()([all_layers[index], prev_layer]))
            prev_layer = all_layers[-1]
        elif section.startswith('upsample'):    # 始于上采样
            stride = int(cfg_parser[section]['stride'])
            assert stride == 2, 'Only stride=2 supported.'
            all_layers.append(UpSampling2D(stride)(prev_layer))
            prev_layer = all_layers[-1]
        elif section.startswith('yolo'):                            # 始于 yolo
            out_index.append(len(all_layers)-1)
            all_layers.append(None)
            prev_layer = all_layers[-1]
        elif section.startswith('net'):                             # 始于 net
            pass
        else:
            raise ValueError(
                'Unsupported section header type: {}'.format(section))
    # 创建并保存模型
    if len(out_index) == 0: out_index.append(len(all_layers)-1)
    model = Model(inputs=input_layer, outputs=[all_layers[i] for i in out_index])
    print(model.summary())
    if args.weights_only:
        model.save_weights('{}'.format(output_path))
        print('Saved Keras weights to {}'.format(output_path))
    else:
        model.save('{}'.format(output_path))
        print('Saved Keras model to {}'.format(output_path))
    # 检查是否已读取所有权重
    remaining_weights = len(weights_file.read()) / 4
    weights_file.close()
```

```
        print('Read {} of {} from Darknet weights.'.format(count, count +
                                                    remaining_weights))
    if remaining_weights > 0:
        print('Warning: {} unused weights'.format(remaining_weights))
    if args.plot_model:
        plot(model, to_file = '{}.png'.format(output_root), show_shapes = True)
        print('Saved model plot to {}.png'.format(output_root))
if __name__ == '__main__':
    _main(parser.parse_args())                   #最后调用解析方法,成功之后即可使用
```

Deep_Sort 下载与特征训练:从 https://github.com/nwojke/deep_sort 下载 Deep_Sort 的代码,这里使用的是 MARS 训练集,因数据量过大,可直接使用训练好的 .pb 文件。

MARS 是 Market-1501 数据集的扩展。视频收集过程中,在清华大学校园内放置了 6 个同步摄像机。有 5 个 1080×1920 高清摄像机和 1 个 640×480 的 SD 摄像机。视频内包含 1261 个不同的行人,每个行人至少被 2 个摄像机捕获。加载训练模型的代码如下:

```
self.model_path = './model_data/yolo.h5'
self.anchors_path = 'model_data/yolo_anchors.txt'
self.classes_path = 'model_data/coco_classes.txt'
model_filename = 'model_data/market1501.pb'
```

5.3.2 数据预处理

输入目标视频后,需要对其进行预处理。

1. 视频读取

视频读取相关代码如下:

```
video_capture = cv2.VideoCapture(args["input"])          #打开路径中的视频
ret, frame = video_capture.read()
#ret(布尔值)表示是否读取到图片,frame 表示截取一帧图片
if ret != True:
    break
```

2. 获取视频流帧的宽度和高度

获取视频流帧的宽度和高度代码如下:

```
w = int(video_capture.get(3))                            #获取视频流帧的宽度
h = int(video_capture.get(4))                            #获取视频流帧的高度
```

3. 指定写入视频帧编码格式为 MJPG

相关代码如下:

```
fourcc = cv2.VideoWriter_fourcc( * 'MJPG')
#OpenCV 转换成 PIL.Image 格式
```

```python
image = Image.fromarray(frame[...,::-1])
```

5.3.3 目标检测

训练模型准备和数据预处理相关工作完成之后，使用 YOLO V3 进行目标检测，寻找到画面中相应的类别。

1. 目标识别

目标识别相关代码如下：

```python
boxs, class_names = yolo.detect_image(image)
features = encoder(frame, boxs)
# 这里的置信分数是1.0
detections = [Detection(bbox, 1.0, feature) for bbox, feature in zip(boxs, features)]
# 运行非最大值抑制
boxes = np.array([d.tlwh for d in detections])
scores = np.array([d.confidence for d in detections])
indices = preprocessing.non_max_suppression(boxes, nms_max_overlap, scores)
# 抑制重复检测
detections = [detections[i] for i in indices]
```

preprocessing.non_max_suppression 用于抑制重复检测，使计数更加精准，函数代码如下：

```python
def non_max_suppression(boxes, max_bbox_overlap, scores=None):  # 非最大抑制
    if len(boxes) == 0:
        return []
    boxes = boxes.astype(np.float)                              # 检测框
    pick = []
    x1 = boxes[:, 0]
    y1 = boxes[:, 1]
    x2 = boxes[:, 2] + boxes[:, 0]
    y2 = boxes[:, 3] + boxes[:, 1]
    area = (x2 - x1 + 1) * (y2 - y1 + 1)
    if scores is not None:
        idxs = np.argsort(scores)                               # 索引
    else:
        idxs = np.argsort(y2)
    while len(idxs) > 0:
        last = len(idxs) - 1
        i = idxs[last]
        pick.append(i)
        xx1 = np.maximum(x1[i], x1[idxs[:last]])
        yy1 = np.maximum(y1[i], y1[idxs[:last]])
        xx2 = np.minimum(x2[i], x2[idxs[:last]])
        yy2 = np.minimum(y2[i], y2[idxs[:last]])
        w = np.maximum(0, xx2 - xx1 + 1)
        h = np.maximum(0, yy2 - yy1 + 1)
        overlap = (w * h) / area[idxs[:last]]                   # 计算重叠区域
```

```
        idxs = np.delete(
            idxs, np.concatenate(
                ([last], np.where(overlap > max_bbox_overlap)[0])))
return pick
```

2. 绘制方框

为使物体能被清晰检测到,需要绘制方框标识物体位置。

```
bbox = track.to_tlbr()                                      #以边界框格式获取当前位置
color = [int(c) for c in COLORS[indexIDs[i] % len(COLORS)]]
cv2.rectangle(frame, (int(bbox[0]), int(bbox[1])), (int(bbox[2]), int(bbox[3])),(color), 3)
cv2.putText(frame,str(track.track_id),(int(bbox[0]), int(bbox[1] -50)),0, 5e - 3 * 150,
(color),2)                                                  #添加文字
if len(class_names) > 0:
    class_name = class_names[0]
    cv2.putText(frame, str(class_names[0]),(int(bbox[0]), int(bbox[1] -20)),0, 5e - 3 * 150,
(color),2)
```

3. 目标计数

目标计数相关代码如下:

```
indexIDs.append(int(track.track_id))
counter.append(int(track.track_id))
```

5.3.4 目标追踪

目标追踪包括两部分,一是对被检测到的物体进行预测,将跟踪状态分布向前传播一步;二是执行测量更新和跟踪管理。

1. predict()

对被检测到的物体进行预测,具体函数定义如下:

```
def predict(self):
    #前进一个时间步长传播轨道状态分布,该函数应该在每次 update 之前调用一次
    for track in self.tracks:
        track.predict(self.kf)
    #由卡尔曼滤波器预测状态分布,记录每个轨迹的均值和方差作为滤波器的输入
```

2. update()

本部分相关代码如下:

```
def update(self, detections):
    #执行测量更新和跟踪管理
    #调用_match进行级联匹配
    matches, unmatched_tracks, unmatched_detections = \
```

```python
            self._match(detections)
        # 根据匹配结果更新轨迹集合
        for track_idx, detection_idx in matches:
            self.tracks[track_idx].update(
                self.kf, detections[detection_idx])
        for track_idx in unmatched_tracks:
            self.tracks[track_idx].mark_missed()
        for detection_idx in unmatched_detections:
            self._initiate_track(detections[detection_idx])
        self.tracks = [t for t in self.tracks if not t.is_deleted()]
        # 传入特征列表及其对应ID,构造一个活跃目标的特征字典
        active_targets = [t.track_id for t in self.tracks if t.is_confirmed()]
        features, targets = [], []
        for track in self.tracks:
            if not track.is_confirmed():
                continue
            features += track.features
            targets += [track.track_id for _ in track.features]
            track.features = []
        self.metric.partial_fit(
            np.asarray(features), np.asarray(targets), active_targets)
```

3. 绘制物体运动路径

绘制物体运动路径的相关代码如下:

```python
for j in range(1, len(pts[track.track_id])):
    if pts[track.track_id][j - 1] is None or pts[track.track_id][j] is None:
        continue
    thickness = int(np.sqrt(64 / float(j + 1)) * 2)           # 线条粗细
    cv2.line(frame,(pts[track.track_id][j-1]),
(pts[track.track_id][j]),(color),thickness)                   # 画线
```

5.3.5 主函数

主函数相关代码如下:

```python
from __future__ import division, print_function, absolute_import   # 导入模块
import os
import datetime
from timeit import import time
import warnings
import cv2
import numpy as np
import argparse
from PIL import Image
from yolo import YOLO
```

```python
from deep_sort import preprocessing
from deep_sort import nn_matching
from deep_sort.detection import Detection
from deep_sort.tracker import Tracker
from tools import generate_detections as gdet
from deep_sort.detection import Detection as ddet
from collections import deque
from keras import backend
backend.clear_session()
#解析命令行参数
ap = argparse.ArgumentParser()
ap.add_argument("-i", "--input",help = "path to input video", default = "E:\项目实战-目标追踪\multi-object-tracking\videos\soccer_01.mp4")
ap.add_argument("-c", "--class",help = "name of class", default = "person")
args = vars(ap.parse_args())
pts = [deque(maxlen = 30) for _ in range(9999)]
warnings.filterwarnings('ignore')
#初始化颜色列表,表示每个可能的类标签
np.random.seed(100)
COLORS = np.random.randint(0, 255, size = (200, 3),         #生成200×3个0~255的随机数
    dtype = "uint8")
def main(yolo):
    start = time.time()
    #参数设定
    max_cosine_distance = 0.5                               #余弦距离的控制阈值
    nn_budget = None
    nms_max_overlap = 0.3                                   #非极大抑制的阈值
    counter = []
    #deep_sort
    model_filename = 'model_data/market1501.pb'
    encoder = gdet.create_box_encoder(model_filename,batch_size = 1)
    #最近邻距离度量,每个目标返回到已观察所有样本的最近距离,由距离度量方法构造Tracker
    metric = nn_matching.NearestNeighborDistanceMetric("cosine", max_cosine_distance, nn_budget)
    tracker = Tracker(metric)
    writeVideo_flag = True
    video_capture = cv2.VideoCapture(args["input"])         #打开路径中的视频
    if writeVideo_flag:
    #定义编解码器并创建VideoWriter对象
        w = int(video_capture.get(3))                       #获取视频流帧的宽度
        h = int(video_capture.get(4))                       #获取视频流帧的高度
        fourcc = cv2.VideoWriter_fourcc( * 'MJPG')
        out = cv2.VideoWriter('./output/' + args["input"][43:57] + "_" + args["class"] + '_output.avi', fourcc, 15, (w, h)) #视频输出
        list_file = open('detection.txt', 'w')
        frame_index = -1
    fps = 0.0
```

```python
while True:
    ret, frame = video_capture.read()          # ret(布尔值)表示是否读取到图
                                                # 片,frame 表示截取到一帧图片
    if ret != True:
        break
    t1 = time.time()
    image = Image.fromarray(frame[...,::-1])   # OpenCV 转换成 PIL.Image 格式
    boxs,class_names = yolo.detect_image(image)
    features = encoder(frame,boxs)
    # 这里的置信分数是 1.0
    detections = [Detection(bbox, 1.0, feature) for bbox, feature in zip(boxs, features)]
    # 运行非最大值抑制
    boxes = np.array([d.tlwh for d in detections])
    scores = np.array([d.confidence for d in detections])
    indices = preprocessing.non_max_suppression(boxes, nms_max_overlap, scores)
                                                # 抑制重复的检测
    detections = [detections[i] for i in indices]
    tracker.predict()                           # 将跟踪状态分布向前传播一步
    tracker.update(detections)                  # 执行测量更新和跟踪管理
    i = int(0)
    indexIDs = []
    c = []
    boxes = []
    for det in detections:
        bbox = det.to_tlbr()                    # 以边界框格式获取当前位置
        cv2.rectangle(frame,(int(bbox[0]), int(bbox[1])), (int(bbox[2]), int(bbox[3])),(255,255,255), 2)        # 画矩形框
    for track in tracker.tracks:
        if not track.is_confirmed() or track.time_since_update > 1:
            continue
        indexIDs.append(int(track.track_id))
        counter.append(int(track.track_id))
        bbox = track.to_tlbr()                  # 以边界框格式获取当前位置
        color = [int(c) for c in COLORS[indexIDs[i] % len(COLORS)]]
        cv2.rectangle(frame, (int(bbox[0]), int(bbox[1])), (int(bbox[2]), int(bbox[3])),(color), 3)
        cv2.putText(frame,str(track.track_id),(int(bbox[0]), int(bbox[1] -50)),0, 5e-3 * 150, (color),2)        # 添加文字
        if len(class_names) > 0:
            class_name = class_names[0]
            cv2.putText(frame, str(class_names[0]),(int(bbox[0]), int(bbox[1] -20)),0, 5e-3 * 150, (color),2)
        i += 1
        # bbox_center_point(x,y)
        center = (int(((bbox[0]) + (bbox[2]))/2), int(((bbox[1]) + (bbox[3]))/2))
        pts[track.track_id].append(center)
        thickness = 5
```

```python
            #中心点
            cv2.circle(frame, (center), 1, color, thickness)
        #绘制运动路径
            for j in range(1, len(pts[track.track_id])):
                if pts[track.track_id][j - 1] is None or pts[track.track_id][j] is None:
                    continue
                thickness = int(np.sqrt(64 / float(j + 1)) * 2)          #线条粗细
                cv2.line(frame,(pts[track.track_id][j-1]), (pts[track.track_id][j]),(color),thickness)                                                            #画线
        count = len(set(counter))
        cv2.putText(frame, "Total Object Counter: " + str(count),(int(20), int(120)),0, 5e-3 * 200,(0,255,0),2)
        cv2.putText(frame, "Current Object Counter: " + str(i),(int(20), int(80)),0, 5e-3 * 200, (0,255,0),2)
        cv2.putText(frame, "FPS: %f"%(fps),(int(20), int(40)),0, 5e-3 * 200, (0,255,0),3)
        cv2.namedWindow("YOLO3_Deep_SORT", 0);
        cv2.resizeWindow('YOLO3_Deep_SORT', 1024, 768);
        cv2.imshow('YOLO3_Deep_SORT', frame)
        if writeVideo_flag:
            #存储一帧
            out.write(frame)
            frame_index = frame_index + 1
            list_file.write(str(frame_index) + ' ')
            if len(boxs) != 0:
                for i in range(0,len(boxs)):
                    list_file.write(str(boxs[i][0]) + ' '+str(boxs[i][1]) + ' '+str(boxs[i][2]) + ' '+str(boxs[i][3]) + ' ')
            list_file.write('\n')
        fps = ( fps + (1./(time.time()-t1)) ) / 2
        #print(set(counter))
        #按下Q键停止
        if cv2.waitKey(1) & 0xFF == ord('q'):
            break
    print(" ")
    print("[Finish]")
    end = time.time()
    if len(pts[track.track_id]) != None:
       print(args["input"][43:57] + ": " + str(count) + " " + str(class_name) + ' Found')
    else:
       print("[No Found]")
    video_capture.release()
    if writeVideo_flag:
        out.release()
        list_file.close()
    cv2.destroyAllWindows()
if __name__ == '__main__':
    main(YOLO())
```

5.4 系统测试

运行效果如图 5-3~图 5-5 所示。

图 5-3　测试图 1　　　　　　　　　图 5-4　测试图 2

图 5-5　测试图 3

项目 6　智能果实采摘指导系统

PROJECT 6

本项目使用 Keras 框架,引入 CNN 进行模型训练,采用 Dropout 梯度下降算法,按比例丢弃部分神经元,实现自动化远程监测果实成熟度的功能,为果农提供采摘指导,有利于节约劳动力,提高生产效率,提升经济效益。

6.1　总体设计

本部分包括系统整体结构和系统流程。

6.1.1　系统整体结构

系统整体结构如图 6-1 所示。

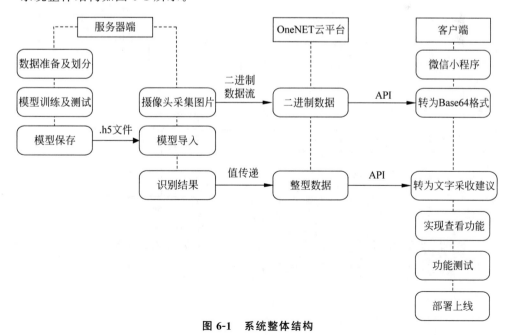

图 6-1　系统整体结构

6.1.2 系统流程

模型训练流程如图 6-2 所示；数据上传流程如图 6-3 所示；小程序流程如图 6-4 所示。

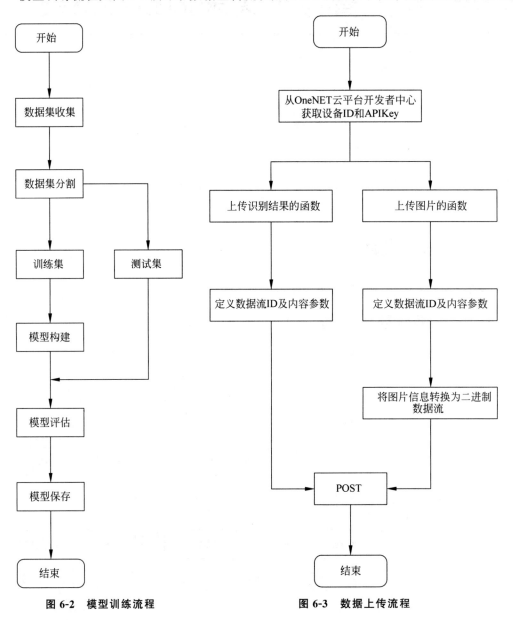

图 6-2　模型训练流程　　　　图 6-3　数据上传流程

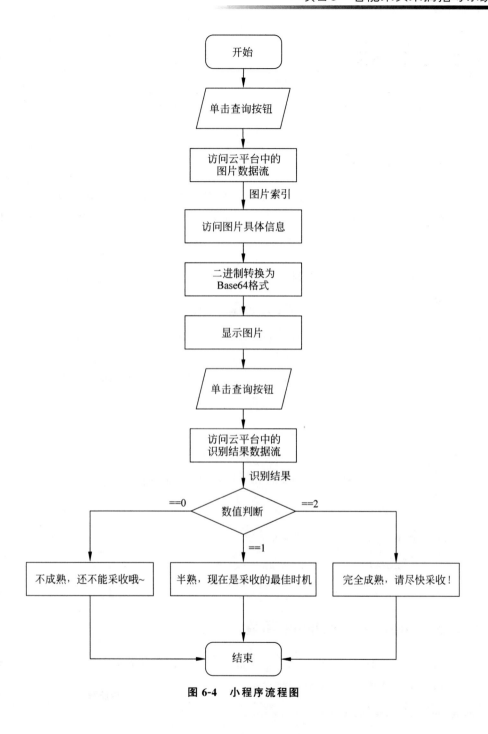

图 6-4 小程序流程图

6.2 运行环境

本部分包括 Python 环境、TensorFlow 环境、Jupyter Notebook 环境、PyCharm 环境、微信开发者工具和 OneNET 云平台。

6.2.1 Python 环境

Python 环境需要 Python 3.6 及以上配置，在 Windows 环境下推荐下载 Anaconda 完成 Python 所需的配置，下载地址为：https://www.anaconda.com/，也可以下载虚拟机在 Linux 环境下运行代码。

6.2.2 TensorFlow 环境

打开 Anaconda Prompt，输入清华仓库镜像，输入命令：

```
conda config -- add channels https://mirrors.tuna.tsinghua.edu.cn/anaconda/pkgs/free/
conda config - set show_channel_urls yes
```

创建 Python 3.6 环境，名称为 TensorFlow，此时 Python 版本和后面 TensorFlow 的版本有匹配问题，此步选择 Python 3.x，输入命令：

```
conda create -n tensorflow python = 3.6
```

有需要确认的地方，都输入 y。
在 Anaconda Prompt 中激活 TensorFlow 环境，输入命令：

```
activate tensorflow
```

安装 CPU 版本的 TensorFlow，输入命令：

```
pip install - upgrade -- ignore-installed tensorflow
```

安装完毕。

6.2.3 Jupyter Notebook 环境

安装 Jupyter Notebook，前提是已安装 Python 2.7 或 Python 3.3 及以上版本。一种方法是使用 Anaconda 安装，在终端输入命令：

```
conda install jupyter notebook
```

另一种方法是使用 pip 命令安装，把 pip 升级到最新版本，输入命令：

```
pip install -upgrade pip
```

再安装 Jupyter Notebook,输入命令:

pip install jupyter

安装完毕。

6.2.4　PyCharm 环境

PyCharm 下载地址为:http://www.jetbrains.com/pycharm/download/#section=windows。选择需要的版本下载,将下载好的 Python 解释器导入 PyCharm 即安装配置完毕。

6.2.5　微信开发者工具

关于微信开发者工具的使用教程,参考用户手册地址如下:

https://developers.weixin.qq.com/miniprogram/dev/devtools/devtools.html。

App ID、小程序账号申请地址为:

https://mp.weixin.qq.com/wxopen/waregister?action=step1。

6.2.6　OneNET 云平台

OneNET 云平台下载地址为:https://open.iot.10086.cn/。注册账号后,可进入开发者中心新建、管理产品。该平台提供多种服务,用户可根据需求选择新建不同类型的产品。本项目使用多协议接入服务,选择 HTTP 协议。

创建产品后,平台会自动分配产品 ID、Master-APIKey 等信息。其中,Master-APIKey 具有最高权限,能够访问产品中的所有设备和数据,查看时需要验证身份,如图 6-5 所示。

图 6-5　产品概况

使用平台进行数据的存储和传输,需要在产品中添加设备,可在"设备列表"中进行添加,如图 6-6 所示。

图 6-6　设备列表

添加设备后,平台会自动分配设备 ID、APIKey 等信息,可在设备详情中查看,外部将通过这些信息访问平台上的数据。当 APIKey 权限不足时,替换为 Master-APIKey,如图 6-7 所示。

图 6-7　设备详情页

同一设备能够上传多个数据流,具体信息在展示页中管理和查看,平台会记录每条数据流的全部历史数据。如果是数值数据,会自动绘制变化曲线图,如图 6-8 所示。

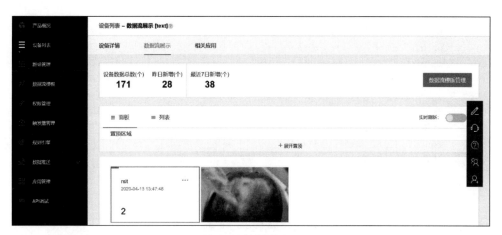

图 6-8　数据流展示

6.3　模块实现

本项目包括 5 个模块：数据预处理、创建模型与编译、模型训练及保存、上传结果、小程序开发。下面分别给出各模块的功能介绍及相关代码。

6.3.1　数据预处理

以红枣为实验对象，在互联网上爬取 1000 张图片作为数据集。

1. 爬取功能

采用 Python 语言自定义爬虫程序，相关代码如下：

```
#定义爬虫函数
def pa():
    try:
        #对输入变量进行控制,预防程序发生错误
        start = 1
        end = int(float(name1.get()))
        #创建下载目录,可以修改成其他目录,也可以下载到现有目录
        if os.path.exists(file_path.get()) is False:
            os.makedirs(file_path.get())
        #打开谷歌浏览器
        browser = webdriver.Chrome()
        browser.get("http://image.baidu.com")
        #print(browser.page_source)
        find_1 = browser.find_element_by_id("kw")
        print('---------------------------- ')
        print(find_1)
```

```python
            find_1.send_keys(keyWord.get())
            btn = browser.find_element_by_class_name("s_search")
            print('----------------------------')
            print(btn)
            btn.click()
            url = browser.find_element_by_name("pn0").get_attribute('href')
            browser.get(url)
            # 设置下载的图片数量
            for i in range(start, end + 1):
                # 获取图片位置
                img = browser.find_elements_by_xpath("//img[@class = 'currentImg']")
                for ele in img:
                    # 获取图片链接
                    target_url = ele.get_attribute("src")
                    # 设置图片名称,以图片链接中的名字为基础选取最后25字节为图片名称
                    img_name = target_url.split('/')[-1]
                    filename = os.path.join(file_path.get(), keyWord.get() + "_" + str(i) + ".jpeg")
                    download(target_url, filename)
                # 下一页
                next_page = browser.find_element_by_class_name("img-next")
                next_page.click()
                time.sleep(3)
                # 显示进度
                print('%d / %d' % (i, end))
            # 关闭浏览器
            browser.quit()
        except ValueError:
            tkinter.messagebox.askokcancel("错误提示","你输入的应该是整数")
```

2. 下载功能

使用自动化测试工具获取第一张图片,将下载链接传给下载模块,然后找到下一张图片的按钮并单击,再次获得新图片的下载链接。

```python
# 定义下载函数
def download(url, filename):
    # 检查下载目录是否存在
    if os.path.exists(filename):
        print('file exists!')
        return
    try:
        # 保存下载图片
        r = requests.get(url, stream = True, timeout = 60)
        r.raise_for_status()
        with open(filename, 'wb') as f:
            for chunk in r.iter_content(chunk_size = 1024):
                if chunk:                        # 过滤保持活动的新块
```

```
                    f.write(chunk)
                    f.flush()
            return filename
        except KeyboardInterrupt:
            if os.path.exists(filename):
                os.remove(filename)
            raise KeyboardInterrupt
        except Exception:
            traceback.print_exc()
            if os.path.exists(filename):
                os.remove(filename)
#获取目录地址
def selectDir():
    global file_path
    file_path.set(filedialog.askdirectory())
```

使用 os.file 下载路径,获取图片 URL,同时按照一定规律命名并进行存储,如图 6-9 所示。

图 6-9　爬取界面

6.3.2　创建模型与编译

数据加载进模型之后,需要定义模型结构并优化损失函数。

1. 定义模型结构

定义架构为二个卷积层,在每个卷积层后都连接一个最大池化层及一个全连接层,进行数据的降维。在每个卷积层上都使用 L2 正则化,并引入 Dropout 和 BN 算法,用以消除模型的过拟合问题。

```python
#第一层使用 L2 正则化,正则化系数为 0.01
X = Conv2D(32, (3, 3), strides = (3, 3), name = 'conv0',kernel_regularizer = regularizers.l2(0.01))(X_input)
X = Dropout(rate = 0.05)(X)                                    #梯度下降,rate = 5%
#归一化
#BN 算法,使该层特征值分布重新符合标准正态分布
X = BatchNormalization(axis = 3, name = 'bn0')(X) #axis 要规范化的轴,通常为特征轴
#非线性激活函数 ReLU
X = Activation('relu')(X)
#最大池化层,用来缩减模型的大小,提高计算速度以及提取特征的健壮性
X = MaxPooling2D((2, 2), name = 'max_pool_1')(X)
#第二层
X = Conv2D(4, (3, 3), strides = (3, 3), name = 'conv1',kernel_regularizer = regularizers.l2(0.01))(X)
X = Dropout(rate = 0.15)(X)
X = BatchNormalization(axis = 3, name = 'bn1')(X)
X = Activation('relu')(X)
#最大池化
X = MaxPooling2D((2, 2), name = 'max_pool_2')(X)
X = Flatten()(X)                                               #降维
X = Dense(3, activation = 'softmax', name = 'dense')(X)        #全连接神经网络层
#使用 Keras 模型中的 API 模型
model = Model(inputs = X_input, outputs = X)
```

2. 优化损失函数

确定模型架构后进行编译。Adam 是常用的梯度下降算法,使用它来优化模型参数。这是多类别的分类问题,因此需要使用交叉熵作为损失函数。由于所有标签都带有相似的权重,因此使用精确度作为性能指标。

```python
#定义损失函数、优化器以及评估模型在训练和测试时的性能指标
model.compile(optimizer = 'Adam',loss = 'categorical_crossentropy',metrics = ['accuracy'])
```

6.3.3 模型训练及保存

定义模型架构和编译之后,通过训练集训练,使模型可以识别红枣的成熟程度。这里将使用训练集和测试集来拟合并保存模型。

1. 模型训练

本部分相关代码如下:

```
#model.fit 函数返回一个 history 的对象
#history 属性记录了损失函数和其他指标的数值随 epoch 变化的情况
hist = model.fit(x = train_data, y = train_label,
                 validation_data = [test_data, test_label],
                 epochs = 500, batch_size = 64)
hist.history['val_acc'][0]                              #记录运行输出
preds = model.evaluate(test_data, test_label)
print ("Loss = " + str(preds[0]))
print ("Test Accuracy = " + str(preds[1]))
```

其中,一个 batch 就是在一次前向/后向传播过程用到的训练样例数量,本项目中每一次用 64 张图片进行训练。预处理数据集后,按照 8∶2 的比例划分训练集和测试集,如图 6-10 所示。

```
Epoch 495/500
735/735 [==============================] - 2s 2ms/step - loss: 0.0651 - accuracy: 0.9878 - val_loss: 0.4812 - val_accuracy: 0.8708
Epoch 496/500
735/735 [==============================] - 2s 2ms/step - loss: 0.0528 - accuracy: 0.9932 - val_loss: 0.5734 - val_accuracy: 0.8202
Epoch 497/500
735/735 [==============================] - 2s 2ms/step - loss: 0.0520 - accuracy: 0.9959 - val_loss: 0.4420 - val_accuracy: 0.8483
Epoch 498/500
735/735 [==============================] - 2s 2ms/step - loss: 0.0589 - accuracy: 0.9878 - val_loss: 1.6353 - val_accuracy: 0.6629
Epoch 499/500
735/735 [==============================] - 2s 2ms/step - loss: 0.0636 - accuracy: 0.9891 - val_loss: 0.4699 - val_accuracy: 0.8764
Epoch 500/500
735/735 [==============================] - 2s 2ms/step - loss: 0.0583 - accuracy: 0.9905 - val_loss: 0.5894 - val_accuracy: 0.8820
178/178 [==============================] - 0s 594us/step
Loss = 0.5893724527251855
Test Accuracy = 0.882022500038147
```

图 6-10 训练结果

通过观察训练集和测试集的损失函数、准确率的大小来评估模型的训练程度,并进行模型训练的进一步决策。一般来说,训练集和测试集的损失函数(或准确率)不变且基本相等为模型训练的最佳状态。

可以将训练过程中保存的准确率和损失函数以图片的形式呈现。

```
import numpy as np
import matplotlib.pyplot as plt
#绘制曲线
#解决中文显示问题
plt.rcParams['font.sans-serif'] = ['KaiTi']             #指定默认字体
plt.rcParams['axes.unicode_minus'] = False              #解决保存图像是"-"显示为方块的问题
fig, ax1 = plt.subplots()
ax2 = ax1.twinx()
lns1 = ax1.plot(np.arange(500), loss, label = "Loss")
#按一定间隔显示实现方法
#ax2.plot(200 * np.arange(len(fig_accuracy)), fig_accuracy, 'r')
lns2 = ax2.plot(np.arange(500), acc, 'r', label = "Accuracy")
ax1.set_xlabel('训练轮次')
ax1.set_ylabel('训练损失值')
ax2.set_ylabel('训练准确率')
#合并图例
lns = lns1 + lns2
```

```
labels = ["损失", "准确率"]
#labels = [l.get_label() for l in lns]
plt.legend(lns, labels, loc = 7)
plt.show()
```

2. 模型保存

为能够被 Python 程序读取，将模型保存为.h5 格式，利用 Keras 中的 Model 模块进行保存。模型被保存后，可以被重用，也可以移植到其他环境中使用。

```
from keras.models import Model
model = HappyModel((IMG_H,IMG_W,3))
#保存为.h5 文件
model.save('C:/Users/SeverusSnape/Desktop/myProject/classifier_3.h5')
```

6.3.4 上传结果

上传结果有两种方法：一是调用计算机摄像头拍摄图片，将图片信息转换为二进制数据流后上传至 OneNET 云平台；二是将数字图片输入 Keras 模型中，获取输出后将识别结果上传至 OneNET 云平台。

1. 图片拍摄

图片拍摄的具体操作如下：

（1）调用摄像头需要引入 cv2 类，对数据进行保护。

```
import cv2
```

（2）调用 cv2 类中的 VideoCapture()函数，实现调用计算机内置摄像头的拍摄功能。

```
#调用计算机内置摄像头,参数为 0,如果有其他摄像头可以调整参数为 1 或 2
cap = cv2.VideoCapture(0)
while True:
    #从摄像头读取图片
    sucess,img = cap.read()
    #显示摄像头
    cv2.imshow("img",img)
    #等待时延为 1ms,保持画面的持续
    k = cv2.waitKey(1)
    if k == 27:
        #按 Esc 键退出摄像
        cv2.destroyAllWindows()
        break
    elif k == 13:
        #按回车键保存图片,并退出
        cv2.imwrite('C:/Users/SeverusSnape/Desktop/myProject/images/try.png',img)
        cv2.destroyAllWindows()
```

```
            break
#关闭摄像头
cap.release()
```

2. 模型导入及调用

模型导入及调用的相关操作如下：

（1）把训练好的.h5文件放入myProject项目目录。

（2）加载Keras模型库，调用Keras模型得到预测结果。

from keras.models import load_model

（3）在xxxtsj.ipynb中声明模型存放路径，调用load_model()函数。

```
#加载模型.h5文件
model = load_model('C:/Users/SeverusSnape/Desktop/myProject/classifier_3.h5')
#定义规范化图片大小和像素值的函数
def get_inputs(src=[]):
    rsltData = []
    for s in src:
        input = cv2.imread(s)                              #读入图像
        input = cv2.resize(input, (IMG_W, IMG_H))          #缩放
        input = cv2.cvtColor(input, cv2.COLOR_BGR2RGB)     #将BGR图片转成RGB图片
pre_y = model.predict(np.reshape(input,[1,IMG_H,IMG_W,3]),batch_size=1)
        print(np.argmax(pre_y, axis=1))                    #打印最大概率对应的标签
        a = np.argmax(pre_y,axis=1)                        #必须使用遍历,否则格式不对
        for i in a:
            rsltData.append(i)                             #将最大概率对应的标签加入rsltData
                                                           #列表尾部
    return rsltData
predict_dir = 'C:/Users/SeverusSnape/Desktop/myProject/images/'
#预测图片所在的文件夹
picName = os.listdir(predict_dir)                          #获得文件夹内的文件名
images = []
for testpath in picName:
    fn = os.path.join(predict_dir, testpath)
    if fn.endswith('png'):  #后缀是png的文件会被存入images列表
        picData = fn
        print(picData)                                     #打印被存入的图片地址
        images.append(picData)                             #地址存入images列表
rsltData = get_inputs(images)                              #调用规范化图片函数得到最大概
                                                           #率对应的标签
```

3. 数据上传至OneNET云平台

本部分包括图片信息上传和识别结果上传。

1) 图片信息上传

将图片信息转换成二进制数据流，使用POST方法上传。

```python
#定义图片上传函数
http_put_pic(data):
    url = "http://api.heclouds.com/bindata"
    headers = {
        "Content-Type": "image/png",                    #格式
        "api-key": "93IlIl2tfXddMN8sgQIInc7qbXs=",
    }
    #device_id 是设备 ID
    #datastream_id 是数据流 ID
    querystring = {"device_id": "586488389", "datastream_id": "pic"}
    #流式上传
    with open(data, 'rb') as f:
        requests.post(url, params=querystring, headers=headers, data=f)
```

2) 识别结果上传

因为识别的结果是整型数据,直接使用 POST 方法上传数值。

```python
#定义识别结果上传函数
def http_put_rslt(data):
    url = "http://api.heclouds.com/devices/" + deviceId + '/datapoints'
    d = time.strftime('%Y-%m-%dT%H:%M:%S')
    data = int(data)                                    #将 int64 类型的 NumPy 数据转换成
                                                        #JSON 可识别的 int 型
    values = {"datastreams": [{"id": "rslt", "datapoints": [{"value": data}]}]}
    jdata = json.dumps(values).encode("utf-8")
    request = urllib.request.Request(url, jdata)        #获取链接数据
    request.add_header('api-key', APIKey)
    request.get_method = lambda: 'POST'                 #POST 方法
    request = urllib.request.urlopen(request)
    return request.read()
```

6.3.5 小程序开发

微信小程序用于查看果实图片、获取采摘建议和查询识别结果。

1. 查询图片

查询图片功能采用两重嵌套回调:第一层通过访问图片数据流获取图片的索引目录,传递给第二层;第二层使用图片索引目录访问图片数据流信息,得到图片的二进制数据流。为使图片能够在界面中显示,将二进制数据转换为 Base64 格式,用 that.setData()函数将值传递给 wxml 文件,并在该函数中修改按钮上的 keyword 为"单击查看采收建议",实现按钮功能的切换。

```
//回调图片
send: function () {
    var that = this
```

```javascript
          if (that.data.keyword == '单击查看你的果园'){
            //多重回调,两次
            const requestPicIndex = wx.request({
              url: 'https://api.heclouds.com/devices/586488389/datapoints?datastream_id=pic',
              header: {
                'content-type': 'application/json',
                'api-key': '93IlIl2tfXddMN8sgQIInc7qbXs='
              },
              success: function (res) {
            var picIndex = res.data.data.datastreams[0].datapoints[0].value.index
                console.log(res.data.data.datastreams[0].datapoints[0].value.index)
//打印图片索引目录;OneNet 上图片的索引
                //嵌套的第二次回调
                const requestTask = wx.request({
                  url: 'http://api.heclouds.com/bindata/' + picIndex,       //图片 URL
                  header: {
                    'content-type': 'application/json',
                    'api-key': 'RSKlDBtVrZ7qDWvK=b6IAyFi=Ow='
                    //master-apikey 可操作 OneNET 上所有信息
                  },
                  responseType: 'arraybuffer',                              //相应类型
                  success: function (res) {
                    console.log(res.data)                                   //打印返回的 data,res 代
                                                                            //表返回数据
                    var data = res.data
                    var base64 = wx.arrayBufferToBase64(res.data)
                    //二进制数据流转换成 Base64 格式
                    base64 = base64.replace(/[\r\n]/g, "")                  //删去换行符
                    that.setData({
                      imgUrl: 'data:image/PNG;base64,' + base64,
                      //能够显示图片的 Base64 格式,传给 wxml
                      keyword: '单击查看采收建议'                            //修改按钮功能为返回采
                                                                            //收建议
                    })
                console.log('http://api.heclouds.com/bindata/' + picIndex)  //打印 URL
                  },
                  fail: function (res) {                                    //异常处理
                    console.log("fail!!!")
                  },
                  complete: function (res) {
                    console.log("end")
                  }
                })
              },
              //回调失败则打印"fail!!!"
              fail: function (res) {
                console.log("fail!!!")
```

```
        },
        //回调完成打印图片URL
        complete: function (res) {
          console.log("end")
        }
      })
    }
  }
```

2. 查询识别结果

得到识别结果后进行一次数值判断:"0"代表未成熟,不适合采收;"1"代表半熟,为最佳采收时机;"2"代表完全成熟,需要尽快采收。通过that.setData()函数赋值给reM,显示采收建议。

```
//回调识别结果
else if (that.data.keyword == '单击查看采收建议'){
        const requestTask = wx.request({
            url: 'https://api.heclouds.com/devices/586488389/datapoints?datastream_id=rslt',
//识别结果的URL
            header: {
                'content-type': 'application/json',
                'api-key': '93IlIl2tfXddMN8sgQIInc7qbXs='
            },
            success: function (res) {
                var app = getApp()
                app.globalData.Zao = res.data.data.datastreams[0]
                var a = app.globalData.Zao.datapoints[0].value
                console.log(app.globalData.Zao)
                //0代表未成熟,不适合采收;1代表半熟,最佳采收时机;2代表完全成熟,尽快采收
                if (a == 2) {
                    console.log(a)
                    that.setData({
                        reM: '完全成熟,请尽快采收!'
                    })
                    console.log('reM:' + that.data.reM)
                }
                else if (a == 1) {
                    console.log(a)
                    that.setData({
                        reM: '半熟,现在是最佳的采收时机'
                    })
                    console.log('reM:' + that.data.reM)
                }
                else if (a == 0) {
                    console.log(a)
                    that.setData({
```

```
          reM: '不成熟,还不能采收哦~'
        })
        console.log('reM:' + that.data.reM)
      }
    },
    //回调失败则打印"fail!!!"
    fail: function (res) {
      console.log("fail!!!")
    },
    //回调完成打印结果
    complete: function (res) {
      console.log("end")
    }
  })
}
```

6.4 系统测试

本部分包括训练准确率、测试效果和外部访问效果。

6.4.1 训练准确率

测试准确率达到 88% 左右,意味着这个预测模型训练比较成功。随着训练轮次的增多,模型在训练数据、测试数据上的损失和准确率逐渐收敛,最终趋于稳定,如图 6-11 所示。

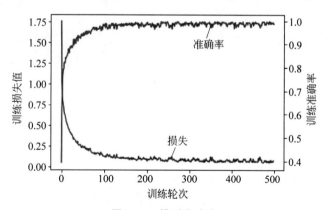

图 6-11　模型准确率

6.4.2 测试效果

将测试集数据代入模型进行测试,并对分类标签与原始数据进行显示和对比,验证了该模型能够实现红枣三类成熟程度的识别。测试结果如图 6-12 所示。

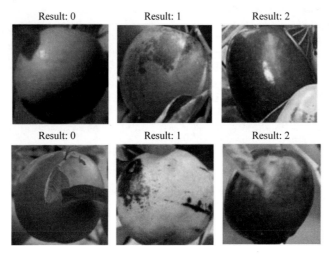

图 6-12　测试结果

6.4.3　外部访问效果

打开小程序，初始界面如图 6-13 所示。

单击界面最上方的"登录"按钮可获得用户微信头像和昵称，如图 6-14 所示。

图 6-13　应用初始界面

图 6-14　登录界面

单击界面中"单击查看你的果园"按钮,在按钮上方会出现果实图片,同时按钮上的文字变成"单击查看采收建议",如图 6-15 所示。

单击"单击查看采收建议"按钮,在按钮下方出现果实成熟度信息和具体采收建议,如图 6-16 所示。

图 6-15　查看采收建议界面　　　　　　图 6-16　查看采收建议界面

移动端测试结果如图 6-17 所示。

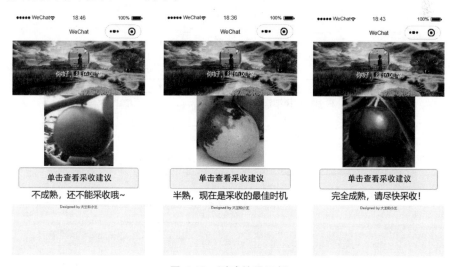

图 6-17　测试结果示例

项目 7 基于 CNN 的猫种类识别

PROJECT 7

本项目基于 CNN 模型训练收集到的数据,采用增强与残差网络结合的方法提高模型性能,实现猫的种类识别。

7.1 总体设计

本部分包括系统整体结构和系统流程。

7.1.1 系统整体结构

系统整体结构如图 7-1 所示。

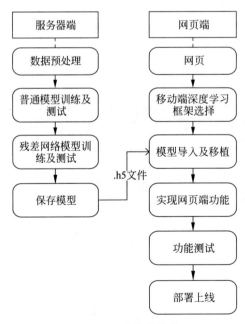

图 7-1 系统整体结构

7.1.2 系统流程

系统流程如图 7-2 所示。

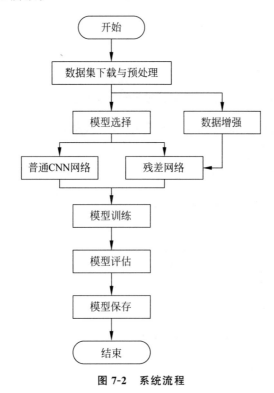

图 7-2 系统流程

7.2 运行环境

本部分包括计算型云服务器、Python 环境、TensorFlow 环境和 MySQL 环境。

7.2.1 计算型云服务器

在阿里云官网注册并充值后,搜索"云服务器 ESC",即可购买计算型云服务器。

付费模式下选择抢占式实例,地域及可用区选择华北 5,类型依次选择异构计算 GPU/FPGA/NPU→GPU 计算型→实例规格:ecs.gn5-c4g1.xlarge。

单台实例规格上限价使用自动出价,数量为 1,镜像选择市场中 CentOS 7.3(预装 NVIDIA GPU 驱动和深度学习框架)V 1.0。

设置密码后,单击"创建实例"即可。远程连接时,输入密码登录。

7.2.2 Python 环境

需要 Python 3.6 及以上配置,以 Linux 环境下安装为例,安装依赖环境,输入命令:

```
yum -y install zlib-devel bzip2-devel openssl-devel ncurses-devel sqlite-devel readline-devel tk-devel gdbm-devel db4-devel libpcap-devel xz-devel
```

下载 Python 3,输入命令:

```
wget https://www.python.org/ftp/python/3.6.1/Python-3.6.1.tgz
```

安装 Python 3,在/usr/local/python3 目录下,输入命令:

```
mkdir -p /usr/local/python3
tar -zxvf Python-3.6.1.tgz
```

进入解压后的目录,编译安装,输入命令:

```
cd Python-3.6.1
./configure -- prefix = /usr/local/python3
```

建立 Python 3 的软链,输入命令:

```
ln -s /usr/local/python3/bin/python3 /usr/bin/python3
```

将/usr/local/python3/bin 加入 PATH,输入命令:

```
vim ~/.bash_profile
.bash_profile
```

获取别名和函数,输入命令:

```
if [ -f ~/.bashrc ]; then
. ~/.bashrc
fi
```

增加新环境的目录,输入命令:

```
PATH = $PATH:$HOME/bin:/usr/local/python3/bin
export PATH
```

按 Esc 键,输入 wq,按回车键退出。使上一步的修改生效,输入命令:

```
source ~/.bash_profile
```

检查 Python 3 及 pip 3 能否正常使用,输入命令:

```
python 3 -V
pip 3 -V
```

7.2.3 TensorFlow 环境

安装 TensorFlow 环境及各种库,升级 pip 3,输入命令:

pip 3 install -- upgrade pip

查询 CUDA 版本,输入命令:

cat /usr/local/cuda/version.txt

查看 cuDNN 版本,输入命令:

cat /usr/local/cuda/include/cudnn.h | grep cuDNN_MAJOR -A 2

安装对应 GPU 版本的 TensorFlow,如图 7-3 所示。

Version:	CPU/GPU:	Python Version:	Compiler:	Build Tools:	cuDNN:	CUDA:
tensorflow-1.6.0	CPU	2.7, 3.3-3.6	GCC 4.8	Bazel 0.9.0	N/A	N/A
tensorflow_gpu-1.6.0	GPU	2.7, 3.3-3.6	GCC 4.8	Bazel 0.9.0	7	9
tensorflow-1.5.0	CPU	2.7, 3.3-3.6	GCC 4.8	Bazel 0.8.0	N/A	N/A
tensorflow_gpu-1.5.0	GPU	2.7, 3.3-3.6	GCC 4.8	Bazel 0.8.0	7	9
tensorflow-1.4.0	CPU	2.7, 3.3-3.6	GCC 4.8	Bazel 0.5.4	N/A	N/A
tensorflow_gpu-1.4.0	GPU	2.7, 3.3-3.6	GCC 4.8	Bazel 0.5.4	6	8
tensorflow-1.3.0	CPU	2.7, 3.3-3.6	GCC 4.8	Bazel 0.4.5	N/A	N/A
tensorflow_gpu-1.3.0	GPU	2.7, 3.3-3.6	GCC 4.8	Bazel 0.4.5	6	8
tensorflow-1.2.0	CPU	2.7, 3.3-3.6	GCC 4.8	Bazel 0.4.5	N/A	N/A
tensorflow_gpu-1.2.0	GPU	2.7, 3.3-3.6	GCC 4.8	Bazel 0.4.5	5.1	8
tensorflow-1.1.0	CPU	2.7, 3.3-3.6	GCC 4.8	Bazel 0.4.2	N/A	N/A
tensorflow_gpu-1.1.0	GPU	2.7, 3.3-3.6	GCC 4.8	Bazel 0.4.2	5.1	8
tensorflow-1.0.0	CPU	2.7, 3.3-3.6	GCC 4.8	Bazel 0.4.2	N/A	N/A
tensorflow_gpu-1.0.0	GPU	2.7, 3.3-3.6	GCC 4.8	Bazel 0.4.2	5.1	8

图 7-3 版本对应图

安装 TensorFlow,输入命令:

pip 3 install tensorflow_gpu == 1.4

安装 TensorFlow 对应的 Keras 库,输入命令:

pip 3 install keras = 2.2.4

安装其他需要使用的库,输入命令:

pip 3 install pillow

```
pip 3 install numpy
pip 3 install h5py
pip 3 install tqdm
```

安装完毕。

7.2.4 MySQL 环境

在 http://www.mysql.com 中下载 MySQL 安装包,选择 Community 版本。

选择 MySQL Community Server,单击 Go to Download Page,打开下载界面,选择本地安装包下载,然后直接下载。

打开下载好的安装包,按照默认设置安装 MySQL(地址可更改)。在 Accounts and Roles 处设置 root 用户名和密码,用于登录数据库。

安装 Navicat for MySQL,便于操作数据库。官网地址为:https://navicat.com.cn/products/navicat-for-mysql,按照默认设置安装即可。

当 Navicat for MySQL 客户端连接到数据库后,鼠标右键"连接名",新建名为 catkind 的数据库,使用 UTF-8 编码。

7.2.5 Django 环境

下载 PyCharm 以及 Anaconda,完成 Python 所需环境的配置,本项目使用 Python 3.6 版本。打开 Anaconda Prompt,输入清华仓库镜像,输入命令:

```
conda config -- add channels https://mirrors.tuna.tsinghua.edu.cn/anaconda/pkgs/free/
conda config - set show_channel_urls yes
```

创建 Python 3.6 的环境,名称为 TensorFlow,输入命令:

```
conda create -n t tensorflow python = 3.6
```

有需要确认的地方,都输入 y。

在 Anaconda Prompt 中激活 TensorFlow 环境,输入命令:

```
activate tensorflow
```

安装 Django,输入命令:

```
pip install django == 1.8.2
pip install pymysql == 0.8.0
```

7.3 模块实现

本项目包括 5 个模块:数据预处理、数据增强、普通 CNN 模型、残差网络模型、模型生成。下面分别给出各模块的功能介绍及相关代码。

7.3.1 数据预处理

打开浏览器,分别搜索布偶猫、孟买猫、暹罗猫和英国短毛猫的图片。用批量下载器下载图片,筛选出特征明显的图片作为数据集。使用的图片包含101张布偶猫、97张孟买猫、101张暹罗猫以及85张英国短毛猫,共计384张图片。

对数据集进行预处理,包括修改图片名、调整格式及大小,将图片按比例划分为训练集和测试集。

```python
import os                                          #导入各种模块
from PIL import Image
import argparse
from tqdm import tqdm
class PrepareData:                                  #准备数据类
    def __init__(self, options):                    #初始化
        self.moudle_name = "prepare data"
        self.options = options
        self.src_images_dir = self.options.src_images_dir
        self.save_img_with = self.options.out_img_size[0]
        self.save_img_height = self.options.out_img_size[1]
        self.save_dir = self.options.save_dir
    #统一图片类型
    def renameJPG(self, filePath, kind):            #图片重命名
        #filePath: 图片文件的路径; kind: 图片的种类标签
        images = os.listdir(filePath)
        for name in images:
            if (name.split('_')[0] in ['0', '1', '2', '3']):
                continue
            else:
                os.rename(filePath + name, filePath + kind + '_' + str(name).split('.')[0] + '.jpg')
    #调用图片处理
    def handle_rename_covert(self):                 #重命名
        save_dir = self.save_dir
        #调用统一图片类型
        list_name = list(os.listdir(self.src_images_dir))
        print(list_name)
        train_dir = os.path.join(save_dir, "train")
        test_dir = os.path.join(save_dir, "test")
        #1.如果已经有存储文件夹,执行后则退出
        if not os.path.exists(save_dir):
            os.mkdir(save_dir)
            os.mkdir(train_dir)
            os.mkdir(test_dir)
```

```python
            list_source = [x for x in os.listdir(self.src_images_dir)]
            #2.获取图片总数
            count_imgs = 0
            for i in range(len(list_name)):
                count_imgs += len(os.listdir(os.path.join(self.src_images_dir, list_name[i])))
            #3.遍历文件夹,并处理每张图片
            for i in range(len(list_name)):
                count = 1
                count_of_each_kind = len(os.listdir(os.path.join(self.src_images_dir, list_name[i])))
                handle_name = os.path.join(self.src_images_dir, list_name[i] + '/')
                self.renameJPG(handle_name, str(i))
                #调用统一图片格式
                img_src_dir = os.path.join(self.src_images_dir, list_source[i])
                for jpgfile in tqdm(os.listdir(handle_name)):
                    img = Image.open(os.path.join(img_src_dir, jpgfile))
                    try:
                        new_img = img.resize((self.save_img_with, self.save_img_height), Image.BILINEAR)
                        if (count > int(count_of_each_kind * self.options.split_rate)):
                            new_img.save(os.path.join(test_dir, os.path.basename(jpgfile)))
                        else:
                            new_img.save(os.path.join(train_dir, os.path.basename(jpgfile)))
                        count += 1
                    except Exception as e:
                        print(e)
#参数设置
def main_args():
    parser = argparse.ArgumentParser()
    parser.add_argument('--src_images_dir', type=str, default='../dataOrig/', help="训练集和测试集的源图片路径")
    parser.add_argument("--split_rate", type=int, default=0.9, help='将训练集二和测试集划分的比例,0.9表示训练集占90%')
    parser.add_argument('--out_img_size', type=tuple, default=(100, 100), help='保存图片的大小,如果使用简单网络结构,大小参数为(100,100);如果使用ResNet,大小参数为(224,224)')
    parser.add_argument("--save_dir", type=str, default='../cat_data_100', help='训练数据的保存位置')
    options = parser.parse_args()
    return options
if __name__ == "__main__":
    #获取参数对象
    options = main_args()
    #获取类对象
    pd_obj = PrepareData(options)
    pd_obj.handle_rename_covert()
```

7.3.2 数据增强

所谓数据增强，是通过翻转、旋转、比例缩放、随机裁剪、移位、添加噪声等操作对现有数据集进行拓展。本项目中数据量较小，无法提取图片的深层特征，使用深层的残差网络时易造成模型过拟合。

```python
from keras.preprocessing.image import ImageDataGenerator, img_to_array, load_img
import argparse, os
from PIL import Image
from tqdm import tqdm                    # 进度条模块
datagen = ImageDataGenerator(
    rotation_range = 40,                 # 整数,数据增强时图片随机转动的角度
    width_shift_range = 0.2,             # 浮点数,图片宽度的某个比例,数据增强时图片水平偏移的幅度
    height_shift_range = 0.2,            # 浮点数,图片高度的某个比例,数据增强时图片竖直偏移的幅度
    rescale = 1. / 255,                  # 重放缩因子,默认为 None
    shear_range = 0.2,                   # 浮点数,剪切强度(逆时针方向的剪切变换角度)
    zoom_range = 0.2,                    # 浮点数或形如[lower,upper]的列表,随机缩放的幅度
                                         # 若为浮点数,则相当于[lower,upper] = [1 - zoom_range, 1 + zoom_range]
    horizontal_flip = True,              # 布尔值,进行随机水平翻转
    vertical_flip = False,               # 布尔值,进行随机竖直翻转
    fill_mode = 'nearest',               # 'constant','nearest','reflect'或'wrap',
                                         # 变换时超出边界的点将根据本参数给定的方法进行处理
    cval = 0,    # 浮点数或整数,当 fill_mode = constant 时,指定向超出边界的点填充值
    channel_shift_range = 0,             # 随机通道转换的范围
)
def data_aug(img_path, save_to_dir, agu_num):
    img = load_img(img_path)
    # 获取被扩充图片的文件名,作为扩充后图片名称的前缀
    save_prefix = os.path.basename(img_path).split('.')[0]
    x = img_to_array(img)
    x = x.reshape((1,) + x.shape)
    i = 0
    for batch in datagen.flow(x, batch_size = 1, save_to_dir = save_to_dir,
                              save_prefix = save_prefix, save_format = 'jpg'):
        i += 1
        # 保存 agu_num 张数据增强图片
        if i >= agu_num:
            break
# 读取文件夹下的图片,并进行数据增强,将结果保存到 dataAug 文件夹下
def handle_muti_aug(options):
    src_images_dir = options.src_images_dir
    save_dir = options.save_dir
    list_name = list(os.listdir(src_images_dir))
    for name in list_name:
        if not os.path.exists(os.path.join(save_dir, name)):
```

```
                    os.mkdir(os.path.join(save_dir, name))
        for i in range(len(list_name)):
            handle_name = os.path.join(src_images_dir, list_name[i] + '/')
            #tqdm()为数据增强添加进度条
            for jpgfile in tqdm(os.listdir(handle_name)):
                #将被扩充的图片保存到增强的文件夹下
Image.open(handle_name + jpgfile).save(save_dir + '/' + list_name[i] + '/' + jpgfile)
                #调用数据增强过程函数
                data_aug(handle_name + jpgfile, os.path.join(options.save_dir, list_name[i]),
options.agu_num)
def main_args():
    parser = argparse.ArgumentParser()
    parser.add_argument('--src_images_dir', type = str, default = '../source_images/', help
= "需要被增强训练集的源图片路径")
    parser.add_argument("--agu_num", type = int, default = 19, help = '每张训练图片需要被增
强的数量,这里设置为19,加上本身的1张,每张图片共计变成20张')
    parser.add_argument("--save_dir", type = str, default = '../dataAug', help = '增强数据的
保存位置')
    options = parser.parse_args()
    return options
if __name__ == "__main__":
    options = main_args()
    handle_muti_aug(options)
```

数据增强进度如图 7-4 所示,数据集拓展为原来的 20 倍,如图 7-5 所示。

图 7-4　数据增强进度

图 7-5　数据增强效果示例

7.3.3 普通 CNN 模型

处理图片数据格式后，转换为数组作为模型的输入，并根据文件名提取标签，定义模型结构、优化器、损失函数和性能指标。本项目使用 Keras 提供类似 VGG 的卷积神经网络。

1. 模型结构

模型结果相关代码如下：

```python
#首先导入相应库
import os
from PIL import Image
import numpy as np
from keras.utils import np_utils
from keras.models import Sequential
from keras.layers.core import Dense, Dropout, Activation, Flatten
from keras.optimizers import SGD, RMSprop, Adam
from keras.layers import Conv2D, MaxPooling2D
import argparse
#将图片转换为数组，并提取标签
def convert_image_array(filename, src_dir):
    img = Image.open(os.path.join(src_dir, filename)).convert('RGB')
    return np.array(img)
def prepare_data(train_or_test_dir):
    x_train_test = []
    #将训练集或者测试集图片转换为数组
    ima1 = os.listdir(train_or_test_dir)
    for i in ima1:
        x_train_test.append(convert_image_array(i, train_or_test_dir))
    x_train_test = np.array(x_train_test)
    #根据文件名提取标签
    y_train_test = []
    for filename in ima1:
        y_train_test.append(int(filename.split('_')[0]))
    y_train_test = np.array(y_train_test)
    #转换标签格式
    y_train_test = np_utils.to_categorical(y_train_test)
    #将特征点从 0~255 转换成 0~1，提高特征提取精度
    x_train_test = x_train_test.astype('float32')
    x_train_test /= 255
    #返回训练和测试数据
    return x_train_test, y_train_test
```

搭建网络模型，定义的架构为 4 个卷积层，每 2 个卷积层后都连接 1 个池化层，进行数据的降维；1 个 Dropout 层，防止模型过拟合；1 个 Flatten 层，把多维输入一维化；最后是全连接层。

```python
def train_model():
    #搭建卷积神经网络
    model = Sequential()
    model.add(Conv2D(32,(3,3),activation = 'relu',input_shape = (100,100,3)))
                                              #提取图像的特征
    model.add(Conv2D(32, (3, 3), activation = 'relu'))
    model.add(MaxPooling2D(pool_size = (2, 2)))
    model.add(Dropout(0.25))              #随机去掉25%的节点权重,防止模型过拟合
    model.add(Conv2D(64, (3, 3), activation = 'relu'))
    model.add(Conv2D(64, (3, 3), activation = 'relu'))
    model.add(MaxPooling2D(pool_size = (2, 2)))
    model.add(Dropout(0.25))
    model.add(Flatten())
    #在Flatten层输入"压平",把多维输入一维化,常用在从卷积层到全连接层的过渡
    model.add(Dense(256, activation = 'relu'))   #Dense是常用的全连接层
    model.add(Dropout(0.5))
    model.add(Dense(4, activation = 'softmax'))
```

2. 模型优化

确定模型架构之后,使用compile()方法对模型进行编译,这是多类别的分类问题,因此需要使用交叉熵作为损失函数。由于所有标签都带有相似的权重,通常使用精确度作为性能指标,使用随机梯度下降算法来优化模型参数。

```python
    sgd = SGD(lr = 0.01, decay = 1e-6, momentum = 0.9, nesterov = True)#SGD优化器
    #完成模型搭建后,使用compile()方法编译
    model.compile(loss = 'categorical_crossentropy',optimizer = sgd,metrics = ['accuracy'])
    return model
```

3. 模型训练

定义模型架构和编译模型后,使用训练集训练模型,使模型可以识别不同种类的猫。这里,将使用训练集和测试集拟合并保存模型。

```python
def main_args():                                              #初始化参数
    parser = argparse.ArgumentParser()
    parser.add_argument('-- train_dir', type = str, default = './cat_data_100/train', help = "the path to the training imgs")
    parser.add_argument('-- test_dir', type = str, default = './cat_data_100/test', help = 'the path to the testing imgs')
    parser.add_argument(" -- save_model", type = str, default = './models/cat_weight.h5', help = 'the path and the model name')
    parser.add_argument(" -- batch_size", type = int, default = 10, help = 'the training batch size of data')
    parser.add_argument(" -- epochs", type = int, default = 32, help = 'the training epochs')
    options = parser.parse_args()
    return options
```

```
#开始模型生成
if __name__ == "__main__":
    #调用函数获取用户参数
    options = main_args()
    #调用函数获取模型
    model = train_model()
    #调用函数获取训练数据和标签
    x_train, y_train = prepare_data(options.train_dir)
    x_test, y_test = prepare_data(options.test_dir)
#使用训练数据按batch进行一定次数的迭代训练网络
model.fit(x_train, y_train, shuffle = True, batch_size = options.batch_size, epochs = options.epochs)
#使用一行代码对模型进行评估,观察模型的指标是否满足要求
score = model.evaluate(x_test, y_test, batch_size = 10)
print("Testing loss:{0},Testing acc:{1}".format(score[0], score[1]))
```

其中,一个batch(批量)就是一次前向/后向传播过程用到的训练样例数量,每次读入10张图片作为一个批量,数据集循环迭代32次。

通过观察训练集和测试集的损失函数、准确率的大小来评估模型的训练程度,进行进一步决策。一般来说,训练集和测试集的损失函数(或准确率)不变且基本相等时,是模型训练的最佳状态。

4. 模型保存

将模型文件保存为.h5格式,以便于移植到其他环境中使用。

```
#保存训练完成的模型文件
save_model = options.save_model
save_model_path = os.path.dirname(save_model)
save_model_name = os.path.basename(save_model)
if not os.path.exists(save_model_path):
    os.mkdir(save_model_path)
model.save_weights(save_model, overwrite = True)
```

7.3.4 残差网络模型

本部分包括残差网络的介绍、模型结构以及模型训练。

1. 残差网络的介绍

网络深度对模型性能至关重要,增加网络层数,可以进行更加复杂的特征提取。但是,深层网络会出现退化问题,即随着网络层数的增加,训练集的损失逐渐下降,然后趋于饱和,当网络深度继续增加时,训练集损失反而会增大。残差网络的思想是把当前层的全部信息映射到下一层,可以有效解决退化问题,优化网络性能。残差网络由一系列残差块组成,残差块分为直接映射部分和残差部分。

2. 模型结构

模型中导入相应库的操作如下：

```python
from __future__ import print_function
import numpy as np
import warnings
from keras.layers import Input
from keras import layers
from keras.layers import Dense
from keras.layers import Activation
from keras.layers import Flatten
from keras.layers import Conv2D
from keras.layers import MaxPooling2D
from keras.layers import GlobalMaxPooling2D
from keras.layers import ZeroPadding2D
from keras.layers import AveragePooling2D
from keras.layers import GlobalAveragePooling2D
from keras.layers import BatchNormalization
from keras.models import Model
from keras.preprocessing import image
import keras.backend as K
from keras.utils import layer_utils
from keras.utils.data_utils import get_file
from keras.applications.imagenet_utils import decode_predictions
from keras.applications.imagenet_utils import preprocess_input
import platform                                #用于平台检测
if platform.system() == "Windows":
    from keras_applications.imagenet_utils import _obtain_input_shape
elif platform.system() == "Linux":
    from keras_applications.imagenet_utils import _obtain_input_shape
from keras.engine.topology import get_source_inputs
```

残差网络模型由 identity_block 和 conv_block 组成。identity_block 与普通的网络相同，包含 3 个卷积层，相关代码如下：

```python
def identity_block(input_tensor, kernel_size, filters, stage, block):
    filters1, filters2, filters3 = filters
    if K.image_data_format() == 'channels_last':
        bn_axis = 3
    else:
        bn_axis = 1
    conv_name_base = 'res' + str(stage) + block + '_branch'
    bn_name_base = 'bn' + str(stage) + block + '_branch'
    #包含 3 个卷积层
    x = Conv2D(filters1, (1, 1), name = conv_name_base + '2a')(input_tensor)
    x = BatchNormalization(axis = bn_axis, name = bn_name_base + '2a')(x)
```

```
    x = Activation('relu')(x)
    x = Conv2D(filters2, kernel_size, padding = 'same', name = conv_name_base + '2b')(x)
    x = BatchNormalization(axis = bn_axis, name = bn_name_base + '2b')(x)
    x = Activation('relu')(x)
    x = Conv2D(filters3, (1, 1), name = conv_name_base + '2c')(x)
    x = BatchNormalization(axis = bn_axis, name = bn_name_base + '2c')(x)
    x = layers.add([x, input_tensor])
    x = Activation('relu')(x)
    return x
```

conv_block 包含 3 个卷积层和 1 个直连,相关代码如下:

```
def conv_block(input_tensor, kernel_size, filters, stage, block, strides = (2, 2)):
                                       #定义卷积块
    filters1, filters2, filters3 = filters
    if K.image_data_format() == 'channels_last':
        bn_axis = 3
    else:
        bn_axis = 1
    conv_name_base = 'res' + str(stage) + block + '_branch'
    bn_name_base = 'bn' + str(stage) + block + '_branch'
    x = Conv2D(filters1, (1, 1), strides = strides, name = conv_name_base + '2a')(input_tensor)
    x = BatchNormalization(axis = bn_axis, name = bn_name_base + '2a')(x)
    x = Activation('relu')(x)
    x = Conv2D(filters2, kernel_size, padding = 'same', name = conv_name_base + '2b')(x)
    x = BatchNormalization(axis = bn_axis, name = bn_name_base + '2b')(x)
    x = Activation('relu')(x)
    x = Conv2D(filters3, (1, 1), name = conv_name_base + '2c')(x)
    x = BatchNormalization(axis = bn_axis, name = bn_name_base + '2c')(x)
    #把输入层的全部信息直接合并到输出
    shortcut = Conv2D(filters3, (1, 1), strides = strides,
                      name = conv_name_base + '1')(input_tensor)
    shortcut = BatchNormalization(axis = bn_axis, name = bn_name_base + '1')(shortcut)
    x = layers.add([x, shortcut])
    x = Activation('relu')(x)
    return x
```

定义了两个模块后,开始搭建残差网络模型,相关代码如下:

```
def ResNet50(include_top = True, weights = 'imagenet',    #定义残差网络
             input_tensor = None, input_shape = None,
             pooling = None,
             classes = 1000):
    if weights not in {'imagenet', 'cat_kind', None}:
        raise ValueError('The `weights` argument should be either '
                         '`None` (random initialization) or `cat_kind` or `imagenet` '
                         '(pre-training on ImageNet).')
    if weights == 'imagenet' and include_top:
```

```python
        classes = 1000
    if weights == 'cat_kind':
        classes = 4
    # 如果在 imagenet 上面微调,并且包含了全连接层,那么类别必须是 1000
    if weights == 'imagenet' and include_top and classes != 1000:
        raise ValueError('If using `weights` as imagenet with `include_top`'
                         ' as true, `classes` should be 1000')
    # 确定合适的输入格式
    input_shape = _obtain_input_shape(input_shape,
                                     default_size = 224,
                                     min_size = 197,
                                     data_format = K.image_data_format(),
                                     # include_top = include_top)
                                     require_flatten = include_top)
    if input_tensor is None:
        img_input = Input(shape = input_shape)
    else:
        if not K.is_keras_tensor(input_tensor):
            img_input = Input(tensor = input_tensor, shape = input_shape)
        else:
            img_input = input_tensor
    if K.image_data_format() == 'channels_last':
        bn_axis = 3
    else:
        bn_axis = 1
    # 构建模型结构
    x = ZeroPadding2D((3, 3))(img_input)
    x = Conv2D(64, (7, 7), strides = (2, 2), name = 'conv1')(x)
    x = BatchNormalization(axis = bn_axis, name = 'bn_conv1')(x)
    x = Activation('relu')(x)
    x = MaxPooling2D((3, 3), strides = (2, 2))(x)
    x = conv_block(x, 3, [64, 64, 256], stage = 2, block = 'a', strides = (1, 1))
    x = identity_block(x, 3, [64, 64, 256], stage = 2, block = 'b')
    x = identity_block(x, 3, [64, 64, 256], stage = 2, block = 'c')
    x = conv_block(x, 3, [128, 128, 512], stage = 3, block = 'a')
    x = identity_block(x, 3, [128, 128, 512], stage = 3, block = 'b')
    x = identity_block(x, 3, [128, 128, 512], stage = 3, block = 'c')
    x = identity_block(x, 3, [128, 128, 512], stage = 3, block = 'd')
    x = conv_block(x, 3, [256, 256, 1024], stage = 4, block = 'a')
    x = identity_block(x, 3, [256, 256, 1024], stage = 4, block = 'b')
    x = identity_block(x, 3, [256, 256, 1024], stage = 4, block = 'c')
    x = identity_block(x, 3, [256, 256, 1024], stage = 4, block = 'd')
    x = identity_block(x, 3, [256, 256, 1024], stage = 4, block = 'e')
    x = identity_block(x, 3, [256, 256, 1024], stage = 4, block = 'f')
    x = conv_block(x, 3, [512, 512, 2048], stage = 5, block = 'a')
    x = identity_block(x, 3, [512, 512, 2048], stage = 5, block = 'b')
    x = identity_block(x, 3, [512, 512, 2048], stage = 5, block = 'c')
```

```python
        x = AveragePooling2D((7, 7), name = 'avg_pool')(x)
    if include_top:
        x = Flatten()(x)
        x = Dense(classes, activation = 'softmax', name = 'fc1000')(x)
    else:
        if pooling == 'avg':
            x = GlobalAveragePooling2D()(x)
        elif pooling == 'max':
            x = GlobalMaxPooling2D()(x)
    #确保模型考虑了 input_tensor 的任何潜在预处理
    if input_tensor is not None:
        inputs = get_source_inputs(input_tensor)
    else:
        inputs = img_input
    #创建模型
    model = Model(inputs, x, name = 'resnet50')
    #加载权重
    if weights == 'imagenet':
        if include_top:
            #调用模型下载,这里由本地提供,所以将其注释
            #weights_path = get_file('resnet50_weights_tf_dim_ordering_tf_kernels.h5',
            #WEIGHTS_PATH,
            #cache_subdir = 'models',
             md5_hash = 'a7b3fe01876f51b976af0dea6bc144eb')
            weights_path = WEIGHTS_PATH_
        else:
            #调用模型下载,这里由本地提供,所以将其注释
            #weights_path = get_file('resnet50_weights_tf_dim_ordering_tf_kernels_notop.h5',
            #WEIGHTS_PATH_NO_TOP,
            #cache_subdir = 'models',
md5_hash = 'a268eb855778b3df3c7506639542a6af')
            weights_path = WEIGHTS_PATH_NO_TOP
        model.load_weights(weights_path)
        if K.backend() == 'theano':
            layer_utils.convert_all_kernels_in_model(model
        if K.image_data_format() == 'channels_first':
            if include_top:
                maxpool = model.get_layer(name = 'avg_pool')
                shape = maxpool.output_shape[1:]
                dense = model.get_layer(name = 'fc1000')
                layer_utils.convert_dense_weights_data_format(dense, shape, 'channels_first')
        if K.backend() == 'tensorflow':
            warnings.warn('You are using the TensorFlow backend, yet you ' 'are using the Theano '
                          'image data format convention '
                          '(`image_data_format = "channels_first"`). '
                          'For best performance, set '
                          '`image_data_format = "channels_last"` in '
```

```python
                            'your Keras config '
                            'at ~/.keras/keras.json.')
    #加载猫种类的权重
    if weights == 'cat_kind':
        WEIGHTS_PATH = '../models/cat_weight_resNet50.h5'
        model.load_weights(WEIGHTS_PATH)
    return model
```

3. 模型训练

用残差网络模型训练数据,相关代码如下:

```python
import os                                    #导入各种模块
from PIL import Image
import numpy as np
from keras.utils import np_utils
from keras.optimizers import SGD, RMSprop, Adam
import argparse
from resnet_example.resnet50 import ResNet50
def convert_image_array(filename, src_dir):      #定义转换图像数组
    img = Image.open(os.path.join(src_dir, filename)).convert('RGB')
    return np.array(img)
def prepare_data(train_or_test_dir):
    x_train_test = []
    #将训练集或者测试集图片转换为数组
    ima1 = os.listdir(train_or_test_dir)
    for i in ima1:
        x_train_test.append(convert_image_array(i, train_or_test_dir))
    x_train_test = np.array(x_train_test)
    #根据文件名提取标签
    y_train_test = []
    for filename in ima1:
        y_train_test.append(int(filename.split('_')[0]))
    y_train_test = np.array(y_train_test)
    #转换标签格式
    y_train_test = np_utils.to_categorical(y_train_test)
    # 将特征点从0~255转换成0~1,提高特征提取精度
    x_train_test = x_train_test.astype('float32')
    x_train_test /= 255
    #返回训练和测试数据
    return x_train_test, y_train_test
def main_args():                              #定义函数参数解析
    parser = argparse.ArgumentParser()
    parser.add_argument('--train_dir', type=str, default='../cat_data_224/train',
                        help="the path to the training imgs")
    parser.add_argument('--test_dir', type=str, default='../cat_data_224/test', help="the path to the testing imgs')
```

```python
    parser.add_argument("--save_model", type=str, default='../models/cat_weight_res.h5'
, help='the path and the model name')
    parser.add_argument("--batch_size", type=int, default=10, help='the training batch size of data')
    parser.add_argument("--epochs", type=int, default=64, help='the training epochs')
    options = parser.parse_args()
    return options
if __name__ == "__main__":
    #调用函数获取用户参数
    options = main_args()
    #搭建卷积神经网络
    #输入大小至少为197x197
    model = ResNet50(weights=None, classes=4)
    #选择在imagenet上进行微调
    #model = ResNet50(include_top=False, weights='imagenet', classes=4)
    sgd = SGD(lr=0.01, decay=1e-6, momentum=0.9, nesterov=True)
    model.compile(loss='categorical_crossentropy', optimizer=sgd, metrics=['accuracy'])
    #调用函数获取训练数据和标签
    x_train, y_train = prepare_data(options.train_dir)
    x_test, y_test = prepare_data(options.test_dir)
    model.fit(x_train, y_train, shuffle=True, batch_size=options.batch_size,
              epochs=options.epochs, validation_data=(x_test, y_test))
```

4. 模型保存

模型保存的相关代码如下:

```python
save_model_path = os.path.dirname(options.save_model)
if not os.path.exists(save_model_path):
    os.mkdir(save_model_path)
#保存模型
model.save_weights(options.save_model, overwrite=True)
score = model.evaluate(x_test, y_test, batch_size=options.batch_size)
print("Testing loss:{0},Testing acc:{1}".format(score[0], score[1]))
```

7.3.5 模型生成

模型应用主要有3部分:一是从本地相册输入猫的图片;二是把输入的图片转换成数据,在输入训练好的模型中进行预测;三是根据预测结果输出数据库中预存的相关信息。

1. 搭建 Django 项目

按快捷键 Win+R,输入 CMD 后回车,打开命令行窗口。

进入虚拟环境:

```
cd H: # Anaconda 的安装路径
cd ProgramData\Anaconda3\Scripts
```

```
activate tensorflow
```

创建 Django 项目：

```
django-admin startproject cat_kind_view
```

进入项目目录，创建一个名为 cat 的应用：

```
cd cat_kind_view
python manage.py startapp cat
```

用 PyCharm 打开创建的项目，如图 7-6 所示。

图 7-6　新建 Django 项目的目录

在 cat_kind_view/cat/views.py 文件中添加网页需要展示的内容，代码如下：

```python
from django.shortcuts import render                    # 导入各种模块
from django.http import HttpResponse
from catclass import settings
from . import models
from .prediction import testcat
import json
# 创建视图
def index(request):
    return render(request, 'index.html')
def catinfo(request):
    if request.method == "POST":
        f1 = request.FILES['pic1']
        # 用于识别
        fname = '%s/pic/%s' % (settings.MEDIA_ROOT, f1.name)
```

```python
            with open(fname, 'wb') as pic:
                for c in f1.chunks():
                    pic.write(c)
            #用于显示
            fname1 = './static/img/%s' % f1.name
            with open(fname1, 'wb') as pic:
                for c in f1.chunks():
                    pic.write(c)
            num = testcat(f1.name)
            if(num == 0):
                num = 4
            #通过ID获取猫的信息
            name = models.Catinfo.objects.get(id = num)
            return render(request, 'info.html', {'nameinfo': name.nameinfo, 'feature': name.feature, 'livemethod': name.livemethod, 'feednn': name.feednn, 'feedmethod': name.feedmethod, 'picname': f1.name})
        else:
            return HttpResponse("上传失败!")
def appupload(request):
    if request.method == "POST":
        f1 = request.FILES['pic1']
        #用于识别
        fname = '%s/pic/%s' % (settings.MEDIA_ROOT, f1.name)
        with open(fname, 'wb') as pic:
            for c in f1.chunks():
                pic.write(c)
        #用于显示
        fname1 = './static/img/%s' % f1.name
        with open(fname1, 'wb') as pic:
            for c in f1.chunks():
                pic.write(c)
        num = testcat(f1.name)
        if (num == 0):
            num = 4
        #通过ID获取猫的信息
        name = models.Catinfo.objects.get(id = num)
        return render(request, 'app.html', {'nameinfo': name.nameinfo, 'feature': name.feature, 'livemethod': name.livemethod, 'feednn': name.feednn, 'feedmethod': name.feedmethod, 'picname': f1.name})
    else:
        return HttpResponse("上传失败!")
#在cat_kind_view/cat目录下新建urls.py
from django.conf.urls import url
from . import views
urlpatterns = [
    url(r'^$', views.index),
    url(r'^info/', views.catinfo),
```

```python
        url(r'^appupload/', views.appupload)
]
# 在 cat_kind_view/cat_kind_view/urls.py 文件中加上新建的 urls
from django.conf.urls import include, url
from django.contrib import admin
urlpatterns = [
        url(r'^admin/', include(admin.site.urls)),
        url(r'^', include('cat.urls')),            # 连接到 cat.urls
]
# 在 cat_kind_view/cat_kind_view/settings.py 中做如下改动
# 定义应用
INSTALLED_APPS = [
        'django.contrib.admin',
        'django.contrib.auth',
        'django.contrib.contenttypes',
        'django.contrib.sessions',
        'django.contrib.messages',
        'django.contrib.staticfiles',
        'cat'                                      # 加上自己的 App
]
# LANGUAGE_CODE = 'en-us'
LANGUAGE_CODE = 'zh-Hans'                          # 修改语言为汉语
# TIME_ZONE = 'UTC'
TIME_ZONE = 'Asia/Shanghai'                        # 修改时区为上海
```

2. 输入图片并预测

创建保存输入图片的文件夹

在 cat_kind_view/cat 目录下添加 prediction.py。此文件的功能是处理输入的图片，并用训练好的模型进行预测。

```python
import os                                          # 导入各种模块
from PIL import Image
import numpy as np
import keras
from keras.models import Sequential
from keras.layers.core import Dense, Dropout, Flatten
from keras.optimizers import SGD
from keras.layers import Conv2D, MaxPooling2D
def prepicture(picname):                           # 定义预处理
    img = Image.open('./media/pic/' + picname)
    new_img = img.resize((100, 100), Image.BILINEAR)
    new_img.save(os.path.join('./media/pic/', os.path.basename(picname)))
def read_image2(filename):
    img = Image.open('./media/pic/' + filename).convert('RGB')
    return np.array(img)
def testcat(picname):
```

```python
#预处理图片大小变成100×100
prepicture(picname)
x_test = []
x_test.append(read_image2(picname))
x_test = np.array(x_test)
x_test = x_test.astype('float32')
x_test /= 255
keras.backend.clear_session()
model = Sequential()
#输入为三通道的100×100图像
#应用32个卷积滤波器,每个大小为3×3
model.add(Conv2D(32,(3,3),activation = 'relu',input_shape = (100,100,3)))
model.add(Conv2D(32, (3, 3), activation = 'relu'))
model.add(MaxPooling2D(pool_size = (2, 2)))
model.add(Dropout(0.25))
model.add(Conv2D(64, (3, 3), activation = 'relu'))
model.add(Conv2D(64, (3, 3), activation = 'relu'))
model.add(MaxPooling2D(pool_size = (2, 2)))
model.add(Dropout(0.25))
model.add(Flatten())
model.add(Dense(256, activation = 'relu'))
model.add(Dropout(0.5))
model.add(Dense(4, activation = 'softmax'))
sgd = SGD(lr = 0.01, decay = 1e-6, momentum = 0.9, nesterov = True)
model.compile(loss = 'categorical_crossentropy', optimizer = sgd, metrics = ['accuracy'])
model.load_weights('./cat/cat_weight.h5') #使用训练好的模型预测
classes = model.predict_classes(x_test)[0]
#target = ['布偶猫', '孟买猫', '暹罗猫', '英国短毛猫']
#print(target[classes])
return classes
```

3. 链接数据库

本部分相关代码如下:

```python
#在cat_kind_view/cat_kind_view/settings.py中修改
DATABASES = {
    'default': {
        'ENGINE': 'django.db.backends.mysql',
        'NAME': 'catkind',                    #数据库名
        'USER': 'root',                       #用户名
        'PASSWORD': 'root',                   #密码
        'HOST': '127.0.0.1',
        'PORT': '3306',
    }
}
```

```
#在 cat_kind_view/cat /__init__.py 中引入数据库
import pymysql
pymysql.version_info = (1, 3, 13, "final", 0)
pymysql.install_as_MySQLdb()
#在 cat_kind_view/cat /__init__.py 中写入数据库的名称和格式
from django.db import models
#创建模型
class Catinfo(models.Model):
    name = models.CharField(max_length=10)
    nameinfo = models.CharField(max_length=1000)
    feature = models.CharField(max_length=1000)
    livemethod = models.CharField(max_length=1000)
    feednn = models.CharField(max_length=1000)
    feedmethod = models.CharField(max_length=1000)
```

在命令行终端输入以下命令,生成迁移文件 0001_initial.py。

`python manage.py makemigrations`

在命令行终端输入以下命令执行迁移,在数据库中生成对应表。

`python manage.py migrate`

在 Navicat for MySQL 中打开数据库,发现由 Django 生成一些列表,如图 7-7 所示,cat_catinfo 是在应用 cat 中由 catinfo 类生成的表。

图 7-7　数据库中的表

在 cat_catinfo 表中添加不同种类猫的相关信息,如图 7-8 所示。

图 7-8　cat_catinfo 表中的信息

这些信息根据预测结果(标签)与主键 ID ——对应,把表中的信息显示在网页上。

4. 美化网页

本部分相关代码如下:

```
# 在 cat_kind_view/cat_kind_view/settings.py 中修改
TEMPLATES = [
    {
        'BACKEND': 'django.template.backends.django.DjangoTemplates',
        'DIRS': [os.path.join(BASE_DIR, 'templates')],   # 加入模板路径
        'APP_DIRS': True,
        'OPTIONS': {
            'context_processors': [
                'django.template.context_processors.debug',
                'django.template.context_processors.request',
                'django.contrib.auth.context_processors.auth',
                'django.contrib.messages.context_processors.messages',
            ],
        },
    },
]
# 静态文件(CSS, JavaScript, Images)
# 参考 https://docs.djangoproject.com/en/3.0/howto/static-files/
STATIC_URL = '/static/'
STATICFILES_DIRS = [
    os.path.join(BASE_DIR, 'static')  # 加上 static 路径,界面美化
]
MEDIA_ROOT = os.path.join(BASE_DIR, "./media")# 加上 media 路径,界面美化
```

美化网页部分代码目录如图 7-9 所示。

图 7-9　美化网页部分代码目录

static/css/style.css 用于设置首页格式，例如背景、字体、各部件的类型、颜色等。

```css
body,ul,li,dl,dt,dd,p,ol,h1,h2,h3,h4,h5,h6,form,img,table,fieldset,legend {
    margin: 0;
    padding: 0;
}
html,body {
    height: 100%;
    width: 100%;
}
ul,li,ol {
    list-style: none;
}
img,fieldset {
    border: 0;
}
img {
    display: block;
}
a {
    text-decoration: none;
    color: #333;
}
h1,h2,h3,h4,h5,h6 {
    font-weight: 100;
}
body {
    font-size: 12px;
    font-family: "微软雅黑";
}
input,a {
    outline: none;
}
* {
    box-sizing: border-box;
}
.bc {
    background: url(../img/image001.jpg) no-repeat;
    background-size: 100% 100%;
}
.main1 {
    width: 100%;
    height: 100%;
    padding-top: 100px;
    padding-left: 100px;
}
.main {
```

```css
        width: 600px;
        height: 500px;
        background-color: rgba(255,255,255,0.3);
        box-shadow: 0px 0px 5px 5px #888888;
}
.title {
        width: 100%;
        height: 55px;
        padding-top: 20px;
}
.title p {
        font-size: 30px;
        text-align: center;
}
.select {
        width: 100%;
        height: 200px;
}
.xfile {
        width: 200px;
        height: 100px;
        display: none;
        border: 0px;
}
.xxfile,.xxsub {
        width: 300px;
        height: 60px;
        background-color: rgba(0,0,0,0.5);
        margin-top: 50px;
        margin-left: 150px;
        font-size: 25px;
        border-radius: 10px;
        border: 0px;
        color: white;
        font-family: "微软雅黑";
        transition: linear 0.2s;
}
.xxfile:hover,.xxsub:hover {
        color: black;
        background-color: rgba(255,255,255,0.5);
}
.select .xxsub {
        margin-top: 30px;
}
.foot {
        width: 100%;
        height: 245px;
```

```css
        padding-top: 30px;
}
.foot ul {
        display: flex;
        justify-content: space-around;
}
.foot dl {
        height: 30px;
        font-size: 16px;
        line-height: 30px;
}
.foot dd {
        height: 40px;
        font-size: 26px;
        line-height: 40px;
}
```

static/css/info.css 用于设置信息页格式,例如背景、字体、各部件的类型、颜色等。

```css
body,ul,li,dl,dt,dd,p,ol,h1,h2,h3,h4,h5,h6,form,img,table,fieldset,legend {
        margin: 0;
        padding: 0;
}
ul,li,ol {
        list-style: none;
}
img,fieldset {
        border: 0;
}
img {
        display: block;
}
a {
        text-decoration: none;
        color: #333;
}
h1,h2,h3,h4,h5,h6 {
        font-weight: 100;
}
body {
        font-size: 12px;
        font-family: "微软雅黑";
        color: black;
}
input,a {
        outline: none;
}
```

```css
* {
    box-sizing: border-box;
}
body {
    background: url(../img/lianzhengaijiujiu003.jpg) no-repeat;
    background-size: cover;
}
.top {
    width: 1000px;
    height: 40px;
    background-color: cornflowerblue;
    margin: 0 auto;
    line-height: 40px;
    font-size: 24px;
    color: white;
    padding-left: 20px;
}
.m1 {
    width: 1000px;
    height: 210px;
    margin: 0 auto;
    margin-top: 20px;
    display: flex;
    justify-content: space-between;
}
.catlogo {
    width: 210px;
    height: 210px;
    padding: 4px;
    border: 1px solid gray;
}
.catlogo img {
    width: 200px;
    height: 200px;
}
.catname {
    width: 770px;
    font-size: 18px;
    padding: 10px;
    text-align: justify;
    background: rgba(255,255,255,0.7);
    border-radius: 10px;
}
.m2,.m3 {
    width: 1000px;
    height: 400px;
    display: flex;
```

```css
            justify-content: space-between;
            margin: 0 auto;
            margin-top: 10px;
        }
        .feature {
            width: 400px;
            height: 400px;
            text-align: justify;
            background: rgba(255,255,255,0.7);
            border-radius: 10px;
        }
        .live {
            width: 560px;
            height: 400px;
            text-align: justify;
            background: rgba(255,255,255,0.7);
            border-radius: 10px;
        }
        .featt,.maintitle,.methodtitle {
            width: 100% ;
            height: 40px;
            line-height: 40px;
            font-size: 24px;
            color: white;
            padding-left: 20px;
            background: cornflowerblue;
        }
        .nn {
            font-size: 18px;
            margin: 8px;
        }
        .m3 {
            height: 450px;
        }
        .feedmethod {
            width: 510px;
            height: 440px;
            text-align: justify;
            background: rgba(255,255,255,0.7);
            border-radius: 10px;
        }
        .feedmain {
            width: 450px;
            height: 440px;
            text-align: justify;
            background: rgba(255,255,255,0.7);
            border-radius: 10px;
```

```css
}
.footcat {
    position: fixed;
    right: 0;
    bottom: 0;
    width: 200px;
    height: 300px;
}
```

static/templates/index.html 设置首页显示的内容。

```html
<!DOCTYPE html>
<html>
<head>
<meta charset = "utf-8" />
<link rel = "stylesheet" href = "/static/css/style.css" />
<script type = "text/javascript" src = "/static/js/cat.js" ></script>
<title>猫的种类识别</title>
</head>
<body class = "bc">
<div class = "main1">
<div class = "main">
    <div class = "title">
<p> The      Kinds      Of      Cats      Identification </p>
    </div>
    <div class = "select">
        <form action = "/info/" method = "post" enctype = "multipart/form-data">
            {% csrf_token %}
            <input type = "file" class = "xfile" id = "btn_file" name = "pic1"/>
            <button type = "button" onclick = "F_Open_dialog()" class = "xxfile">选择图片</button>
            <br />
            <input type = "submit" class = "xxsub" value = "种类识别"/>
        </form>
    </div>
    <div class = "foot">
        <ul>
        <li>
            <ol>
                <dd>识别的种类</dd>
                <dl>布偶猫</dl>
                <dl>孟买猫</dl>
                <dl>暹罗猫</dl>
                <dl>英国短毛猫</dl>
            </ol>
        </li>
```

```html
        <li>
            <ol>
                <dd>应用说明</dd>
                <dl>上传图片均为 jpg</dl>
                <dl>识别精度达 90% 以上</dl>
            </ol>
        </li>
    </ul>
</div>
</div>
</div>
</body>
</html>
```

static/templates/app.html 设置信息页显示的内容。

```html
<!DOCTYPE html>
<html lang="en">
<head>
<meta name="viewport" content="width=device-width, initial-scale=1, minimum-scale=1, maximum-scale=1,user-scalable=no" />
<title></title>
<script src="/static/js/mui.min.js"></script>
<link href="/static/css/mui.min.css" rel="stylesheet"/>
<link rel="stylesheet" type="text/css" href="/static/css/appinfo.css"/>
</head>
<body>
<header class="mui-bar mui-bar-nav ihead">
    <a class="mui-action-back mui-icon mui-icon-left-nav mui-pull-left iha"></a>
    <h1 class="mui-title">识别结果</h1>
</header>
<div class="mui-scroll-wrapper">
<div class="mui-scroll">
<div class="ifhead ihh">种类</div>
<div class="ipic">
    <img src="/static/img/{{ picname }}" />
    <div class="iname">
        <p>       {{ nameinfo }}</p>
    </div>
</div>
<div class="ifhead">外形特征</div>
<div class="ifeature">
    <p>       {{ feature }}</p>
</div>
<div class="ifhead">生活习性</div>
<div class="ifeature">
    <p>       {{ livemethod }}</p>
```

```
        </div>
        <div class = "ifhead">饲养须知</div>
        <div class = "ifeature">
            <p>       {{ feednn }}</p>
        </div>
        <div class = "ifhead">饲养方法</div>
        <div class = "ifeature">
            <p>       {{ feedmethod }}</p>
        </div>
    </div>
</div>
<script type = "text/javascript" charset = "utf-8">
    mui.init();
    mui('.mui-scroll-wrapper').scroll({
        deceleration: 0.0005 //flick 为减速系数,系数越大,滚动速度越慢,滚动距离越小,默认值为
                             //0.0006
    });
</script>
</body>
</html>
```

7.4 系统测试

本部分包括训练准确率、测试效果及模型应用。

7.4.1 训练准确率

经过训练,普通网络在原始数据集上的准确率为 73.2%;残差在原始数据集上的准确率为 85.4%;残差在数据增强后的准确率为 99.3%。

1. 普通网络模型准确率

普通网络模型训练结果如图 7-10 所示,准确率为 73%,预测模型效果不够理想,其原因是数据集太小且图片背景复杂,无法提供更多的信息,模型的深度不够,无法提取更深层次的信息。普通网络模型在测试集上的准确率如图 7-11 所示。

2. 残差网络模型准确率

用 ResNet50 模型在原始数据集上训练 100 次,训练结果如图 7-12 所示。残差网络模型在测试集上的准确率如图 7-13 所示。

3. 数据增强后残差网络模型准确率

使用数据增强将数据集拓展 20 倍后,用残差网络模型训练,训练结果如图 7-14 所示,准确率如图 7-15 所示。

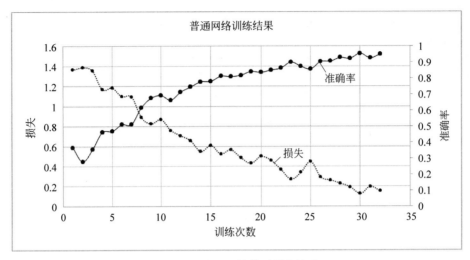

图 7-10　普通网络模型训练结果

Testing loss:0.7071030750507261,Testing acc:0.7317073112580834

图 7-11　普通网络模型在测试集上的准确率

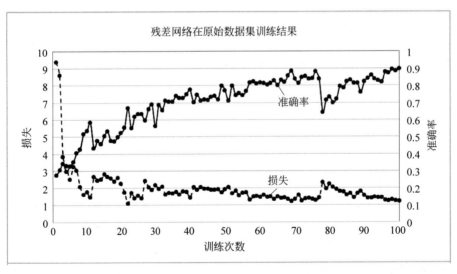

图 7-12　残差网络模型训练结果

Testing loss:0.34894240593037956,Testing acc:0.8536585307702785

图 7-13　残差网络模型在测试集上的准确率

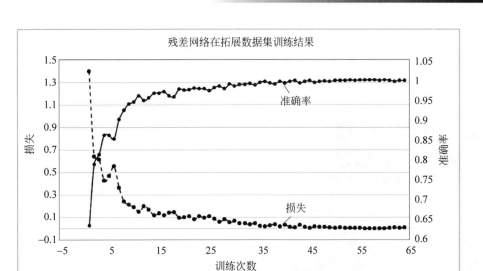

图 7-14　数据增强后残差网络模型训练结果

```
Testing loss:0.017364256225164354,Testing acc:0.9934895833333334
```

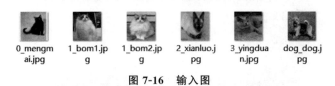

图 7-15　数据增强后残差网络模型准确率

7.4.2　测试效果

如图 7-16 所示，将测试集数据输入模型进行测试，分类的标签与原始数据进行显示和对比。如图 7-17 所示，可得到验证：模型可以识别四种猫。

图 7-16　输入图

```
Ground Truth: 1,   class: 1,   name: 布偶猫
Ground Truth: dog,   class: 1,   name: 布偶猫
Ground Truth: 2,   class: 2,   name: 暹罗猫
Ground Truth: 1,   class: 1,   name: 布偶猫
Ground Truth: 0,   class: 0,   name: 孟买猫
Ground Truth: 3,   class: 3,   name: 英国短毛猫
```

图 7-17　模型训练效果

7.4.3　模型应用

页面项目编译成功后，在命令行终端输入 python manage. py runserver，即可在

http://127.0.0.1:8000/预览网页。打开网页，初始界面如图 7-18 所示。

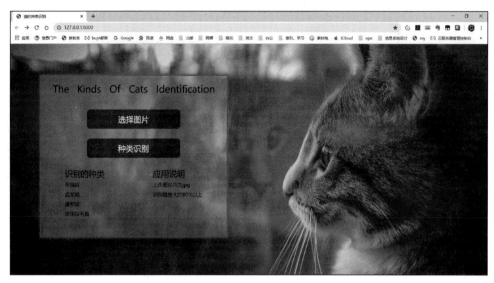

图 7-18　应用初始界面

界面左侧有两个按钮。单击"选择图片"按钮，即可从本地文件中选取要测试的猫图片。单击"种类识别"按钮，识别出猫的种类，并显示相关信息。

在网页上用孟买猫测试结果如图 7-19 所示。

图 7-19　测试结果示例

项目 8 基于 VGG-16 的驾驶行为分析

PROJECT 8

本项目使用 VGG-16 网络模型,通过 Kaggle 开源数据集,提取图片中的用户特征,在移动端实现识别不良驾驶行为的功能。

8.1 总体设计

本部分包括系统整体结构和系统流程。

8.1.1 系统整体结构

系统整体结构如图 8-1 所示。

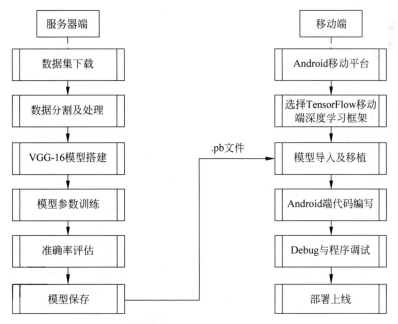

图 8-1 系统整体结构

8.1.2 系统流程

系统流程如图 8-2 所示,网络结构如图 8-3 所示。

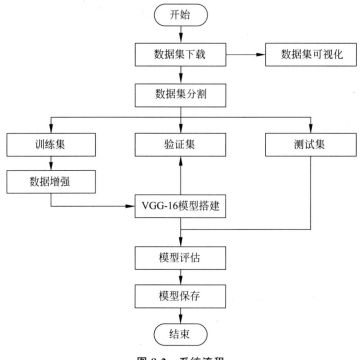

图 8-2 系统流程

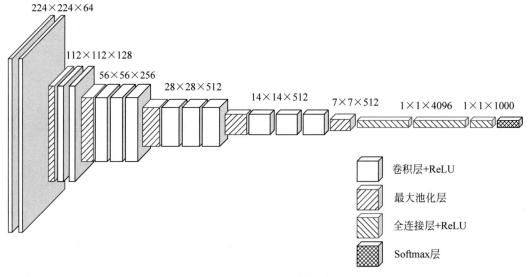

图 8-3 VGG-16 网络结构

8.2 运行环境

本部分包括 Python 环境、TensorFlow 环境和 Android 环境。

8.2.1 Python 环境

选择 Python 3.7.4 版本,在 Python 官网下载对应版本的安装文件,下载地址为:https://www.python.org。下载 Anaconda 完成 Python 所需的环境配置,下载地址为:https://www.anaconda.com/。在 Windows 操作系统下进行环境搭建。

8.2.2 TensorFlow 环境

执行 conda 命令完成 TensorFlow 环境配置,在 PyCharm 软件中完成其余具体库的安装配置,输入清华仓库镜像,输入命令:

```
conda config -- add channels https://mirrors.tuna.tsinghua.edu.cn/anaconda/pkgs/free/
conda config - set show_channel_urls yes
```

在 Anaconda Prompt 中激活 TensorFlow 环境,输入命令:

```
activate tensorflow
```

安装 CPU 版本的 TensorFlow,输入命令:

```
pip install - upgrade -- ignore-installed tensorflow
```

安装完毕。

在 PyCharm 中新建项目,使用构建的 Anaconda 环境,在 File→Settings→Project:XXX→Project Interpreter 中添加其余需要用到的库,包括 OpenCV、NumPy、Keras、OS 等。

8.2.3 Android 环境

安装 Android Studio,新建 Android 项目。在 app/build.gradle 文件中添加 TensorFlow mobile 依赖项,命令如下:

```
implementation 'org.tensorflow:tensorflow-android: +
```

完整的 app/build.gradle 配置代码如下:

```
apply plugin: 'com.android.application'
android {
    compileSdkVersion 26
    defaultConfig {
        applicationId "com.specpal.mobileai"
```

```
            minSdkVersion 21
            targetSdkVersion 26
            versionCode 1
            versionName "1.0"
            testInstrumentationRunner "android.support.test.runner.AndroidJUnitRunner"
        }
        buildTypes {
            release {
                minifyEnabled false
                proguardFiles getDefaultProguardFile('proguard-android.txt'), 'proguard-rules.pro'
            }
        }
    }
    dependencies {
        implementation fileTree(dir: 'libs', include: ['*.jar'])
        implementation 'com.android.support:appcompat-v7:26.1.0'
        implementation 'com.android.support.constraint:constraint-layout:1.0.2'
        implementation 'com.android.support:design:26.1.0'
        testImplementation 'junit:junit:4.12'
        androidTestImplementation 'com.android.support.test:runner:1.0.2'
        androidTestImplementation 'com.android.support.test.espresso:espresso-core:3.0.2'
        implementation 'org.tensorflow:tensorflow-android: + '
    }
```

app/build.gradle 里的内容有任何改动，Android Studio 会弹出如图 8-4 所示的提示。

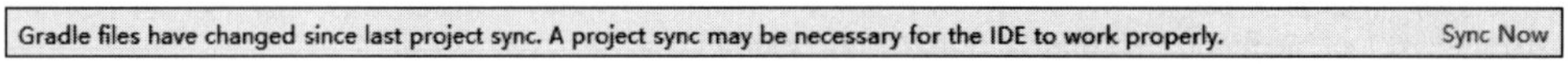

> Gradle files have changed since last project sync. A project sync may be necessary for the IDE to work properly.　　Sync Now

图 8-4　改动后提示

单击 Sync Now 或 图标，同步该配置，配置完成。

8.3　模块实现

本项目包括 4 个模块：数据预处理、模型构建、模型训练及保存、模型生成。下面分别给出各模块的功能介绍及相关代码。

8.3.1　数据预处理

本部分包括数据集来源、内容和预处理。

1. 数据集来源

使用开源数据集 state-farm-distracted-driver-detection，下载地址为：https://www.kaggle.com/c/state-farm-distracted-driver-detection/data，直接在网页上单击 Download 按

钮下载。也可以调用 API 下载，下载地址为：kaggle competitions download -c state-farm-distracted-driver-detection。数据集包括 22424 张训练集图片，若干测试集图片。由于是竞赛数据集，测试集的图片没有标签，故不使用。

2. 数据集内容

数据集共包括 10 个驾驶状态：安全驾驶、右手使用手机、右手打电话、左手使用手机、左手打电话、调广播、喝水、向后拿东西、整理头发或化妆、与乘客交流。除安全驾驶外都是危险状态，如图 8-5 所示。

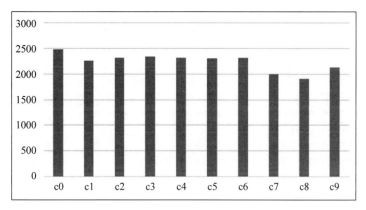

图 8-5　数据集中各驾驶状态

3. 数据集预处理

数据集中每张图片大小都是 640×480 像素，从同一角度拍摄的车载监控图片，数据集十分规范，无噪声，不需要进行过多的预处理，在训练前通过图像增广进行预处理产生相似但又不同的训练样本。随机改变训练样本可以降低模型对某些属性的依赖，提高模型的泛化能力。

```
train_datagen = ImageDataGenerator(rescale = 1.0/255, shear_range = 0.2, zoom_range = 0.2, horizontal_flip = True)
#采用图像缩放、随机水平翻转、随机剪切、随机缩放进行图像增广
```

8.3.2　模型构建

数据加载进模型之后，需要定义模型结构，并优化损失函数。

1. 定义模型结构

使用 VGG-16 网络结构，共 13 个卷积层、3 个全连接层和 5 个池化层，使用 Keras 库搭建模型，构建 Sequential 顺序模型，它由多个网络层线性堆叠而成。

```
model = Sequential()
#快速开始顺序模型 Sequential,通过 add()方法将各层逐个加入模型中
```

```python
model.add(ZeroPadding2D((1, 1), input_shape = (150, 150, 3)))
# 图像上下左右都补 0,输入图像
model.add(Convolution2D(64, 3, 3, activation = 'relu'))
# 使用 64 个 3×3 的卷积核,激活 ReLU 函数
model.add(ZeroPadding2D((1, 1)))
model.add(Convolution2D(64, 3, 3, activation = 'relu'))
model.add(MaxPooling2D((2, 2), strides = (2, 2)))
# 最大值池化窗口大小为 2×2,步长两个方向都是 2
model.add(ZeroPadding2D((1, 1)))
model.add(Convolution2D(128, 3, 3, activation = 'relu'))
model.add(ZeroPadding2D((1, 1)))
model.add(Convolution2D(128, 3, 3, activation = 'relu'))
model.add(MaxPooling2D((2, 2), strides = (2, 2)))
model.add(ZeroPadding2D((1, 1)))
model.add(Convolution2D(256, 3, 3, activation = 'relu'))
model.add(ZeroPadding2D((1, 1)))
model.add(Convolution2D(256, 3, 3, activation = 'relu'))
model.add(ZeroPadding2D((1, 1)))
model.add(Convolution2D(256, 3, 3, activation = 'relu'))
model.add(MaxPooling2D((2, 2), strides = (2, 2)))
model.add(ZeroPadding2D((1, 1)))
model.add(Convolution2D(512, 3, 3, activation = 'relu'))
model.add(ZeroPadding2D((1, 1)))
model.add(Convolution2D(512, 3, 3, activation = 'relu'))
model.add(ZeroPadding2D((1, 1)))
model.add(Convolution2D(512, 3, 3, activation = 'relu'))
model.add(MaxPooling2D((2, 2), strides = (2, 2)))
model.add(ZeroPadding2D((1, 1)))
model.add(Convolution2D(512, 3, 3, activation = 'relu'))
model.add(ZeroPadding2D((1, 1)))
model.add(Convolution2D(512, 3, 3, activation = 'relu'))
model.add(ZeroPadding2D((1, 1)))
model.add(Convolution2D(512, 3, 3, activation = 'relu'))
model.add(MaxPooling2D((2, 2), strides = (2, 2)))
# 卷积层以及池化层搭建完毕
model.load_weights('vgg16_weights_tf_dim_ordering_tf_kernels_notop.h5')
# 加载上述模型中预训练的权重参数,不含 top model
top_model = Sequential()
# 构建多层感知机 MLP 作为 top model
top_model.add(Flatten(input_shape = model.output_shape[1:]))
# Flatten 用于将多维数据变成一维,卷积层到全连接层的过渡
top_model.add(Dense(256, activation = 'relu'))
# 加入全连接层,输出维度 256 个神经元,激活函数为 ReLU 函数
top_model.add(Dropout(0.5))
# 加入 Dropout 层防止训练过拟合,控制需要断开的神经元比例为 0.5
top_model.add(Dense(256, activation = 'relu'))
top_model.add(Dropout(0.5))
```

```python
top_model.add(Dense(10, activation = 'softmax'))
#输出层,Softmax 分类,共 10 类
model.add(top_model)
#将 top_model 加入之前的模型中
for layer in model.layers[:25]:
    layer.trainable = False
#将卷积神经网络权重参数设置为不可改变,减少训练量
#VGG-16 模型搭建完毕
```

2. 优化损失函数

优化损失函数的相关代码如下:

```python
model.compile(loss = 'categorical_crossentropy', optimizer = 'adadelta', metrics = ['accuracy'])
#因为是多分类问题,标签为 One-Hot 编码,损失函数设置为类别交叉熵,指标列表 metrics 设置为
#accuracy,使用 adadelta 优化器,加快收敛速度,避免出现局部最优解
```

8.3.3 模型训练及保存

在定义模型架构和编译后,通过训练集训练,使模型可以识别数据集中图像的特征。

1. 模型训练

模型训练相关代码如下:

```python
train_generator = train_datagen.flow_from_directory(train_data_dir, target_size = (img_height, img_width), batch_size = 32, class_mode = 'categorical')
#读取训练集
validation_generator = train_datagen.flow_from_directory(validation_data_dir, target_size = (img_height, img_width), batch_size = 32, class_mode = 'categorical')
#读取验证集
model.fit_generator(train_generator, samples_per_epoch = nb_train_samples, epochs = nb_epoch, validation_data = validation_generator, nb_val_samples = nb_validation_samples)
#训练模型
model.save('model + weights.h5')
#保存模型及权重
```

2. 模型保存

上述由 Keras 库生成的模型及权重文件为 .h5 格式,为了能够被 Android 程序读取,需要将 .h5 文件转换为 .pb 格式的文件,模型被保存后,可以被重用,也可以移植到其他环境中使用。

```python
def h5_to_pb(h5_model, output_dir, model_name, out_prefix = "output_", log_tensorboard = True):
#.h5 模型文件转换成 .pb 模型文件
    if os.path.exists(output_dir) == False:
        os.mkdir(output_dir)
```

```
out_nodes = []
for i in range(len(h5_model.outputs)):
    out_nodes.append(out_prefix + str(i + 1))
    tf.identity(h5_model.output[i], out_prefix + str(i + 1))
sess = backend.get_session()
from tensorflow.python.framework import graph_util, graph_io
#写入.pb模型文件
init_graph = sess.graph.as_graph_def()
main_graph = graph_util.convert_variables_to_constants(sess, init_graph, out_nodes)
graph_io.write_graph(main_graph, output_dir, name = model_name, as_text = False)
#输出日志文件
if log_tensorboard:
    from tensorflow.python.tools import import_pb_to_tensorboard
    import_pb_to_tensorboard.import_to_tensorboard(os.path.join(output_dir, model_name), output_dir)
```

8.3.4 模型生成

将图片转化为数据,输入 TensorFlow 的模型中并获取输出。

1. 模型导入及调用

本部分包括模型导入及调用的操作方法。

(1) 编写代码进行实际预测之前,需要将转换后的模型添加到应用程序的资源文件夹中。在 Android Studio 中,鼠标右键"项目",跳转至 Add Folder(添加文件夹)部分,并选择 Assets Folder(资源文件夹)。在应用程序目录中创建一个资源文件夹,将模型复制到其中,如图 8-6 所示。

图 8-6 添加模型

(2) 将新的 Java 类添加到项目的主程序包中,并命名为 ImageUtils,ImageUtils 为图片工具类,可用于 Bitmap、byte、array、Drawable 图片类型之间进行转换以及缩放。相关代码如下:

```
package com.example.doremi.testkeras2tensorflow;
import android.content.res.AssetManager;
import android.graphics.Bitmap;
import android.graphics.Canvas;
```

```java
import android.graphics.Matrix;
import android.os.Environment;
import java.io.File;
import java.io.FileOutputStream;
import java.io.InputStream;
import org.json.*;
//用于处理图像的实用程序类
public class ImageUtils {
    /*
     * 返回转换矩阵,处理裁切(如果需要保持宽高比)和旋转
     * 参数 srcWidth 为源帧的宽度
     * 参数 srcHeight 为源帧的高度
     * 参数 dstWidth 为目标帧的宽度
     * 参数 dstHeight 为目标帧的高度
     * 参数 applyRotation 为旋转角度,为 90°的整数倍
     * 参数 maintainAspectRatio 为是否维持缩放比例
     * 返回满足所需要求的转换
     */
    public static Matrix getTransformationMatrix(
            final int srcWidth,
            final int srcHeight,
            final int dstWidth,
            final int dstHeight,
            final int applyRotation,
            final boolean maintainAspectRatio) {
        final Matrix matrix = new Matrix();
        if (applyRotation != 0) {
            //进行平移,使图像中心在原点
            matrix.postTranslate(-srcWidth / 2.0f, -srcHeight / 2.0f);
            //绕原点旋转
            matrix.postRotate(applyRotation);
        }
        //考虑已经应用的旋转(如果有),然后确定每个轴需要多少缩放
        final boolean transpose = (Math.abs(applyRotation) + 90) % 180 == 0;
        final int inWidth = transpose ? srcHeight : srcWidth;
        final int inHeight = transpose ? srcWidth : srcHeight;
        //必要时应用缩放
        if (inWidth != dstWidth || inHeight != dstHeight) {
            final float scaleFactorX = dstWidth / (float) inWidth;
            final float scaleFactorY = dstHeight / (float) inHeight;
            if (maintainAspectRatio) {
                //按最小比例缩放,以便在保持宽高比的同时完全填充,某些图像可能会被截掉边缘
                final float scaleFactor = Math.max(scaleFactorX, scaleFactorY);
                matrix.postScale(scaleFactor, scaleFactor);
            } else {
                //精确缩放
                matrix.postScale(scaleFactorX, scaleFactorY);
```

```java
            }
        }
        if (applyRotation != 0) {
    //从以原点为中心的参考点转换回目标帧
            matrix.postTranslate(dstWidth / 2.0f, dstHeight / 2.0f);
        }
        return matrix;
    }
    public static Bitmap processBitmap(Bitmap source,int size){
        int image_height = source.getHeight();
        int image_width = source.getWidth();
        Bitmap croppedBitmap = Bitmap.createBitmap(size, size, Bitmap.Config.ARGB_8888);
        Matrix frameToCropTransformations = getTransformationMatrix(image_width, image_height,size,size,0,false);
        Matrix cropToFrameTransformations = new Matrix();
        frameToCropTransformations.invert(cropToFrameTransformations);
        final Canvas canvas = new Canvas(croppedBitmap);
        canvas.drawBitmap(source, frameToCropTransformations, null);
        return croppedBitmap;
    }
    public static float[] normalizeBitmap(Bitmap source,int size,float mean,float std){
        float[] output = new float[size * size * 3];
        int[] intValues = new int[source.getHeight() * source.getWidth()];
        source.getPixels(intValues, 0, source.getWidth(), 0, 0, source.getWidth(), source.getHeight());
        for (int i = 0; i < intValues.length; ++i) {
            final int val = intValues[i];
            output[i * 3] = (((val >> 16) & 0xFF) - mean)/std;
            output[i * 3 + 1] = (((val >> 8) & 0xFF) - mean)/std;
            output[i * 3 + 2] = ((val & 0xFF) - mean)/std;
        }
        return output;
    }
    public static Object[] argmax(float[] array){
        int best = -1;
        float best_confidence = 0.0f;
        for(int i = 0;i < array.length;i++){
            float value = array[i];
            if (value > best_confidence){
                best_confidence = value;
                best = i;
            }
        }
        return new Object[]{best,best_confidence};
    }
    public static String getLabel( InputStream jsonStream,int index){
        String label = "";
```

```java
        try {
            byte[] jsonData = new byte[jsonStream.available()];
            jsonStream.read(jsonData);
            jsonStream.close();
            String jsonString = new String(jsonData,"utf-8");
            JSONObject object = new JSONObject(jsonString);
            label = object.getString(String.valueOf(index));
        }
        catch (Exception e){
        }
        return label;
    }
}
```

(3) 在主活动(Main Activity)中添加代码,用于显示图像和预测结果。

```java
public void predict(final Bitmap bitmap){
    //在后台线程中运行预测
    new AsyncTask<Integer,Integer,Integer>(){
        @Override
        protected Integer doInBackground(Integer...params){
            //将图像大小调整为 150×150
            Bitmap resized_image = ImageUtils.processBitmap(bitmap,150);
            //归一化像素

            floatValues = ImageUtils.normalizeBitmap(resized_image,150,127.5f,1.0f);
            //将输入图片传到 TensorFlow
            tf.feed(INPUT_NAME,floatValues,1,150,150,3);
            //计算预测
            tf.run(new String[]{OUTPUT_NAME});
            //将输出复制到预测数组中
            tf.fetch(OUTPUT_NAME,PREDICTIONS);
            //获得最高预测
            Object[] results = argmax(PREDICTIONS);
            int class_index = (Integer) results[0];
            float confidence = (Float) results[1];
            try{
                final String conf = String.valueOf(confidence * 100).substring(0,5);
                //将预测的类别索引转换为实际的标签名称
                final String label = ImageUtils.getLabel(getAssets().open("labels.json"),
class_index);
                //展示结果
                runOnUiThread(new Runnable() {
                    @Override
                    public void run() {
                        progressBar.dismiss();
                        resultView.setText(label + " : " + conf + "%");
```

```
                }
            });
        }
        catch (Exception e){
        }
        return 0;
    }
}.execute(0);
```

2. 相关代码

本部分包括布局文件和主活动类。

1）布局文件

布局文件相关代码如下：

```
/res/layout/activity_main.xml
<?xml version = "1.0" encoding = "utf-8"?>
< android.support.design.widget.CoordinatorLayout xmlns:android = "http://schemas.android.com/apk/res/android"
    xmlns:app = "http://schemas.android.com/apk/res-auto"
    xmlns:tools = "http://schemas.android.com/tools"
    android:layout_width = "match_parent"
    android:layout_height = "match_parent"
    tools:context = ".MainActivity">
    < android.support.design.widget.AppBarLayout
        android:layout_width = "match_parent"
        android:layout_height = "wrap_content"
        android:theme = "@style/AppTheme.AppBarOverlay">
        < android.support.v7.widget.Toolbar
            android:id = "@ + id/toolbar"
            android:layout_width = "match_parent"
            android:layout_height = "?attr/actionBarSize"
            android:background = "?attr/colorPrimary"
            app:popupTheme = "@style/AppTheme.PopupOverlay" />
    </android.support.design.widget.AppBarLayout >
    < include layout = "@layout/content_main" />
    < android.support.design.widget.FloatingActionButton
        android:id = "@ + id/predict"
        android:layout_width = "wrap_content"
        android:layout_height = "wrap_content"
        android:layout_gravity = "bottom|end"
        android:layout_margin = "@dimen/fab_margin"
        app:srcCompat = "@android:drawable/ic_media_play" />
</android.support.design.widget.CoordinatorLayout >
/res/layout/content_main.xml
<?xml version = "1.0" encoding = "utf-8"?>
```

```xml
<android.support.constraint.ConstraintLayout xmlns:android = "http://schemas.android.com/apk/res/android"
    xmlns:app = "http://schemas.android.com/apk/res-auto"
    xmlns:tools = "http://schemas.android.com/tools"
    android:layout_width = "match_parent"
    android:layout_height = "match_parent"
    app:layout_behavior = "@string/appbar_scrolling_view_behavior"
    tools:context = ".MainActivity"
    tools:showIn = "@layout/activity_main">
    <ScrollView
        android:layout_width = "match_parent"
        android:layout_height = "match_parent">
        <LinearLayout
            android:layout_width = "match_parent"
            android:layout_height = "wrap_content"
            android:orientation = "vertical"
            >
            <TextView
                android:layout_width = "match_parent"
                android:layout_height = "wrap_content"
                android:textSize = "30dp"
                android:layout_marginBottom = "30dp"
                android:text = "Click the Red-Colored floating button below to show and predict the image"
                />
            <ImageView
                android:layout_width = "match_parent"
                android:layout_height = "wrap_content"
                android:adjustViewBounds = "true"
                android:scaleType = "fitCenter"
                android:id = "@+id/imageview"
                android:layout_marginBottom = "10dp"
                />
            <TextView
                android:layout_width = "wrap_content"
                android:layout_height = "wrap_content"
                android:id = "@+id/results"
                />
        </LinearLayout>
    </ScrollView>
</android.support.constraint.ConstraintLayout>
```

2) 主活动类

主活动类相关代码如下：

```
package com.specpal.mobileai;
import android.graphics.Bitmap;
```

```java
import android.graphics.BitmapFactory;
import android.os.AsyncTask;
import android.os.Bundle;
import android.renderscript.ScriptGroup;
import android.support.design.widget.FloatingActionButton;
import android.support.design.widget.Snackbar;
import android.support.v7.app.AppCompatActivity;
import android.support.v7.widget.Toolbar;
import android.util.JsonReader;
import android.view.View;
import android.widget.ImageView;
import android.widget.TextView;
import android.widget.Toast;
import org.json.*;
import org.tensorflow.contrib.android.TensorFlowInferenceInterface;
import java.io.FileInputStream;
import java.io.InputStream;
public class MainActivity extends AppCompatActivity {
    //加载流推理库
    static {
        System.loadLibrary("tensorflow_inference");
    }
    //模型存放路径和输入/输出节点名称
    private String MODEL_PATH = "file:///android_asset/model+weights_22424.pb";
    private String INPUT_NAME = "zero_padding2d_1_input";
    private String OUTPUT_NAME = "output_1";
    private TensorFlowInferenceInterface tf;
    //保存预测的数组和图像数据的浮点值
    float[] PREDICTIONS = new float[10];
    private float[] floatValues;
    private int[] INPUT_SIZE = {150,150,3};
    ImageView imageView;
    TextView resultView;
    Snackbar progressBar;
    @Override
    protected void onCreate(Bundle savedInstanceState) {
        super.onCreate(savedInstanceState);
        setContentView(R.layout.activity_main);
        Toolbar toolbar = (Toolbar) findViewById(R.id.toolbar);
        setSupportActionBar(toolbar);
        //初始化TensorFlow
        tf = new TensorFlowInferenceInterface(getAssets(),MODEL_PATH);
        imageView = (ImageView) findViewById(R.id.imageview);
        resultView = (TextView) findViewById(R.id.results);
        progressBar = Snackbar.make(imageView,"PROCESSING IMAGE",Snackbar.LENGTH_INDEFINITE);
        final FloatingActionButton predict = (FloatingActionButton) findViewById(R.id.predict);
        predict.setOnClickListener(new View.OnClickListener() {
```

```java
            @Override
            public void onClick(View view) {
                try{
                    //从 assets 文件夹读取图片
                    InputStream imageStream = getAssets().open("testimage4.gif");
                    Bitmap bitmap = BitmapFactory.decodeStream(imageStream);
                    imageView.setImageBitmap(bitmap);
                    progressBar.show();
                    predict(bitmap);
                }
                catch (Exception e){
                }
            }
        });
    }
    //计算最大预测及其置信度的函数
    public Object[] argmax(float[] array){
        int best = -1;
        float best_confidence = 0.0f;
        for(int i = 0;i < array.length;i++){
            float value = array[i];
            if (value > best_confidence){
                best_confidence = value;
                best = i;
            }
        }
        return new Object[]{best,best_confidence};
    }
    public void predict(final Bitmap bitmap){
        //在后台线程中运行预测
        new AsyncTask< Integer,Integer,Integer >(){
            @Override
            protected Integer doInBackground(Integer ...params){
                //将图像大小调整为 150×150
                Bitmap resized_image = ImageUtils.processBitmap(bitmap,150);
                //归一化像素
floatValues = ImageUtils.normalizeBitmap(resized_image,150,127.5f,1.0f);
                //将输入图像传到 TensorFlow
                tf.feed(INPUT_NAME,floatValues,1,150,150,3);
                //计算预测
                tf.run(new String[]{OUTPUT_NAME});
                //将输出复制到预测数组中
                tf.fetch(OUTPUT_NAME,PREDICTIONS);
                //获得最高预测
                Object[] results = argmax(PREDICTIONS);
                int class_index = (Integer) results[0];
```

```
                float confidence = (Float) results[1];
                try{
                    final String conf = String.valueOf(confidence * 100).substring(0,5);
                    //将预测的类别索引转换为实际的标签名称
                    final String label = ImageUtils.getLabel(getAssets().open("labels.json"),class_index);
                    //展示结果
                    runOnUiThread(new Runnable() {
                        @Override
                        public void run() {
                            progressBar.dismiss();
                            resultView.setText(label + " : " + conf + "%");
                        }
                    });
                }
                catch (Exception e){
                }
                return 0;
            }
        }.execute(0);
    }
}
```

8.4 系统测试

本部分包括训练准确率、测试效果及模型应用。

8.4.1 训练准确率

训练准确率在98.2%左右,损失率在10.6%左右,可见整个预测模型的训练比较成功,如图8-7所示。

8.4.2 测试效果

使用OpenCV库读取图片,将测试集中的数据代入模型中进行预测,如图8-8和图8-9所示。

```
img = cv2.imread('img_test.jpg')
img = cv2.resize(img, (150, 150))
images = [img]
output = model.predict(np.array(images), batch_size=1)
pro = output.max()
index = output.argmax()
print(names[index], pro)
print(output)
```

项目8　基于VGG-16的驾驶行为分析

```
635/653 [============================>.] - ETA: 1:56 - loss: 0.1052 - accuracy: 0.9820
636/653 [============================>.] - ETA: 1:50 - loss: 0.1050 - accuracy: 0.9821
637/653 [============================>.] - ETA: 1:43 - loss: 0.1049 - accuracy: 0.9821
638/653 [============================>.] - ETA: 1:37 - loss: 0.1055 - accuracy: 0.9821
639/653 [============================>.] - ETA: 1:30 - loss: 0.1054 - accuracy: 0.9821
640/653 [============================>.] - ETA: 1:24 - loss: 0.1059 - accuracy: 0.9821
641/653 [============================>.] - ETA: 1:17 - loss: 0.1061 - accuracy: 0.9821
642/653 [============================>.] - ETA: 1:11 - loss: 0.1059 - accuracy: 0.9821
643/653 [============================>.] - ETA: 1:04 - loss: 0.1063 - accuracy: 0.9820
644/653 [============================>.] - ETA: 58s - loss: 0.1062 - accuracy: 0.9820
645/653 [============================>.] - ETA: 51s - loss: 0.1061 - accuracy: 0.9820
646/653 [============================>.] - ETA: 45s - loss: 0.1059 - accuracy: 0.9820
647/653 [============================>.] - ETA: 38s - loss: 0.1057 - accuracy: 0.9821
648/653 [============================>.] - ETA: 32s - loss: 0.1056 - accuracy: 0.9821
649/653 [============================>.] - ETA: 25s - loss: 0.1054 - accuracy: 0.9821
650/653 [============================>.] - ETA: 19s - loss: 0.1053 - accuracy: 0.9822
651/653 [============================>.] - ETA: 12s - loss: 0.1051 - accuracy: 0.9822
652/653 [============================>.] - ETA: 6s - loss: 0.1049 - accuracy: 0.9822
```

图 8-7　准确率与损失率

图 8-8　模型测试（左手）

图 8-9　模型测试（右手）

8.4.3 模型应用

Android 项目编译成功后,建议将项目部署到真机上进行测试。模拟器运行较慢,不建议使用。部署到真机的方法如下:

将手机数据线连接到计算机,开启开发者模式,打开 USB 调试,单击 Android 项目的"运行"按钮,出现"连接手机"的选项,单击该选项即可。

Android Studio 生成 apk,发送到手机,在手机上下载 apk,安装即可。打开 App,初始界面如图 8-10 所示。

单击右下角按钮,显示测试照片结果,如图 8-11 所示,移动端测试结果如图 8-12 所示。

图 8-10　应用初始界面

图 8-11　预测结果显示界面

图 8-12　移动端测试结果示例

项目 9 基于 Mask R-CNN 的娱乐视频生成器

PROJECT 9

本项目借助 Facebook 的 Detectron2 目标检测平台，基于 Mask R-CNN 网络和 PointRend 优化方法的预训练模型以及 Python 的音视频编辑库，实现将视频中动态元素叠加到另一段视频中的功能，并提供音视频剪辑功能和自定义特效。

9.1 总体设计

本部分包括系统整体结构和系统流程。

9.1.1 系统整体结构

系统整体结构如图 9-1 所示。

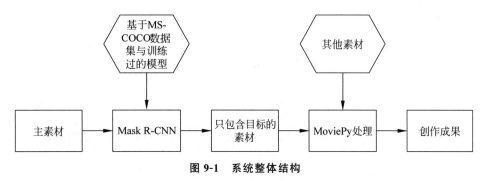

图 9-1 系统整体结构

9.1.2 系统流程

系统流程如图 9-2 所示。

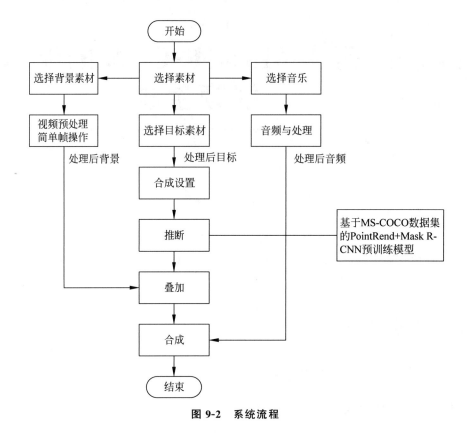

图 9-2　系统流程

9.2　运行环境

本部分包括 Python 环境、PyTorch 环境、Detectron2 平台和 MoviePy 的安装。

9.2.1　Python 环境

需要 Python 3.6 及以上配置，推荐在 Anaconda 等软件中建立新的虚拟环境。

9.2.2　PyTorch 环境

由于一些兼容性问题，在 Windows 系统下运行 PyTorch 1.4 版本时会出现问题，因此建议使用 PyTorch 1.3 版本，torchvision 选择 0.4.1 版本。在 Anaconda 环境下已配置 CUDA 10.1 和 cuDNN 7.0＋的计算机上，运行以下命令安装：

```
conda install pytorch==1.3.0 torchvision==0.4.1 cudatoolkit=10.1 -c pytorch
```

9.2.3 Detectron2 平台

Detectron2 平台分为 CPU 版本和 GPU 版本。其中,CPU 版本通用性较好,但由于并行处理能力弱于 CPU,建议使用 GPU 版本,它需要 CUDA 驱动,以 CUDA 10.1 版本为例,CPU 版本安装命令如下:

```
python -m pip install detectron2 -f https://dl.fbaipublicfiles.com/detectron2/wheels/cpu/index.html
```

GPU 版本安装命令如下:

```
python -m pip install detectron2 -f https://dl.fbaipublicfiles.com/detectron2/wheels/cu101/index.html
```

Detectron2 平台还有一些其他依赖,例如 fvcore,在安装时会自动检测依赖并安装。

上述命令适用于 MacOS 或 Linux 系统的 Pytorch 1.4 版本,无须手动编译。但在 Windows 系统下只能手动编译。步骤如下:

(1) 在 GitHub 页面 https://github.com/facebookresearch/detectron2 中下载 Master 分支或者在 Release 页面 https://github.com/facebookresearch/detectron2/release 中下载 0.1.1 版本。

(2) 安装 fvcore,输入命令:

```
pip install git+https://github.com/facebookresearch/fvcore
```

(3) 安装 pycocotools,输入命令:

```
pip install cython;
pip install git+https://github.com/philferriere/cocoapi.git#subdirectory=PythonAPI
```

(4) 安装 Visual Studio 2019。

(5) 在 PyTorch 环境和上述依赖安装完成后修改 Python 环境下的 argument_spec.h 文件(路径{your evn path}\Lib\site-packages\torch\include\torch\csrc\jit\argument_spec.h),将 static constexpr size_t DEPTH_LIMIT = 128 修改为 static const size_t DEPTH_LIMIT = 128。

(6) 修改 cast.h(路径{your evn path}\Lib\site-packages\torch\include\pybind11\cast.h),将 explicit operator type&() { return *(this->value); }修改为 explicit operator type&() { return *((type*)this->value); }。

(7) 在命令行中激活 Visual Studio 2019 准备编译,形如 E:\Program Files (x86)\Microsoft Visual Studio\2019\Community\VC\Auxiliary\Build\vcvars64.bat。

(8) 在根目录中运行命令: python setup.py build develop。

9.2.4 MoviePy 的安装

MoviePy 是一个较为成熟的视频处理 Python 库,通过输入命令:pip install moviepy,即可安装,相关依赖项会自动补足,包括叠加、拼接、RGB 通道处理、音频处理和视频输出等各类功能。

9.2.5 PyQt 的安装

PyQt 可以制作具有图形用户界面的程序,提高程序的易用性和用户友好性。安装命令如下:

```
pip install PyQt5-tools
```

9.3 模块实现

本项目包括 3 个模块:数据处理、视频处理和 PyQt 界面。下面分别给出各模块的功能介绍及相关代码。

9.3.1 数据处理

将目标素材叠加到背景素材,通过 Mask R-CNN 网络和 PointRend 优化方法的预训练模型,对目标素材的每一帧进行推断,并根据结果对目标素材的像素进行处理,通过 MoviePy 叠加时明确哪部分在背景素材上可见。下载地址为:https://github.com/facebookresearch/detectron2/tree/master/projects/PointRend,路径为:datasets/model_final_3c3198.pkl。

1. 读取参数设置和模型权重

本部分相关代码如下:

```
#Detectron2 默认设置
cfg = get_cfg()
#PointRend 设置
add_pointrend_config(cfg)
#从该配置文件读取 PointRend 的参数设置
cfg.merge_from_file("projects/PointRend/configs/InstanceSegmentation/pointrend_rcnn_R_50_FPN_3x_coco.yaml")
    #若阈值过低预测速度会很慢
    cfg.MODEL.ROI_HEADS.SCORE_THRESH_TEST = 0.5
    #读取预训练模型的权重参数
    cfg.MODEL.WEIGHTS = "datasets/model_final_3c3198.pkl"
```

```python
    if not torch.cuda.is_available():
        cfg.MODEL.DEVICE = 'cpu'
# 默认预测
predictor = DefaultPredictor(cfg)
```

2. 根据预测输出结果处理每帧像素

逐帧输入大小为 $W \times H \times 3$ 的 RGB 视频帧数据，假设从中检测出 N 个物体，输出结果 output 包含：classes 表示识别出来的 N 个物体类别 ID(ID 是 MS-COCO 数据集中 80 种物体类别标签的编号，为 $0 \sim 79$，其中人类编号为 0)；boxes 包含 N 个检测框 4 个顶点的位置数据；scores 是对于 N 个物体的预测评分；masks 是二进制帧数据，大小为 $N \times W \times H$，属于目标类别的像素值为 True，不属于的像素值为 False。

```python
# 逐帧过滤不需要的类别数据，为 to_mask 转换做准备
def custom_frame(frame):
    _frame = frame.copy()
    output = predictor(_frame)
    instances = output['instances'].to('cpu')
    data = {'classes':instances.pred_classes.numpy(),'boxes':instances.pred_boxes.tensor.numpy(),
            'masks':instances.pred_masks.numpy(),'scores':instances.scores.numpy()}
    # 设定接收人类的数据
    data = process(data, target_class = [0])
    result = custom_show(_frame, data['masks'])
    return result
# process()接收输出数据和想要的物体类别 ID 集合，滤除不属于该集合的数据类别
def process(data, target_class = None):
    if target_class is not None:
        filter = []
        for i in range(len(data['classes'])):
            if data['classes'][i] in target_class:
                filter.append(i)
        data['classes'] = data['classes'][filter]
        data['masks'] = data['masks'][filter, :, :]
        data['scores'] = data['scores'][filter]
        data['boxes'] = data['boxes'][filter]
    return data
```

过滤完类别以后对输入的视频帧数据进行处理，masks 值为 True 时设为白色，否则设为黑色，因为 MoviePy 可以通过黑白帧决定视频叠加时上层画面数据的透明度。

```python
# 目标以外设为黑色，否则设为白色
def custom_show(image, masks):
    alpha = 1
    if not len(masks):
        for i in range(3):
```

```python
            image[:, :, i] = image[:, :, i] * (1 - alpha) + alpha * 0
    return image
final_mask = masks[0]
for i in range(1, masks.shape[0]):
    final_mask = np.bitwise_or(final_mask, masks[i])
for j in range(3):
    image[:, :, j] = np.where(
        final_mask == 1,
        image[:, :, j] * (1 - alpha) + alpha * 255,
        image[:, :, j] * (1 - alpha) + alpha * 0
    )
return image
```

9.3.2 视频处理

本部分包括音视频的存储与导入、预处理、生成参数设置以及处理结果。

1. 音视频的存储与导入

MoviePy 中的两个主要类 VideoFileClip 和 AudioFileClip 可接收路径文件名为参数，调用底层 ffmpeg 读取视频、音频文件的元信息并得到一个与该文件关联的对象。该对象存储对应文件里一小部分信息，在逐帧读取或写入新文件时进行实际的帧数据处理，而中间的变换只是改写了读取与写入规则，可以用 Clip 对象对应视频、音频片段。

生成视频首先要导入目标素材、背景素材和音频素材，相关代码如下：

```python
#打开目标文件
    def openTarget(self):
        fileName, _ = QFileDialog.getOpenFileName(self, "Open Video", QDir.homePath(),
"Video( *.mp4; *.wmv; *.rmvb; *.avi; *.mkv; *.mov; *.mpeg; *.flv; *.rm; *.mpeg; *.mpg)")
        if fileName != '':
            if self.state != State.IDLE:
                self.stop.emit()
            v = VideoFileClip(fileName)
            item = MyListWidgetItem(fileName.split("/")[-1], v, v.audio)
            self.listWidget.addItem(item)
            self.listWidget.setCurrentItem(item)
            self.cur_listWidget = self.listWidget
            self.setupMedia(item)
#打开背景文件
    def openBackground(self):
        fileName, _ = QFileDialog.getOpenFileName(self, "Open Video", QDir.homePath(),
"Video( *.mp4; *.wmv; *.rmvb; *.avi; *.mkv; *.mov; *.mpeg; *.flv; *.rm; *.mpeg; *.mpg)")
        if fileName != '':
            if self.state != State.IDLE:
                self.stop.emit()
            v = VideoFileClip(fileName)
```

```python
                item = MyListWidgetItem(fileName.split("/")[-1], v, v.audio)
                self.listWidget_2.addItem(item)
                self.listWidget_2.setCurrentItem(item)
                self.cur_listWidget = self.listWidget_2
                self.setupMedia(item)
    # 打开音频文件
    def openAudio(self):
        fileName, _ = QFileDialog.getOpenFileName(self, "Open Audio", QDir.homePath(),"
Audio(*.mp3;*.wav;*.ogg;*.wma;)")
        if fileName != '':
            if self.state != State.IDLE:
                self.stop.emit()
            a = AudioFileClip(fileName)
            item = MyListWidgetItem(clipName = fileName.split("/")[-1], video = None, audio = a)
            self.listWidget_3.addItem(item)
            self.listWidget_3.setCurrentItem(item)
            self.cur_listWidget = self.listWidget_3
            self.setupMedia(item)
# 设置要播放的视频、音频信息
    def setupMedia(self, item):
      duration = item.video.duration if item.video else item.audio.duration
        self.video_backend = item.video
        self.audio_backend = item.audio
        # 还没有播放过的片段,建立子线程并初始化相应对象
        if self.flag == 0:
            self.video = Video(item.video)
            self.durationChanged(duration)
            self.calthread = QThread()
            self.video.moveToThread(self.calthread)
            self.video.t_changed.connect(self.positionChanged)
            self.stop.connect(self.video.stop)
            self.pause.connect(self.video.pause)
            self.resume.connect(self.video.resume)
            self.video.finish.connect(self.processFinish)
            self.calthread.started.connect(self.video.work)
            self.set_video.connect(self.video.setClip)
            self.audioThread = QThread()
            if self.audio_backend:
                self.audio = Audio(clip = item.audio, fps = item.audio.fps, buffersize = int
(1.0 / self.video.fps * item.audio.fps), nbytes = 2)
            else:
                self.audio = Audio(clip = None, fps = 0, buffersize = 0, nbytes = 0)
            self.audio.moveToThread(self.audioThread)
            self.audioThread.started.connect(self.audio.work)
            self.set_audio.connect(self.audio.setClip)
        # 已经播放过,此时视频线程与音频线程均存在,设置相应对象信息即可
        else:
```

```python
            v_fps = item.video.fps if item.video else 30
            if item.video and item.audio:
                self.durationChanged(duration)
                self.set_video.emit(item.video)
                self.set_audio.emit(item.audio, item.audio.fps,
                                    int(1.0 / v_fps * item.audio.fps), 2)
            elif item.video and not item.audio:
                self.set_video.emit(item.video)
                self.set_audio.emit(None, 44100, 0, 2)
                self.durationChanged(duration)
            else:
                self.set_video.emit(item.video)
                self.durationChanged(item.audio.duration)
                self.set_audio.emit(item.audio, item.audio.fps,
                                    int(1.0 / v_fps * item.audio.fps), 2)
        self.state = State.READY
        self.fitState()
```

2. 音视频的预处理

本部分相关代码如下:

```python
    #设置切分起始时间并显示
    def setStart(self):
        self.cutter_start = self.video.t
        duration = self.video_backend.duration if self.video_backend else self.audio_backend.duration
        currentTime = QTime(int(self.cutter_start/3600) % 60, int(self.cutter_start/60) % 60, self.cutter_start % 60, (self.cutter_start * 1000) % 1000)
        format = 'hh:mm:ss' if duration > 3600 else 'mm:ss'
        tStr = currentTime.toString(format)
        self.statusbar.showMessage("设置起始时间 % s" % tStr, 1000)
    #若当前进度大于切分起始时间,则对当前播放片段原地切分
    def cut(self):
        if self.video.t <= self.cutter_start:
            QMessageBox.critical(self, "警告", "请在起始时间之后切分")
        elif self.video.t > self.cutter_start:
            if self.video_backend and self.audio_backend:
                self.video_backend = self.video_backend.subclip(self.cutter_start, self.video.t)
                self.audio_backend = self.video_backend.audio
                self.set_video.emit(self.video_backend)
                self.set_audio.emit(self.audio_backend, self.audio_backend.fps, int(1.0 / self.video_backend.fps * self.audio_backend.fps), 2)
                self.cur_listWidget.currentItem().setClip(self.video_backend, self.audio_backend)
                self.durationChanged(self.video_backend.duration)
            elif self.video_backend and not self.audio_backend:
                self.video_backend = self.video_backend.subclip(self.cutter_start, self.
```

```python
                    video.t)
                    self.set_video.emit(self.video_backend)
            self.cur_listWidget.currentItem().setClip(self.video_backend, self.audio_backend)
                    self.durationChanged(self.video_backend.duration)
                elif not self.video_backend and self.audio_backend:
            self.audio_backend = self.audio_backend.subclip(self.cutter_start, self.video.t)
                    self.set_video.emit(self.video_backend)
                    self.set_audio.emit(self.audio_backend, self.audio_backend.fps, int(1.0 /
            self.video.fps * self.audio_backend.fps), 2)
            self.cur_listWidget.currentItem().setClip(self.video_backend, self.audio_backend)
                    self.durationChanged(self.audio_backend.duration)
            self.cutter_start = 0
    # 将目标音频或背景音频添加为音频行的item
    def addToAudio(self):
            newName = "(Audio)" + self.cur_listWidget.currentItem().text
        item = MyListWidgetItem(clipName = newName, video = None, audio = self.cur_listWidget.
        currentItem().audio)
            self.listWidget_3.addItem(item)
            self.setupMedia(item)
    # 重复添加选中的item
    def copySelected(self):
        self.cur_listWidget.addItem(self.cur_listWidget.currentItem().copy())
    # 某一片段的抖音效果
    def tiktok(self):
            if isinstance(self.clip, VideoFileClip):
                self.clip = self.clip.fl_image(tiktok_effect)
            ui.cur_listWidget.currentItem().setClip(self.clip, self.clip.audio)
                ui.openGLWidget.update()
    # 删除item,若删除的是正在播放的,会退到初始状态以防止错误
    def deleteItem(self):
            if self.video_backend:
                if self.cur_listWidget.currentItem().video == self.video_backend:
                    self.returnIDLE()
            elif self.audio_backend:
                if self.cur_listWidget.currentItem().audio == self.audio_backend:
                    self.returnIDLE()
            ditem = self.cur_listWidget.takeItem(self.cur_listWidget.row(self.cur_listWidget.
        currentItem()))
            del ditem
    # 保存该片段到本地
    def saveItem(self):
            self.returnIDLE()
            item = self.cur_listWidget.currentItem()
            self.saveitem = item
            if item.video:
                self.saveObject = SaveTemp(item.video.copy())
```

```python
            else:
                self.saveObject = SaveTemp(item.audio.copy())
            item.setFlags(item.flags() & ((Qt.ItemIsSelectable | Qt.ItemIsEnabled) ^ 0xff))
            self.saveThread = QThread()
            self.saveObject.moveToThread(self.saveThread)
            self.saveObject.finish_process.connect(self.finishSave)
            self.saveObject.message.connect(self.thread_message)
            self.saveObject.progress.connect(self.thread_progress)
            self.saveThread.started.connect(self.saveObject.process)
            self.progressBar.setVisible(True)
            self.saveThread.start()
# 保存临时片段
@check_outpath
class SaveTemp(QObject):
    finish_process = pyqtSignal()
    progress = pyqtSignal(int)
    message = pyqtSignal(str)
    def __init__(self, clip, parent = None):                  # 初始化
        super(SaveTemp, self).__init__(parent)
        self.clip = clip
    def process(self):                                         # 处理函数
        myLogger = MyBarLogger(self.message, self.progress)
        if isinstance(self.clip, VideoClip):
self.clip.write_videofile(f'./output/{time.strftime("%Y-%m-%d_%H-%M-%S", time.localtime())}.mp4', codec = 'mpeg4',
                audio_codec = "libmp3lame",
                bitrate = "8000k",
                threads = 4, logger = myLogger)
        elif isinstance(self.clip, AudioClip):
self.clip.write_audiofile(f'./output/{time.strftime("%Y-%m-%d_%H-%M-%S", time.localtime())}.mp3', logger = myLogger)
        self.finish_process.emit()
```

3. 生成参数设置

将所有素材预处理后设置需要添加的全局特效，可以选择抖音效果、分身效果和高斯模糊效果。

```python
        # 在对话框里设置处理参数
        def setupPara(self):
            def choose():
                if self.gauss.isChecked():
                    self.gauss_target.setEnabled(True)
                    self.gauss_background.setEnabled(True)
                else:
                    self.gauss_target.setEnabled(False)
                    self.gauss_background.setEnabled(False)
```

项目9 基于Mask R-CNN的娱乐视频生成器

```python
        #至少选择目标和背景才能在对话框里设置处理参数
        if self.listWidget.count() and self.listWidget_2.count():
            self.dialog = QtWidgets.QDialog(self)
            self.dialog.setAttribute(Qt.WA_DeleteOnClose)
            self.dialog.setWindowModality(Qt.WindowModal)
            self.dialog.setWindowTitle("准备合成")
            gridLayout = QtWidgets.QGridLayout(self.dialog)
            self.dialog.setLayout(gridLayout)
            self.check = QtWidgets.QCheckBox("全局分身效果", self.dialog)
            self.gauss = QtWidgets.QCheckBox("高斯模糊", self.dialog)
            self.gauss_target = QtWidgets.QCheckBox("目标", self.dialog)
            self.gauss_background = QtWidgets.QCheckBox("背景", self.dialog)
            self.tiktok = QtWidgets.QCheckBox("抖音效果", self.dialog)
            sizePolicy = QtWidgets.QSizePolicy(QtWidgets.QSizePolicy.Fixed, QtWidgets.QSizePolicy.Fixed)
            self.gauss_target.setSizePolicy(sizePolicy)
            self.gauss_background.setSizePolicy(sizePolicy)
            self.check.setSizePolicy(sizePolicy)
            self.gauss.setSizePolicy(sizePolicy)
            self.tiktok.setSizePolicy(sizePolicy)
            self.gauss.stateChanged.connect(choose)
            self.gauss_target.setEnabled(False)
            self.gauss_background.setEnabled(False)
            dialogbtns = QDialogButtonBox(QDialogButtonBox.Cancel | QDialogButtonBox.Ok)
            dialogbtns.button(QDialogButtonBox.Cancel).setText("取消")
            dialogbtns.button(QDialogButtonBox.Ok).setText("开始")
            dialogbtns.rejected.connect(self.dialog.reject)
            dialogbtns.accepted.connect(self.dialog.accept)
            dialogbtns.accepted.connect(self.composite)
            dialogbtns.setSizePolicy(sizePolicy)
            label = QtWidgets.QLabel(self.dialog)
            label.setText("默认分辨率 1280x720")
            gridLayout.addWidget(self.check, 0, 0, 1, 1)
            gridLayout.addWidget(dialogbtns, 4, 0, 1, 3)
            gridLayout.addWidget(label, 3, 0, 1, 1)
            gridLayout.addWidget(self.gauss, 1, 0, 1, 1)
            gridLayout.addWidget(self.tiktok, 2, 0, 1, 1)
            gridLayout.addWidget(self.gauss_target, 1, 1, 1, 1)
            gridLayout.addWidget(self.gauss_background, 1, 2, 1, 1)
            self.dialog.resize(400, 300)
            self.dialog.show()
    #特效相关代码
    #抖音特效
    def tiktok_effect(frame):
        #单独抽取去掉红色通道的图像
        gb_channel_frame = frame.copy()
        gb_channel_frame[:, :, 0].fill(0)
```

```python
        #单独抽取红色通道图像
        r_channel_frame = frame.copy()
        r_channel_frame[:, :, 1].fill(0)
        r_channel_frame[:, :, 2].fill(0)
        #错位合并图像,形成抖音效果
        result = frame.copy()
        result[:-5, :-5, :] = r_channel_frame[:-5, :-5, :] + gb_channel_frame[5:, 5:, :]
    return result
#高斯模糊
def blur(image):
    return gaussian(image.astype(float), sigma = 10)
#分身效果,构造红蓝两种颜色的纯色片段,设置在中央人物的左右两侧,通过mask显示出与人物相
#同的轮廓
    def triple_effect(self, clip, mask_clip, width, height):
        red_clip = ColorClip(clip.size, (255, 0, 0), duration = clip.duration)
        blue_clip = ColorClip(clip.size, (0, 0, 255), duration = clip.duration)
        center_person_clip = clip.set_mask(mask_clip).set_position("center", "center")
        left_person_clip = red_clip.set_mask(mask_clip).set_opacity(0.5)
        right_person_clip = blue_clip.set_mask(mask_clip).set_opacity(0.5)
        left_person_clip_x = (width / 2 - left_person_clip.w / 2) - int(left_person_clip.w * 0.3)
        right_person_clip_x = (width / 2 - left_person_clip.w / 2) + int(left_person_clip.w * 0.3)
        person_clip_y = height / 2 - left_person_clip.h / 2
        left_person_clip = left_person_clip.set_position((left_person_clip_x, person_clip_y))
        right_person_clip = right_person_clip.set_position((right_person_clip_x, person_clip_y))
        return [left_person_clip, right_person_clip, center_person_clip]
#处理的片段与参数
    def __init__(self, targets, backgrounds, audios = None, triple = False, gauss = False, tiktok = False,
                 gauss_target = False, gauss_background = False,
                 parent = None):
        super(CompositeObject, self).__init__(parent)
        self.targets = targets
        self.backgrounds = backgrounds
        self.audios = audios
        self.triple = triple
        self.gauss = gauss
        self.tiktok = tiktok
        self.gauss_target = gauss_target
        self.gauss_background = gauss_background
        self.width = 1280
        self.height = 720
```

4. 处理结果

参数设置完毕后对所有修剪后的音视频进行最终处理。

```python
        #以最终帧高度为准,使所有目标素材的帧高度相同,不因拉伸而失真
```

```python
        for i in range(len(self.targets)):
            self.targets[i] = self.targets[i].fx(vfx.resize, height = self.height)
        for i in range(len(self.backgrounds)):
            self.backgrounds[i] = self.backgrounds[i].fx(vfx.resize, (self.width, self.height))
        #简单拼接
        target = concatenate_videoclips(self.targets, method = "compose").without_audio()
        background = concatenate_videoclips(self.backgrounds).without_audio()
        #计算总时长,若有音频则拼接音频
        audio = None
        duration = min(target.duration, background.duration)
        if self.audios:
            audio = concatenate_audioclips(self.audios)
            duration = min(target.duration, background.duration, audio.duration)
        #把目标的识别结果转换为mask,表明它所属的片段哪些部分在背景上可见
        mask_clip = target.fl_image(custom_frame).to_mask()
        #在目标或背景上进行高斯模糊
        if self.gauss_target:
            target = target.fl_image(blur)
        if self.gauss_background:
            background = background.fl_image(blur)
        #在目标上添加抖音效果
        if self.tiktok:
            target = target.fl_image(tiktok_effect)
        #分身效果
        if self.triple:
            temp = self.triple_effect(target, mask_clip, width = self.width, height = self.height)
            temp.insert(0, background)
        else:
            #set_mask 使得被识别为 True 的部分在背景上可见
            target = target.set_mask(mask_clip).set_position("center","center")
            temp = [background, target]
        #拼接所有目标素材
        final_clip = CompositeVideoClip(temp).set_audio(audio). \
            set_duration(duration) if audio else CompositeVideoClip(temp).set_duration(duration)
        #导出为文件
        final_clip.write_videofile(
            f'./output/{time.strftime("%Y-%m-%d_%H-%M-%S", time.localtime())}.mp4',
            fps = 30,
            codec = 'mpeg4',
            bitrate = "8000k",
            audio_codec = "libmp3lame",
            threads = 4,
            logger = my_logger
        )
```

9.3.3 PyQt 界面

本软件通过 PyQt 提供对视频和音频简单处理和生成参数设置的可视化交互，且在使用过程中可播放所有未在处理中的视频、音频片段。

1. 片段对象可视化

通过自定义 PyQt 的 ListWidgetItem 来表现、处理各片段的拼接顺序。

```python
listHeight = 60
#存储界面上各片段音视频信息,与用户交互的item
class MyListWidgetItem(QListWidgetItem):
    def __init__(self, clipName = None, video = None, audio = None, parent = None):   #定义初始化
        super(MyListWidgetItem, self).__init__(clipName, parent)
        self.video = video
        self.audio = audio
        self.text = clipName
        self.duration = self.video.duration if self.video else self.audio.duration
        self.setSizeHint(QSize(self.duration * 15, listHeight))
    def setClip(self, video = None, audio = None):                                    #设置片段
        self.video = video
        self.audio = audio
        self.duration = self.video.duration if self.video else self.audio.duration
        self.setSizeHint(QSize(self.duration * 15, listHeight))
    def copy(self):                                                                   #复制
        return MyListWidgetItem(self.text, self.video, self.audio)
```

2. 片段播放

第一次播放片段时,开启两个子线程,借助相应的对象来同步音视频播放的进度,切换要播放的片段时,向这两个对象传递新片段的参数即可。

```python
#记录应用状态的枚举变量
@unique
class State(Enum):
    IDLE = 0
    READY = 1
    PLAYING = 2
    PAUSE = 3
    FINISHED = 4
#在子线程中定时读取视频数据的对象
class Video(QObject):
    finish = pyqtSignal()
    t_changed = pyqtSignal(int)
    #初始化视频信息
    def __init__(self, clip):
        super(Video, self).__init__()
```

```python
        self.clip = clip
        self.t = 0
        self.fps = self.clip.fps if self.clip else 30
        self.stride = 1 / self.fps
        self.duration = self.clip.duration if self.clip else ui.audio_backend.duration
    # 设置定时器,定时视频帧并提醒同步读取音频对象
    def work(self):
        self.timer = QTimer()
        self.timer.setTimerType(Qt.PreciseTimer)
        if self.clip:
            self.timer.timeout.connect(ui.openGLWidget.update)
        self.timer.timeout.connect(self.updatetime)
        self.timer.timeout.connect(ui.audio.update)
        self.timer.setInterval(1000 * self.stride)
        self.timer.start()
    # 根据进度入队新视频数据,若超时则结束
    def updatetime(self):
        if self.t < self.duration:
            if self.clip:
                im = self.clip.get_frame(self.t)
                img = QImage(im.data, im.shape[1], im.shape[0], im.shape[1] * 3, QImage.Format_RGB888)
                q.put(img)
            self.setT(self.t + self.stride)
        else:
            self.stop()
            self.finish.emit()
    def pause(self):                                         # 暂停
        self.timer.stop()
        q.queue.clear()
        self.t -= (2 * self.stride)
    def resume(self):                                        # 继续
        self.timer.start(1000 * self.stride)
    def stop(self):                                          # 停止
        self.timer.stop()
        self.setT(0)
        q.queue.clear()
    # 某一片段的抖音效果
    def tiktok(self):                                        # 抖音效果
        if isinstance(self.clip, VideoFileClip):
            self.clip = self.clip.fl_image(tiktok_effect)
            ui.cur_listWidget.currentItem().setClip(self.clip, self.clip.audio)
            ui.openGLWidget.update()
    # 设置要播放的新视频
    def setClip(self, clip):                                 # 设置片段
        self.clip = clip
        self.setT(0)
```

```python
            self.fps = self.clip.fps if self.clip else 30
            self.stride = 1 / self.fps
            self.duration = self.clip.duration if self.clip else ui.audio_backend.duration
            self.timer = QTimer()
            self.timer.setTimerType(Qt.PreciseTimer)
            if self.clip:
                self.timer.timeout.connect(ui.openGLWidget.update)
            self.timer.timeout.connect(self.updatetime)
            self.timer.timeout.connect(ui.audio.update)
            self.timer.setInterval(1000 * self.stride)
    #调整播放进度
    def setT(self, t):
        self.t = t
        self.t_changed.emit(self.t * 10)
#播放音频类,独占子线程播放音频
class Audio(QObject):
    #初始化音频片段
    def __init__(self, clip, fps = 44100, buffersize = 4000, nbytes = 2):
        super(Audio, self).__init__()
        self.clip = clip
        self.fps = fps
        self.buffersize = buffersize
        self.nbytes = nbytes
        self.index = 0
        if self.clip:
            self.totalsize = int(self.fps * self.clip.duration)
            self.pospos = np.array(list(range(0, self.totalsize, self.buffersize)) + [self.totalsize])
        else:
            self.totalsize = 0
            self.pospos = None
    #音频流开启
    def work(self):
        if self.clip:
            self.stream = sd.OutputStream(
                samplerate = self.fps, blocksize = self.buffersize,
                dtype = 'int16')
            self.stream.start()
    #逐段写入音频流至结束
    def update(self):
        if self.clip and self.index < len(self.pospos) - 1:
            t = (1.0 / self.fps) * np.arange(self.pospos[self.index], self.pospos[self.index + 1])
            data = self.clip.to_soundarray(t, nbytes = self.nbytes, quantize = True)
            self.stream.write(data)
            self.index += 1
        else:
```

```python
        self.index = 0
    # 设置新音频片段
    def setClip(self, clip, fps = 44100, buffersize = 4000, nbytes = 2):
        self.clip = clip
        self.index = 0
        self.fps = fps
        self.buffersize = buffersize
        self.nbytes = nbytes
        if self.clip:
            self.totalsize = int(self.fps * self.clip.duration)
            self.pospos = np.array(list(range(0, self.totalsize, self.buffersize)) +
[self.totalsize])
        else:
            self.totalsize = 0
            self.pospos = None
        self.work()
    # 控制音频播放进度
    def setIndex(self, i):
        self.index = i
# 用于播放视频的自定义控件
class MyOpenGLWidget(QOpenGLWidget):
    def __init__(self, parent = None):
        super(MyOpenGLWidget, self).__init__(parent)
    # 播放状态下不断从队列获得要播放的视频数据,而暂停状态下只在需要时重绘当前进度的画面
    def paintEvent(self, e: QtGui.QPaintEvent) -> None:
        if not q.empty() and ui.state == State.PLAYING:
            painter = QPainter()
            painter.begin(self)
            painter.setRenderHint(QPainter.SmoothPixmapTransform)
            painter.drawImage(QRect(0,0,self.width(),self.height()), q.get())
            painter.end()
        elif ui.state == State.PAUSE and ui.video_backend:
            painter = QPainter()
            painter.begin(self)
            painter.setRenderHint(QPainter.SmoothPixmapTransform)
            im = ui.video.clip.get_frame(ui.video.t)
            img = QImage(im.data, im.shape[1], im.shape[0], im.shape[1] * 3, QImage.Format_RGB888)
            painter.drawImage(QRect(0,0,self.width(),self.height()), img)
            painter.end()
    # 单击播放时根据状态实施相应行为
    def play(self):
        self.flag += 1
        if self.state in (State.FINISHED, State.READY):
            self.state = State.PLAYING
            self.fitState()
            if self.flag == 1:
                self.calthread.start()
```

```python
                self.audioThread.start()
            else:
                self.resume.emit()
        elif self.state == State.PLAYING:
            self.state = State.PAUSE
            self.pause.emit()
            self.fitState()
        elif self.state == State.PAUSE:
            self.resume.emit()
            self.state = State.PLAYING
            self.fitState()
    # 播放结束
    def processFinish(self):
        self.state = State.FINISHED
        self.fitState()
    # 播放进度改变时移动进度条滑块并记录时间
    def positionChanged(self, position):
        self.horizontalSlider.setValue(position)
        self.updateDurationInfo(position / 10)
    # 播放对象改变时重设时间信息
    def durationChanged(self, duration):
        self.horizontalSlider.setRange(0, duration * 10)
        self.updateDurationInfo(0)
    # 控制视频播放进度
    def setPosition(self, position):
        self.video.setT(position / 10)
    # 控制音频播放进度
    def setApos(self, position):
        if self.audio_backend:
            self.audio.setIndex(int(position / 10 * self.audio.fps / self.audio.buffersize))
    # 更新时间信息
    def updateDurationInfo(self, currentInfo):
        # 正确计算时长
        if self.video_backend:
            duration = self.video_backend.duration
        elif self.audio_backend:
            duration = self.audio_backend.duration
        else:
            duration = 0
        # 计算当前时间和总时间并显示
        if currentInfo or duration:
            currentTime = QTime((currentInfo / 3600) % 60, (currentInfo / 60) % 60,
                               currentInfo % 60, (currentInfo * 1000) % 1000)
            totalTime = QTime((duration / 3600) % 60, (duration / 60) % 60,
                              duration % 60, (duration * 1000) % 1000)
            format = 'hh:mm:ss' if duration > 3600 else 'mm:ss'
            tStr = currentTime.toString(format) + " / " + totalTime.toString(format)
```

```
else:
    tStr = ""
self.label.setText(tStr)
```

9.4 系统测试

本部分包括训练准确率、运行效率和应用使用说明。

9.4.1 训练准确率

以打篮球的视频片段进行验证。该片段共 287 帧,每帧中人与球都存在。某一帧检测、分割的可视化结果如图 9-3 所示。

图 9-3 某一帧检测、分割结果

从图中可以看出人物和球都被检测到,分割出来的物体轮廓误差较小。运动中的人物和球比较模糊,人物的双腿、身体与球在平面上是重叠的,因而这一帧球未成功被检测成 Sports Ball,分割出来的轮廓误差相对较大。

视频片段的总体检测结果中,287 帧中每帧都检测出了人物,但只有 166 帧检测出有球,如图 9-4 所示。

图 9-4 另一帧检测、分割结果

9.4.2 运行效率

对一段目标视频进行检测、分割后与背景视频画面叠加作测试。运行相关参数为：CPU，英特尔酷睿，i5-7300HQ，四核，2.50GHz；GPU，NVIDIA GeForce GTX 1050 Ti 4GB；驱动版本 425.31；CUDA 版本 10.1。

如图 9-5 所示，本机 CUDA 环境下，平均每秒可处理 2 帧画面。

```
t:  3%|▊         | 10/287 [00:04<02:02,  2.27it/s, now=None]10
t:  4%|▊         | 11/287 [00:04<02:02,  2.26it/s, now=None]11
t:  4%|▊         | 12/287 [00:04<02:02,  2.25it/s, now=None]12
t:  5%|▊         | 13/287 [00:05<02:02,  2.25it/s, now=None]13
t:  5%|▊         | 14/287 [00:05<02:02,  2.23it/s, now=None]14
t:  5%|▊         | 15/287 [00:06<02:02,  2.22it/s, now=None]15
t:  6%|▊         | 16/287 [00:06<02:02,  2.21it/s, now=None]16
t:  6%|▊         | 17/287 [00:07<02:02,  2.21it/s, now=None]17
t:  6%|▊         | 18/287 [00:07<02:02,  2.19it/s, now=None]18
t:  7%|▊         | 19/287 [00:08<02:02,  2.18it/s, now=None]19
t:  7%|▊         | 20/287 [00:08<02:02,  2.18it/s, now=None]20
t:  7%|▊         | 21/287 [00:09<02:01,  2.19it/s, now=None]21
t:  8%|▊         | 22/287 [00:09<02:01,  2.18it/s, now=None]22
t:  8%|▊         | 23/287 [00:09<02:00,  2.18it/s, now=None]23
t:  8%|▊         | 24/287 [00:10<02:00,  2.19it/s, now=None]24
t:  9%|▊         | 25/287 [00:10<01:59,  2.19it/s, now=None]25
t:  9%|▊         | 26/287 [00:11<01:59,  2.19it/s, now=None]26
t:  9%|▊         | 27/287 [00:11<01:58,  2.20it/s, now=None]27
t: 10%|▊         | 28/287 [00:12<01:57,  2.20it/s, now=None]28
t: 10%|▊         | 29/287 [00:12<01:57,  2.19it/s, now=None]29
```

图 9-5 某一时段的运行进度

CUDA 环境下运行时的资源消耗如图 9-6 所示。

名称	状态	87% CPU	68% 内存	4% 磁盘	1% 网络	1% GPU	GPU 引擎
∨ ▣ PyCharm (10)		74.8%	3,418.7 ...	0.1 MB/秒	0 Mbps	0.7%	GPU 1 - Copy
▣ Python		71.5%	1,926.6 ...	0 MB/秒	0 Mbps	0.7%	GPU 1 - Copy
▣ ffmpeg-win64-v4.1.exe		2.3%	63.0 MB	0.1 MB/秒	0 Mbps	0%	
▣ ffmpeg-win64-v4.1.exe		1.0%	57.3 MB	0.1 MB/秒	0 Mbps	0%	

图 9-6 CUDA 环境下运行时的资源消耗

如图 9-7 所示，本机 CPU 环境下，平均每 10s 处理 1 帧画面。

图 9-8 表示本机 CUDA 环境下运行时的资源消耗。

图 9-7 运行进度

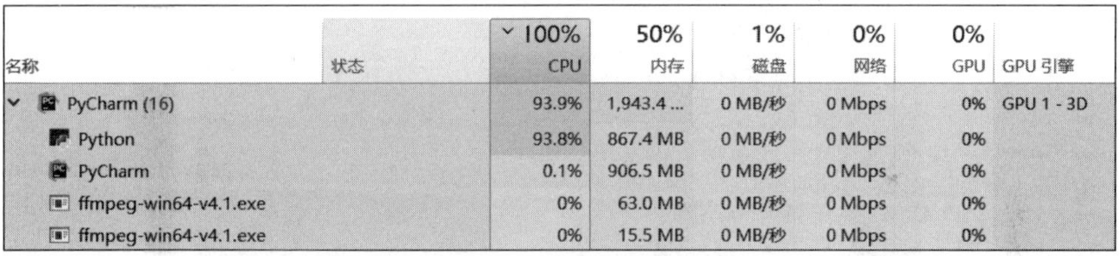

图 9-8 资源消耗

9.4.3 应用使用说明

应用主程序初始界面如图 9-9 所示。

视频导入后如图 9-10 所示,可对素材进行简单预处理(切分、换序、重复、删除),也可以对其中的每一片段进行预览。

单击"生成"按钮设定合成参数。图 9-11 选择了分身效果和背景高斯模糊,单击"开始"按钮即可生成。

这些片段会按放置顺序拼接、叠加,处理过程中它们无法交互,但可导入其他素材进行操作,如图 9-12 所示。

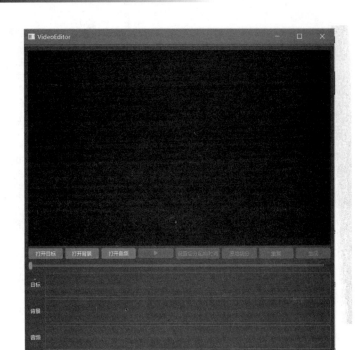

图 9-9　应用初始界面

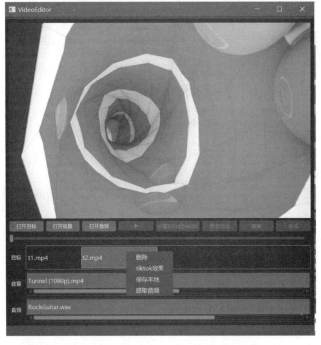

图 9-10　视频导入后可进行的操作

图 9-11 合成参数设置

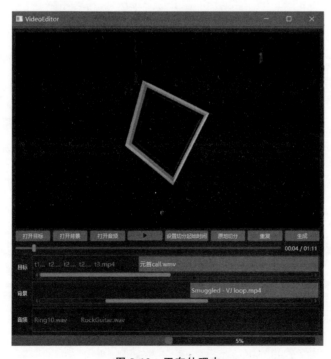

图 9-12 正在处理中

成品效果如图 9-13 所示。

图 9-13　成品效果

项目 10　基于 CycleGAN 的图像转换

PROJECT 10

本项目采用对抗网络 CycleGAN 对一批油画图片和现实风景图片进行训练，在微信小程序上实现图像转换。

10.1　总体设计

本部分包括系统整体结构和系统流程。

10.1.1　系统整体结构

系统整体结构如图 10-1 所示。

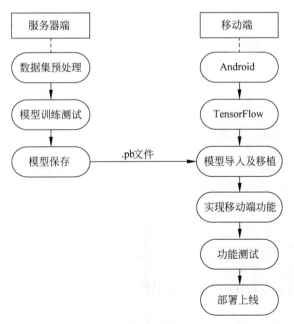

图 10-1　系统整体结构

10.1.2 系统流程

系统流程如图 10-2 所示。

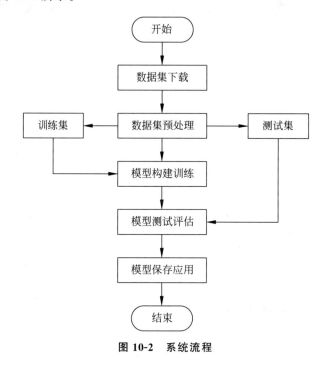

图 10-2 系统流程

10.2 运行环境

本部分包括 Python 环境、TensorFlow GPU 环境和 Android 环境配置。

10.2.1 Python 环境

安装 Python 3.6.8 或以上版本，下载地址为：https://www.python.org/downloads/，选择相应操作系统的安装包。也可在 Anaconda 上配置 Python，下载地址为：https://www.anaconda.com/。

集成开发环境为 PyCharm，下载地址为：http://www.jetbrains.com/pycharm/download/#section=windows，安装、配置环境变量即可使用。

10.2.2 TensorFlow GPU 环境

为提高下载速度，需要添加国内镜像源，打开 Anaconda Prompt，输入命令：

```
conda config -- add channels https://mirrors.tuna.tsinghua.edu.cn/anaconda/pkgs/free/
conda config -- add channels https://mirrors.tuna.tsinghua.edu.cn/anaconda/pkgs/main/
conda config -- set show_channel_urls yes
```

成功添加清华仓库镜像源后，创建 Python 3.6 环境，名称为 tensorflow-gpu，输入命令：

```
conda create -n tensorflow-gpu python = 3.6
```

激活环境，输入命令：

```
conda activate tensorflow-gpu
```

安装 tensorflow-gpu，输入命令：

```
conda install tensorflow-gpu == 1.14
```

在激活的环境中安装 TensorFlow 的 GPU 版本，后续安装的 CUDA 和 cuDNN 的版本号要与之匹配，如图 10-3 所示。

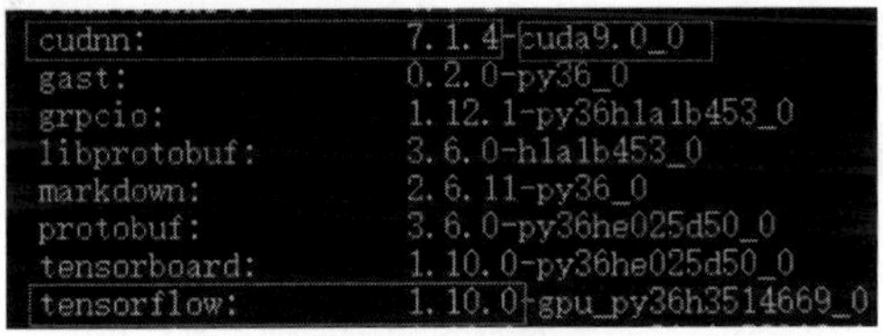

图 10-3　版本匹配

如果三者版本号不对应，会导致安装失败。本次使用的版本为 tensorflow-gpu-1.14、cuDNN 7.6.4、CUDA 10.1。

CUDA 安装，下载地址为：https://developer.nvidia.com/cuda-toolkit-archive，版本选择如图 10-4 所示。

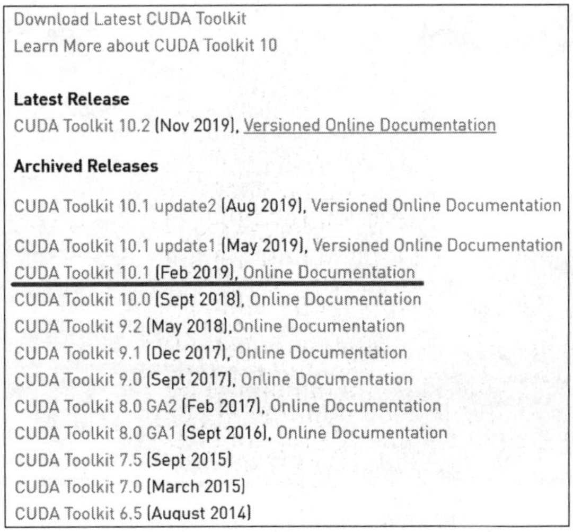

图 10-4　CUDA 版本

cuDNN 安装，下载地址为：https://developer.nvidia.com/rdp/cudnn-archive，下载之前需要注册英伟达账号，版本选择如图 10-5 所示。

图 10-5　cuDNN 版本

下载完成后解压文件，将文件全部复制到 NVIDIA GPU Computing Toolkit\CUDA\v10.1 文件夹中。

10.2.3　Android 环境

安装 Android Studio，下载地址为：https://developer.android.google.cn/studio/install.html。新建 Android 项目，打开 Android Studio，选择 File→New→New Project→Empty Activity→Next，出现如图 10-6 所示的项目配置界面。

图 10-6　Android 项目配置界面

导入 TensorFlow 的 jar 包和 so 库,可以使用 bazel 工具自行编译,难度较大;也可以在网上下载 libtensorflow_inference.so 和 libandroid_tensorflow_inference_java.jar,下载地址为: https://github.com/PanJinquan/Mnist-tensorFlow-AndroidDemo/tree/master/app/libs。

将 libandroid_tensorflow_inference_java.jar 文件放到 app/libs 目录下,libtensorflow_inference.so 放到 app/libs/armeabi-v7a 目录下。app/build.gradle 进行如下配置:

在 defaultConfig 中添加以下命令:

```
multiDexEnabled true
    ndk {
        abiFilters "armeabi-v7a"
    }
```

在 android 节点下添加 sourceSets,用于制定 jniLibs 的路径:

```
sourceSets {
    main {
        jniLibs.srcDirs = ['libs']
    }
}
```

配置完成情况如图 10-7 所示。

图 10-7　添加部分配置

在 dependencies 中(若没有)增加 TensorFlow 编译的 jar 文件：

```
implementation files('libs/libandroid_tensorflow_inference_java.jar')
```

导入 jar 包位置如图 10-8 所示。

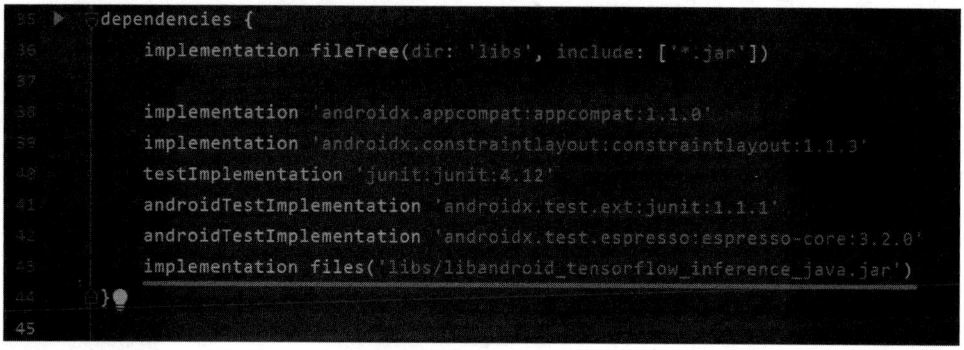

图 10-8 导入 jar 包位置

app/build.gradle 中的内容有任何改动，Android Studio 都会弹出如图 10-9 所示的提示信息。

图 10-9 弹窗同步

直接单击 Sync Now 即可配置成功。

10.3 模块实现

本项目包括 5 个模块：数据集预处理、模型构建、模块分析、模型训练及保存、模型生成。下面分别给出各模块的功能介绍及相关代码。

10.3.1 数据集预处理

数据集下载地址为：https://people.eecs.berkeley.edu/~taesung_park/CycleGAN/datasets/，选择数据集 vangogh2photo。

由于数据集中包含上千张图片，占用内存较大，因此，转为 TensorFlow 中专用的 tfrecords 格式以供训练。TensorFlow 训练数据需要接收 tfrecords 格式的文件。

训练用到 trainA 和 trainB 两个文件，其中 trainA 有 400 张彩色图片，trainB 有 6287 张图片，所有图片的尺寸均为 256×256×3，训练集部分图片如图 10-10 和图 10-11 所示。

图 10-10　训练集 trainA 部分图片

图 10-11　训练集 trainB 部分图片

鉴于训练集 trainB 图片数量较多，若计算机运行内存只有 8GB，在训练模型时将 trainB 全部数据读入，会出现内存不足的情况。因此，保存前 500 张图片进行模型训练。删减数据集虽然可以保证模型的正常训练，但会使模型训练效果变差。

10.3.2 模型构建

本部分包括定义模型结构及优化损失函数。

1. 定义模型结构

该模型从域 X 获取输入图像,将输入的图像传到生成器 G 中,转换到目标域 Y 中,生成的图像传递到生成器 F 中,转换回原始域 X 中。这样一个循环后,生成的图像与原始图像之间差距被定义为循环一致损失。

两个输入图像传递到判别器中,以判别器 Y 为例,输入的图像一部分是 Y 域中原图,一部分是通过生成器生成的假图,判别器的目的是尽可能多地识别出生成的图像,相对应的生成器尽可能生成以假乱真的假图。

```python
def model(self):
    X_reader = Reader(self.X_train_file, name = 'X',          #读取数据
                     image_size = self.image_size, batch_size = self.batch_size)
    Y_reader = Reader(self.Y_train_file, name = 'Y',
                     image_size = self.image_size, batch_size = self.batch_size)
    x = X_reader.feed()
    y = Y_reader.feed()
    #定义循环一致损失
    cycle_loss = self.cycle_consistency_loss(self.G, self.F, x, y)
    #X-> Y
    #生成器 G 生成的图片
    fake_y = self.G(x)
    #fake_y损失:判别器 D_Y 对其生成图片的判别值
    G_gan_loss = self.generator_loss(self.D_Y, fake_y, use_lsgan = self.use_lsgan)
    #生成器 G 损失 = fake_y 损失 + 循环一致损失
    G_loss = G_gan_loss + cycle_loss
    #判别器 Y 的损失:判别器 D_Y 对 Y 域真实图片的判别值
    D_Y_loss = self.discriminator_loss(self.D_Y, y, self.fake_y, use_lsgan = self.use_lsgan)
    #Y-> X
    fake_x = self.F(y)
    F_gan_loss = self.generator_loss(self.D_X, fake_x, use_lsgan = self.use_lsgan)
    F_loss = F_gan_loss + cycle_loss
    D_X_loss = self.discriminator_loss(self.D_X, x, self.fake_x, use_lsgan = self.use_lsgan)
```

2. 优化损失函数

损失函数由生成器损失、判别器损失和重建损失构成,求损失时使用最小二乘法,并用 L2 范式约束损失,即矩阵中每个元素相减后求平方。模型优化部分需要定义初始学习率、衰减速度,规定每迭代 100000 次后学习率衰减至 0,选择 Adam 优化器损失最小。

```python
#定义损失函数和优化器
cross_entropy = -tf.reduce_sum(y_ * tf.log(y_conv))
```

```python
train_step = tf.train.AdamOptimizer(1e-4).minimize(cross_entropy)
correct_predict = tf.equal(tf.argmax(y_conv, 1), tf.argmax(y_, 1))
    def generator_loss(self, D, fake_y, use_lsgan = True):
        #使用最小二乘法
        if use_lsgan:
            #tf.reduce_mean:求矩阵均值,求生成 y 与真实 y 的损失
            #f.squared_difference:两个矩阵的每个元素相减后求平方,形成新矩阵输出
            loss = tf.reduce_mean(tf.squared_difference(D(fake_y), REAL_LABEL))
        #均方误差
        else:
            #表达式:-log(D(G(x)))
            loss = -tf.reduce_mean(ops.safe_log(D(fake_y))) / 2
        return loss
    def discriminator_loss(self, D, y, fake_y, use_lsgan = True):
        #使用最小二乘法
        if use_lsgan:
            #计算 y-x 为真实样本的均方误差
            error_real = tf.reduce_mean(tf.squared_difference(D(y), REAL_LABEL))
            #均方误差
            #计算 x-y-x 的均方误差 fake_y:x-y
            error_fake = tf.reduce_mean(tf.square(D(fake_y)))          #均方误差
        else:
            #表达式:-(log(D(y)) + log(1 - D(G(x))))/2
            error_real = -tf.reduce_mean(ops.safe_log(D(y)))           #-log(D(y))
            error_fake = -tf.reduce_mean(ops.safe_log(1-D(fake_y)))    #-log(1-D(G(x)))
        loss = (error_real + error_fake) / 2
        #-(log(D(y)) + log(1-D(G(x))))/2
        return loss
    def cycle_consistency_loss(self, G, F, x, y):
        #x -> G(x) -> F(G(x)) ≈ x,前向损失
        forward_loss = tf.reduce_mean(tf.abs(F(G(x))-x))
        #y -> F(y) -> G(F(y)) ≈ y,后向损失
        backward_loss = tf.reduce_mean(tf.abs(G(F(y))-y))
        loss = self.lambda1 * forward_loss + self.lambda2 * backward_loss
        return loss
    def optimize(self, G_loss, D_Y_loss, F_loss, D_X_loss):
        def make_optimizer(loss, variables, name = 'Adam'):
            global_step = tf.Variable(0, trainable = False)        #迭代次数
            starter_learning_rate = self.learning_rate             #初始学习率
            end_learning_rate = 0.0                                #最后的学习率,每 100000 次
                                                                   #之后线性衰减到 0
            start_decay_step = 100000                              #初始迭代数
            decay_steps = 100000#衰减速度,在迭代到该次数时学习率衰减为:学习率×衰减系数
            beta1 = self.beta1
            #学习率
            learning_rate = (
                    tf.where(
```

```
                    #返回一个布尔值的张量
                    tf.greater_equal(global_step, start_decay_step),
                    #实现指数衰减学习率
                    tf.train.polynomial_decay(starter_learning_rate, #初始学习率
                        global_step-start_decay_step,         #当前迭代次数-开始迭代数
                                        decay_steps,
                    #衰减速度(在迭代到该次数时学习率衰减为 earning_rate * decay_rate)
                                        end_learning_rate,    #最终学习率
                                        power = 1.0),
                    #初始学习率
                    starter_learning_rate
            )
    )
    #以标量形式输出学习率
    tf.summary.scalar('learning_rate/{}'.format(name), learning_rate)
    learning_step = (
        # 选择 Adam 优化器,让损失最小
        tf.train.AdamOptimizer(learning_rate, beta1 = beta1, name = name)
                .minimize(loss,                         #最小化的目标变量
            global_step = global_step,                  #梯度下降一次加 1,一般用
                                                        #于记录迭代优化的次数
                        var_list = variables            #每次要迭代更新的参数集合
                    ))
    return learning_step
```

10.3.3 模块分析

本部分包括数据预处理、生成器模块和判别器模块。

1. 数据预处理

data_reader()函数将解压后的数据集读入并打乱图片的顺序,data_writer()函数将图片写成 tfrecords 格式文件。函数中使用官方提供的 tfrecord 生成器,直接遍历数据集中所有图片并写入 tfrecords 格式输出。

```
def data_writer(input_dir, output_file):
    file_paths = data_reader(input_dir)                 #读数据
    #遍历输出路径文件夹是否存在,没有则新建
    output_dir = os.path.dirname(output_file)
    try:
        os.makedirs(output_dir)
    except os.error as e:
        pass
    images_num = len(file_paths)                        #获取文件数量
    writer = tf.python_io.TFRecordWriter(output_file)   #创建 tf 文件
    #遍历每一个图片文件
    for i in range(len(file_paths)):
```

```
file_path = file_paths[i]
# 读取图片
with tf.gfile.FastGFile(file_path, 'rb') as f:
    image_data = f.read()
example = _convert_to_example(file_path, image_data)   # 方法实例化
writer.write(example.SerializeToString())
# 将实例序列转化为一个字符串,写入 tfrecords
```

2. 生成器模块

生成器主要由编码器、转换器、解码器构成。

(1) 编码部分使用 3 层卷积神经网络从输入图像中提取特征,并压缩成 128 个 64×64 的特征向量。

```
# 卷积 1
c7s1_32 = ops.c7s1_k(input, self.ngf, is_training = self.is_training, norm = self.norm,
reuse = self.reuse, name = 'c7s1_32')
# 卷积 2
d64 = ops.dk(c7s1_32, 2 * self.ngf, is_training = self.is_training, norm = self.norm,
reuse = self.reuse, name = 'd64')
# 卷积 3
d128 = ops.dk(d64, 4 * self.ngf, is_training = self.is_training, norm = self.norm,
reuse = self.reuse, name = 'd128')
```

(2) 转换部分通过组合图像的不相近特征,将图像在 DA 域中的特征向量转换为 DB 域中的特征向量。使用 6 层或者 9 层 ResNet 模块,每个 ResNet 模块是由 2 个卷积层构成的神经网络层,能够在转换时保留原始图像特征。

```
# 若是输出的图片小于 128,使用 6 层残差网络
if self.image_size <= 128:
    res_output = ops.n_res_blocks(d128, reuse = self.reuse, n = 6)
# 大于 128,则使用 9 层残差网络
else:
    res_output = ops.n_res_blocks(d128, reuse = self.reuse, n = 9)
```

(3) 解码部分使用反卷积层完成从特征向量中还原出低级特征的工作,最后生成图像。

```
# 反卷积
u64 = ops.uk(res_output, 2 * self.ngf, is_training = self.is_training, norm = self.norm, reuse
= self.reuse, name = 'u64')
# 反卷积
u32 = ops.uk(u64, self.ngf, is_training = self.is_training, norm = self.norm, reuse = self.
reuse, name = 'u32', output_size = self.image_size)
```

(4) 生成器模块相关代码如下:

```
class Generator:
```

```python
    # 定义实例(Generator)的私有属性,外部无法访问,实例内可以传递
    def __init__(self, name, is_training, ngf = 64, norm = 'instance', image_size = 128):
        self.name = name                              # 卷积层名称
        self.reuse = False                            # 直接创建新变量
        self.ngf = ngf                                # 滤波器数量:32
        self.norm = norm                              # 归一化
        self.is_training = is_training                # 是否训练中
        self.image_size = image_size                  # 输出图片大小
    # 编码(3层卷积)-转换(残差网络)-解码(2层反卷积、1层卷积、1个输出层)
    def __call__(self, input):
        # 上下文变量管理
        with tf.variable_scope(self.name):
            # 卷积1
            c7s1_32 = ops.c7s1_k(input, self.ngf, is_training = self.is_training, norm = self.norm, reuse = self.reuse, name = 'c7s1_32')
            # 卷积2
            d64 = ops.dk(c7s1_32, 2 * self.ngf, is_training = self.is_training, norm = self.norm, reuse = self.reuse, name = 'd64')
            # 卷积3
            d128 = ops.dk(d64, 4 * self.ngf, is_training = self.is_training, norm = self.norm,
                reuse = self.reuse, name = 'd128')
            # 若是输出的图片小于128,使用6层残差网络
            if self.image_size <= 128:
                res_output = ops.n_res_blocks(d128, reuse = self.reuse, n = 6)
            # 大于128,则使用9层残差网络
            else:
                res_output = ops.n_res_blocks(d128, reuse = self.reuse, n = 9)
            # 反卷积
            u64 = ops.uk(res_output, 2 * self.ngf, is_training = self.is_training, norm = self.norm,
                reuse = self.reuse, name = 'u64')
            # 反卷积
            u32 = ops.uk(u64, self.ngf, is_training = self.is_training, norm = self.norm,
                reuse = self.reuse, name = 'u32', output_size = self.image_size)
            # 输出层
            output = ops.c7s1_k(u32, 3, norm = None,
                activation = 'tanh', reuse = self.reuse, name = 'output')
            self.reuse = True
        # 获取优化器训练的变量参数
        self.variables = tf.get_collection(tf.GraphKeys.TRAINABLE_VARIABLES, scope = self.name)
        return output
    def sample(self, input):
        # 把浮点型张量转换为 int 型,0~1 转换为 0~255
        image = utils.batch_convert2int(self.__call__(input))
        # 图片解码,即把图片还原成三维矩阵的形式
        image = tf.image.encode_jpeg(tf.squeeze(image, [0]))
        # tf.squeeze()用于压缩张量中形状为 1 的轴,除去张量中形状为 1 的轴
        return image
```

3. 判别器模块

判别器对输入的图片进行判决,确定其为原始图片或生成图片。判别器使用卷积网络结构实现,通过添加产生一维输出的卷积层提取特征。

```
class Discriminator:
    #定义实例(Discriminator)的私有属性,外部无法访问,实例内可以传递
    def __init__(self, name, is_training, norm = 'instance', use_sigmoid = False):
        self.name = name                              #卷积层名称
        self.is_training = is_training                #是否训练中
        self.norm = norm                              #归一化
        self.reuse = False                            #直接创建新变量
        self.use_sigmoid = use_sigmoid                #使用sigmoid阶跃函数
    def __call__(self, input):
        #调用ops.py的CK()函数,对图像做4层卷积
        with tf.variable_scope(self.name):
            C64  = ops.Ck(input, 64, reuse = self.reuse, norm = None,
                is_training = self.is_training, name = 'C64')
            C128 = ops.Ck(C64, 128, reuse = self.reuse, norm = self.norm,
                is_training = self.is_training, name = 'C128')
            C256 = ops.Ck(C128, 256, reuse = self.reuse, norm = self.norm,
                is_training = self.is_training, name = 'C256')
            C512 = ops.Ck(C256, 512, reuse = self.reuse, norm = self.norm,
                is_training = self.is_training, name = 'C512')
            #最后一层卷积,加一个偏量,使用sigmoid阶跃函数
            output = ops.last_conv(C512, reuse = self.reuse, use_sigmoid = self.use_sigmoid, name = 'output')
            self.reuse = True
            #获取优化器训练的变量参数
            self.variables = tf.get_collection(tf.GraphKeys.TRAINABLE_VARIABLES, scope = self.name)
            return output
```

10.3.4 模型训练及保存

本部分包括模型训练和模型保存。

1. 模型训练

在模型训练之前先定义 tf.Session 的运行方式为 GPU,若指定设备不存在或无法启动可允许 TensorFlow 自动分配设备(GPU 不可用,训练任务会分配给 CPU),使用 GPU 时最大利用率为 70%。

```
#定义选用0号GPU
os.environ["CUDA_VISIBLE_DEVICES"] = "0"
#若指定的设备不存在,则允许TensorFlow自动分配设备
config = tf.ConfigProto(allow_soft_placement = True)
```

```python
        # 选择 GPU 的运行方式
        config.gpu_options.allow_growth = True
        # 限制 GPU 的使用率,最大 70%
        gpu_options = tf.GPUOptions(per_process_gpu_memory_fraction = 0.7)
        # 确定会话为以上配置
        sess = tf.Session(config = config)
    with tf.Session(graph = graph) as sess:
        if FLAGS.load_model is not None:
            checkpoint = tf.train.get_checkpoint_state(checkpoints_dir)
            # 模型文件路径
            meta_graph_path = checkpoint.model_checkpoint_path + ".meta"
            # meta 文件路径
            restore = tf.train.import_meta_graph(meta_graph_path)          # 导入 meta
            restore.restore(sess, tf.train.latest_checkpoint(checkpoints_dir))  # 恢复图片并
                                                                                # 得到数据
            step = int(meta_graph_path.split("-")[2].split(".")[0])
        else:
            sess.run(tf.global_variables_initializer())                    # 初始化
            step = 0
        coord = tf.train.Coordinator()
        threads = tf.train.start_queue_runners(sess = sess, coord = coord)
        try:
            # 调用 unit.py 中的 ImagePool()方法
            fake_Y_pool = ImagePool(FLAGS.pool_size)
            fake_X_pool = ImagePool(FLAGS.pool_size)
            while not coord.should_stop():
                # 获取之前生成的图片
                fake_y_val, fake_x_val = sess.run([fake_y, fake_x])
                # 训练
          _, G_loss_val,D_Y_loss_val, F_loss_val, D_X_loss_val, summary = (
                    sess.run(
                [optimizers, G_loss, D_Y_loss, F_loss, D_X_loss, summary_op],
                    feed_dict = {cycle_gan.fake_y: fake_Y_pool.query(fake_y_val),
                        cycle_gan.fake_x: fake_X_pool.query(fake_x_val)}
                )
            )
            train_writer.add_summary(summary, step)   # 存入训练的汇总参数和步长
                train_writer.flush()
                if step % 100 == 0:                   # 每过 100 步打印生成器损失和判别器损失
                    logging.info('----------- Step %d: ------------- ' % step)
                    logging.info('G_loss : {}'.format(G_loss_val))
                    logging.info('D_Y_loss : {}'.format(D_Y_loss_val))
                    logging.info('F_loss : {}'.format(F_loss_val))
                    logging.info('D_X_loss : {}'.format(D_X_loss_val))
                if step % 10000 == 0:                 # 每 10000 步保存 ckpt 模型
                    save_path = saver.save(sess, checkpoints_dir + "/model.ckpt", global_
step = step)                                                            # 保存路径
```

```python
            logging.info("Model saved in file: %s" % save_path)    # 打印日志信息
                step += 1
        # 异常处理
        except KeyboardInterrupt:                                   # 键盘按键中断
            logging.info('Interrupted')                             # 输出中止信息
            coord.request_stop()
        except Exception as e:                                      # 其他中断
            coord.request_stop(e)
        finally:
            save_path = saver.save(sess, checkpoints_dir + "/model.ckpt", global_step=step)
                                                                    # 使用全局变量 step,在指
                                                                    # 定位置保存 ckpt 模型
            logging.info("Model saved in file: %s" % save_path)     # 打印模型已保存信息
            coord.request_stop()                                    # 完成时,可中止
            coord.join(threads)
```

2. 模型保存

使用 TensorFlow 中自带的 graph_util 模型将保存的 ckpt 导出为 .pb 文件,以便在 Android 程序中调用该模型。

```python
def export_graph(model_name, XtoY=True):
    graph = tf.Graph()
    with graph.as_default():
        cycle_gan = CycleGAN(ngf=FLAGS.ngf, norm=FLAGS.norm, image_size=FLAGS.image_size)
                                                                    # 调用 CycleGAN 方法
        # 输入图片
        input_image = tf.placeholder(tf.float32, shape=[FLAGS.image_size, FLAGS.image_size, 3], name='input_image')
        cycle_gan.model()
        if XtoY:
            output_image = cycle_gan.G.sample(tf.expand_dims(input_image, 0))
                                                                    # 生成器 G 生成图片
        else:
            output_image = cycle_gan.F.sample(tf.expand_dims(input_image, 0))
                                                                    # 生成器 F 生成图片
        output_image = tf.identity(output_image, name='output_image')
        restore_saver = tf.train.Saver()
        export_saver = tf.train.Saver()
    with tf.Session(graph=graph) as sess:
        sess.run(tf.global_variables_initializer())
        latest_ckpt = tf.train.latest_checkpoint(FLAGS.checkpoint_dir)
        restore_saver.restore(sess, latest_ckpt)
        output_graph_def = tf.graph_util.convert_variables_to_constants(
            sess, graph.as_graph_def(), [output_image.op.name])
        # 导出模型 graph 结构,即 .pb 模型
        tf.train.write_graph(output_graph_def, 'pretrained', model_name, as_text=False)
```

10.3.5 模型生成

本部分包括模型导入和 Android 应用。

1. 模型导入

将训练好的.pb 模型放入 Android 项目 app/src/main/assets 下，若不存在 assets 目录，鼠标右键选择 main→New→Directory，输入 assets，如图 10-12 所示。

图 10-12　模型位置

新建类 Transfer_image.java，在用到模型之处，首先加载 libtensorflow_inference.so 库并初始化 TensorFlowInferenceInterface 对象，再调用导入的.pb 模型得到预测结果。

```
TensorFlowInferenceInterface inferenceInterface;
    static {
        //加载 libtensorflow_inference.so 库文件
        System.loadLibrary("tensorflow_inference");
        Log.e("tensorflow","libtensorflow_inference.so 库加载成功");
    }
    Classifier(AssetManager assetManager, String modePath) {
        //初始化 TensorFlowInferenceInterface 对象
        inferenceInterface = new TensorFlowInferenceInterface(assetManager,modePath);
        Log.e("tf","TensorFlow 模型文件加载成功");
    }
```

2. Android 应用

该部分主要由界面布局、调用模型转换、主活动控制三部分构成。

1) 界面布局

界面布局相关代码如下：

```
/app/res/layout/activity_main.xml
    <?xml version = "1.0" encoding = "utf-8"?>
<androidx.constraintlayout.widget.ConstraintLayout xmlns:android = "http://schemas.android.com/apk/res/android"
    xmlns:app = "http://schemas.android.com/apk/res-auto"
    xmlns:tools = "http://schemas.android.com/tools"
    android:layout_width = "match_parent"
    android:layout_height = "match_parent"
    tools:context = ".MainActivity">
    <Button
        android:id = "@+id/button1"
        android:layout_width = "114dp"
        android:layout_height = "48dp"
        android:layout_marginStart = "52dp"
        android:layout_marginEnd = "82dp"
        android:layout_marginBottom = "54dp"
        android:text = "从相册中选择"
        app:layout_constraintBottom_toBottomOf = "parent"
        app:layout_constraintEnd_toStartOf = "@+id/button2"
        app:layout_constraintStart_toStartOf = "parent"
        app:layout_constraintTop_toBottomOf = "@+id/textView2" />
    <Button
        android:id = "@+id/button2"
        android:layout_width = "114dp"
        android:layout_height = "48dp"
        android:layout_marginEnd = "49dp"
        android:layout_marginBottom = "68dp"
        android:text = "输出结果"
        app:layout_constraintBottom_toBottomOf = "parent"
        app:layout_constraintEnd_toEndOf = "parent"
        app:layout_constraintStart_toEndOf = "@+id/button1"
        app:layout_constraintTop_toBottomOf = "@+id/imageView4" />
    <ImageView
        android:id = "@+id/imageView3"
        android:layout_width = "180dp"
        android:layout_height = "180dp"
        android:layout_marginTop = "130dp"
        android:layout_marginBottom = "90dp"
        app:layout_constraintBottom_toTopOf = "@+id/imageView4"
        app:layout_constraintEnd_toEndOf = "@+id/imageView4"
        app:layout_constraintStart_toStartOf = "@+id/imageView4"
```

```
            app:layout_constraintTop_toTopOf = "parent"
            app:srcCompat = "@drawable/start_image" />
    <ImageView
            android:id = "@+id/imageView4"
            android:layout_width = "180dp"
            android:layout_height = "180dp"
            android:layout_marginStart = "65dp"
            android:layout_marginEnd = "65dp"
            android:layout_marginBottom = "43dp"
            app:layout_constraintBottom_toTopOf = "@+id/button2"
            app:layout_constraintEnd_toEndOf = "@+id/button2"
            app:layout_constraintStart_toStartOf = "@+id/button1"
            app:layout_constraintTop_toBottomOf = "@+id/imageView3"
            app:srcCompat = "@drawable/start_image0" />
    <TextView
            android:id = "@+id/textView"
            android:layout_width = "90dp"
            android:layout_height = "wrap_content"
            android:layout_marginTop = "70dp"
            android:layout_marginBottom = "230dp"
            android:text = "请输入图片:"
            app:layout_constraintBottom_toTopOf = "@+id/textView2"
            app:layout_constraintEnd_toEndOf = "@+id/textView2"
            app:layout_constraintStart_toStartOf = "@+id/textView2"
            app:layout_constraintTop_toTopOf = "parent" />
    <TextView
            android:id = "@+id/textView2"
            android:layout_width = "90dp"
            android:layout_height = "wrap_content"
            android:layout_marginStart = "52dp"
            android:layout_marginTop = "10dp"
            android:layout_marginBottom = "234dp"
            android:text = "输出结果:"
            app:layout_constraintBottom_toTopOf = "@+id/button1"
            app:layout_constraintStart_toStartOf = "parent"
            app:layout_constraintTop_toBottomOf = "@+id/textView" />
</androidx.constraintlayout.widget.ConstraintLayout>
```

2) 调用模型转换

使用 TensorFlowInferenceInterface 中提供的 feed() 方法获取输入数据，使用 run() 方法运行导入的模型，使用 fetch() 方法取出输出节点的数据。

```
package com.example.cyclegandemo;
import android.content.res.AssetManager;
import android.graphics.Bitmap;
import android.graphics.Color;
import android.graphics.Matrix;
```

```java
import android.util.Log;
import org.tensorflow.contrib.android.TensorFlowInferenceInterface;
public class Transfer_image {
    private static final String TAG = "Transfer_image";
    //设置模型输入/输出节点的数据维度
    private static final int IN_COL = 3;
    private static final int IN_ROW = 256 * 256;
    private static final int OUT_COL = 3;
    private static final int OUT_ROW = 256 * 256;
    //模型中输入变量的名称
    private static final String inputName = "input_image";
    //模型中输出变量的名称
    private static final String outputName = "output_image";
    TensorFlowInferenceInterface inferenceInterface;
    static {
        //加载 libtensorflow_inference.so 库文件
        System.loadLibrary("tensorflow_inference");
        Log.e(TAG,"libtensorflow_inference.so 库加载成功");
    }
    Transfer_image(AssetManager assetManager, String modePath) {
        //初始化 TensorFlowInferenceInterface 对象
        inferenceInterface = new TensorFlowInferenceInterface(assetManager,modePath);
        Log.e(TAG,"TensorFlow 模型文件加载成功");
    }
    /*
     * 使用训练好的 TensorFlow 模型预测结果
     * bitmap 为输入被测试的 Bitmap 图
     */
    public int[] get_transfer(Bitmap bitmap) {
        float[] inputdata = bitmapToFloatArray(bitmap,256, 256);
        //需要将图片缩放到 28×28
        //将数据 feed 给 TensorFlow 的输入节点
        inferenceInterface.feed(inputName, inputdata, IN_COL, IN_ROW);
        //运行 TensorFlow
        String[] outputNames = new String[] {outputName};
        inferenceInterface.run(outputNames);
        ///获取输出节点的信息
        int[] outputs = new int[OUT_COL * OUT_ROW]; //用于存储模型的输出数据
        inferenceInterface.fetch(outputName, outputs);
        return outputs; //返回输出结果,int 数组
    }
    /*
     * 将 bitmap 转为(按行优先)一个 float 数组,并且每个像素点都归一化到 0~1
     * bitmap 为输入被测试的 Bitmap 图片
     */
    public static float[] bitmapToFloatArray(Bitmap bitmap, int rx, int ry){
        int height = bitmap.getHeight();
```

```
            int width = bitmap.getWidth();
            //计算缩放比例
            float scaleWidth = ((float) rx);
            float scaleHeight = ((float) ry);
            Matrix matrix = new Matrix();
            matrix.postScale(scaleWidth, scaleHeight);
            bitmap = Bitmap.createBitmap(bitmap,0,0,width,height,matrix,true);
            Log.i(TAG,"bitmap width:" + bitmap.getWidth() + ",height:" + bitmap.getHeight());
            Log.i(TAG,"bitmap.getConfig():" + bitmap.getConfig());
            height = bitmap.getHeight();
            width = bitmap.getWidth();
            float[] result = new float[height * width * 3];
            return result;
        }
}
```

3) 主活动控制

在 MainActivity.java 中声明模型存放路径,调用 Transfer_image 类。

```
package com.example.cyclegandemo;
import androidx.appcompat.app.AppCompatActivity;
import android.graphics.Bitmap;
import android.graphics.BitmapFactory;
//import android.support.v7.app.AppCompatActivity;
import android.os.Bundle;
import android.util.Log;
import android.view.View;
import android.widget.ImageView;
import android.widget.TextView;
public class MainActivity extends AppCompatActivity {
    private static final String TAG = "MainActivity";
    private static final String MODEL_FILE = "file:///android_asset/photo_to_vangogh.pb";
//模型存放路径
    TextView txt;
    TextView tv;
    ImageView imageView;
    Bitmap bitmap;
    Transfer_image transfer;
    @Override
    protected void onCreate(Bundle savedInstanceState) {
        super.onCreate(savedInstanceState);
        setContentView(R.layout.activity_main);
        //布局控件 ID 绑定
        tv = (TextView) findViewById(R.id.textView);
        txt = (TextView)findViewById(R.id.textView2);
        imageView = (ImageView)findViewById(R.id.imageView4);
        bitmap = BitmapFactory.decodeResource(getResources(), R.drawable.start_image);
```

```java
        imageView.setImageBitmap(bitmap);
        transfer = new Transfer_image(getAssets(),MODEL_FILE);
        //输入模型存放路径,并加载 TensorFlow 模型
    }
    //按钮触发事件
    public void click01(View v){
        String res = "输出结果";
        int[] result =  transfer.get_transfer(bitmap);
        for (int i = 0;i < result.length;i++){
            Log.i(TAG, res + result[i] );
            res = res + String.valueOf(result[i]) + " ";
        }
        txt.setText(res);
        tv.setText(stringFromJNI());
    }
    /**
     * A native method that is implemented by the 'native-lib' native library,
     * which is packaged with this application.
     */
    public native String stringFromJNI();                    //可以去掉
}
```

10.4 系统测试

生成器损失和判别器损失越小,代表训练的效果越好,如图 10-13 所示。

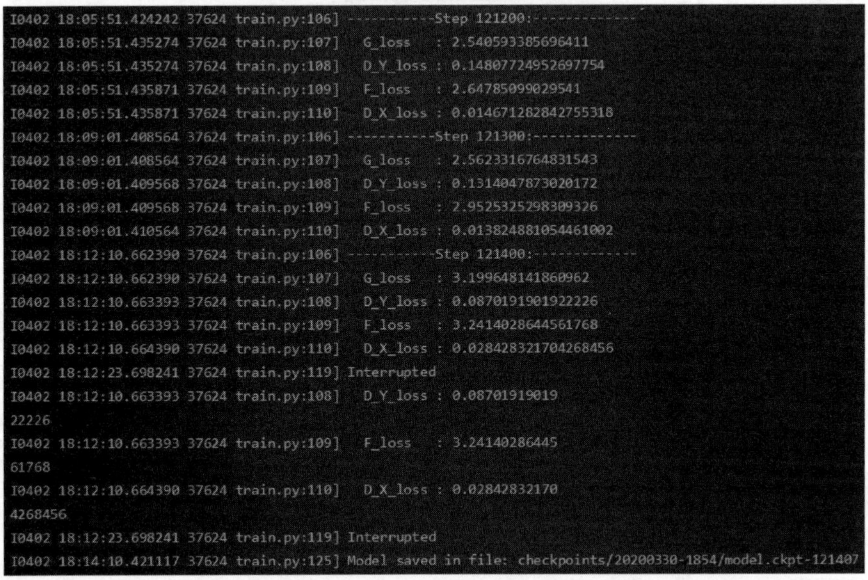

```
I0402 18:05:51.424242 37624 train.py:106] -----------Step 121200:------------
I0402 18:05:51.435274 37624 train.py:107]   G_loss    : 2.540593385696411
I0402 18:05:51.435274 37624 train.py:108]   D_Y_loss : 0.14807724952697754
I0402 18:05:51.435871 37624 train.py:109]   F_loss    : 2.64785099029541
I0402 18:05:51.435871 37624 train.py:110]   D_X_loss : 0.014671282842755318
I0402 18:09:01.408564 37624 train.py:106] -----------Step 121300:------------
I0402 18:09:01.408564 37624 train.py:107]   G_loss    : 2.5623316764831543
I0402 18:09:01.409568 37624 train.py:108]   D_Y_loss : 0.1314047873020172
I0402 18:09:01.409568 37624 train.py:109]   F_loss    : 2.9525325298309326
I0402 18:09:01.410564 37624 train.py:110]   D_X_loss : 0.013824881054461002
I0402 18:12:10.662390 37624 train.py:106] -----------Step 121400:------------
I0402 18:12:10.662390 37624 train.py:107]   G_loss    : 3.199648141860962
I0402 18:12:10.663393 37624 train.py:108]   D_Y_loss : 0.0870191901922226
I0402 18:12:10.663393 37624 train.py:109]   F_loss    : 3.2414028644561768
I0402 18:12:10.664390 37624 train.py:110]   D_X_loss : 0.028428321704268456
I0402 18:12:23.698241 37624 train.py:119] Interrupted
I0402 18:12:10.663393 37624 train.py:108]   D_Y_loss : 0.08701919019
22226
I0402 18:12:10.663393 37624 train.py:109]   F_loss    : 3.24140286445
61768
I0402 18:12:10.664390 37624 train.py:110]   D_X_loss : 0.02842832170
4268456
I0402 18:12:23.698241 37624 train.py:119] Interrupted
I0402 18:14:10.421117 37624 train.py:125] Model saved in file: checkpoints/20200330-1854/model.ckpt-121407
```

图 10-13 训练损失

训练 120000 次,保存的模型效果,重建损失保持在 3% 以下,如图 10-14 所示。

图 10-14　模型效果

将模型移植到 Android 的效果如图 10-15 所示。

图 10-15　移植效果

项目 11 交通警察——车辆监控系统

PROJECT 11

本项目调用 TensorFlow 的 Object Detection API，采用 SSD＋MobileNet V2 算法，对车辆进行识别，实现 Windows 系统下可交互的车辆违章行为记录系统。

11.1 总体设计

本部分包括系统整体结构和系统流程。

11.1.1 系统整体结构

系统整体结构如图 11-1 所示。

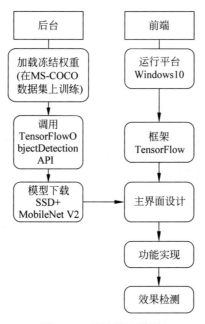

图 11-1 系统整体结构

11.1.2 系统流程

系统流程如图 11-2 所示。

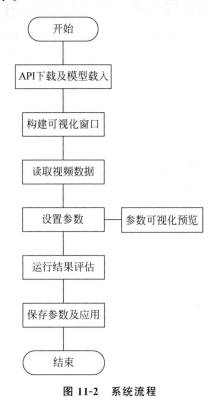

图 11-2 系统流程

11.2 运行环境

本部分包括 Python 环境、TensorFlow 环境、PyCharm IDE 配置和 Protoc 配置。

11.2.1 Python 环境

需要 Python 3.6 及以上配置，在 Windows 环境下推荐下载 Anaconda 完成 Python 所需的配置，下载地址为：https://www.anaconda.com/，也可以下载虚拟机在 Linux 环境下运行代码。

11.2.2 TensorFlow 环境

打开 Anaconda Prompt，输入清华仓库镜像，输入命令：

```
conda config -- add channels https://mirrors.tuna.tsinghua.edu.cn/anaconda/pkgs/free/
conda config - set show_channel_urls yes
```

创建 Python 3.7 的环境,名称为 TensorFlow,此时 Python 版本和后面 TensorFlow 的版本有匹配问题,此步选择 Python 3.x,输入命令:

```
conda create -n tensorflow python = 3.7
```

有需要确认的地方,都输入 y。

在 Anaconda Prompt 中激活 TensorFlow 环境,输入命令:

```
activate tensorflow
```

安装 CPU 版本的 TensorFlow,输入命令:

```
pip install - upgrade -- ignore-installed tensorflow
```

安装完毕。

11.2.3 PyCharm IDE 配置

PyCharm IDE 下载地址为:

https://www.jetbrains.com/pycharm/download/#section=windows,安装最新的 PyCharm 2020.1 版本。打开 PyCharm,选择 File → Settings → Project → Python interpreter,翻译器选择 Anaconda 下的 Python.exe,运行环境设置到当前目录下的/evn 文件夹(自己创建),单击安装按钮完成安装。

11.2.4 Protoc 配置

Protoc 下载地址为:https://github.com/protocolbuffers/protobuf/releases,下载对应编译好的 zip 包。解压后在环境变量 path 中添加.exe 文件路径,并新建一个 proto_path,路径为.exe。调用 CMD,输入命令:

```
Protoc
```

提示 missing input file,安装成功。添加环境变量 PYTHONPATH,输入命令:

```
PYTHONPATH    G:\TensorFlow\models\slim
```

11.3 模块实现

本项目包括 4 个模块:API 下载及导入、识别训练、导入模型并编译、模型生成。下面分别给出各模块的功能介绍及相关代码。

11.3.1 API 下载及载入

本部分包括 TensorFlow Object Detection API 中的 MobileNet V2 COCO 模型的介绍及相关代码。从 GitHub 下载 API 源码，下载地址为：https://github.com/tensorflow/models.git，进入 pipeline.config 的相关代码如下：

```
model {
  ssd {
    #识别的类别数
    num_classes: 90
    image_resizer {
      fixed_shape_resizer {
        height: 300
        width: 300
      }
    }
#定位加入训练集和测试集的位置
train_input_reader {
  label_map_path: "PATH_TO_BE_CONFIGURED/mscoco_label_map.pbtxt"
  tf_record_input_reader {
    input_path: "PATH_TO_BE_CONFIGURED/mscoco_train.record"
  }
}
eval_config {                              #测试配置
  num_examples: 8000
  max_evals: 10
  use_moving_averages: false
}
eval_input_reader {                        #测试集读取
  label_map_path: "PATH_TO_BE_CONFIGURED/mscoco_label_map.pbtxt"
  shuffle: false
  num_readers: 1
  tf_record_input_reader {
    input_path: "PATH_TO_BE_CONFIGURED/mscoco_val.record"
  }
}
#进入 vehicle_detection_main.py
#导入相应数据包
import numpy as np
import os
import six.moves.urllib as urllib
import sys
import tarfile
import tensorflow as tf
import zipfile
```

```
import cv2
import numpy as np
import csv
import time
import _thread
import threading
import tkinter.filedialog, tkinter.messagebox
import tkinter as tk
from collections import defaultdict
from io import StringIO
from matplotlib import pyplot as plt
import sqlite3
import os
from PIL import ImageTk
import PIL.Image
#导入 Object detection
from utils import label_map_util
from utils import visualization_utils as vis_util
from utils.color_recognition_module import color_recognition_api
from utils.vehicle_detection_module import vehicle_detection_api
from utils.vehicle_detection_module.vehicle_detection_api import *
from utils.image_utils import image_saver
from tkinter import *
from tkinter import ttk, Entry, Label, LabelFrame
#API 和 frozen_inference_ graph
MODEL_NAME = 'ssdlite_mobilenet_v2_coco_2018_05_09'
MODEL_FILE = MODEL_NAME + '.tar.gz'
DOWNLOAD_BASE = 'http://download.tensorflow.org/models/object_detection/'
PATH_TO_CKPT = MODEL_NAME + '/frozen_inference_graph.pb'
```

11.3.2 识别训练

为使交通监控自动化,加入对交通灯颜色的监控识别,相关代码如下:

```
from PIL import Image                  #导入各种模块
import os
import cv2
import numpy as np
import matplotlib.pyplot as plt
from scipy.stats import itemfreq
from utils.color_recognition_module import knn_classifier as knn_classifier
current_path = os.getcwd()
def color_histogram_of_test_image(test_src_image):
    #加载图像
    image = test_src_image
    chans = cv2.split(image)
```

```python
        colors = ('b', 'g', 'r')
        features = []
        feature_data = ''
        counter = 0
        for (chan, color) in zip(chans, colors):
            counter = counter + 1
            hist = cv2.calcHist([chan], [0], None, [256], [0, 256])  #print(hist)
            features.extend(hist)
            #查找 R、G 和 B 的峰值像素值
            elem = np.argmax(hist)
            #print(elem)
            if counter == 1:
                blue = str(elem)
            elif counter == 2:
                green = str(elem)
            elif counter == 3:
                red = str(elem)
                feature_data = red + ',' + green + ',' + blue
        with open(current_path + '/utils/color_recognition_module/'
                  + 'test.data', 'w') as myfile:
            myfile.write(feature_data)
def color_histogram_of_training_image(img_name):
    if 'red' in img_name:
        data_source = 'red'
    elif 'yellow' in img_name:
        data_source = 'yellow'
    elif 'green' in img_name:
        data_source = 'green'
    elif 'black' in img_name:
        data_source = 'black'
    #加载图片
    image = cv2.imread(img_name)
    chans = cv2.split(image)
    colors = ('b', 'g', 'r')
    features = []
    feature_data = ''
    counter = 0
    for (chan, color) in zip(chans, colors):
        counter = counter + 1
        hist = cv2.calcHist([chan], [0], None, [256], [0, 256])
        features.extend(hist)
        elem = np.argmax(hist)
        if counter == 1:
            blue = str(elem)
        elif counter == 2:
            green = str(elem)
        elif counter == 3:
```

```python
            red = str(elem)
            feature_data = red + ',' + green + ',' + blue
    with open('training.data', 'a') as myfile:
        myfile.write(feature_data + ',' + data_source + '\n')
def training():
    #训练红色
    for f in os.listdir('./training_dataset/red'):
        color_histogram_of_training_image('./training_dataset/red/' + f)
    # 训练黄色
    for f in os.listdir('./training_dataset/yellow'):
        color_histogram_of_training_image('./training_dataset/yellow/' + f)
    #训练绿色
    for f in os.listdir('./training_dataset/green'):
        color_histogram_of_training_image('./training_dataset/green/' + f)
    #训练黑色
    for f in os.listdir('./training_dataset/black'):
        color_histogram_of_training_image('./training_dataset/black/' + f)
```

11.3.3 导入模型与编译

将模型文件保存为.pb 格式,用 TensorFlow 中的 graph_util 模块进行模型保存。

```python
def object_detection_function():                #目标检测函数
    cap = cv2.VideoCapture(VIDEO_FILE_PATH)
vis_util.set_detection_area_value(INTEREST_AREA_Y_START, INTEREST_AREA_Y_END, SPEED_LIMIT,
VIDEO_FILE_NAME)
    reset_vehicle_count()
    image_saver.reset_stored_value()
    vehicle_detection_api.reset_stored_value()
vehicle_detection_api.set_roi_value(LINE_CROSSING_DETECTION_POS_TOP, LINE_CROSSING_
DETECTION_POS_BOTTOM,
IS_LANE_FIRST_AVAILABLE, IS_LANE_SECOND_AVAILABLE, IS_LANE_THIRD_AVAILABLE,
LEFT_DETECTION_POSITION_LANE_FIRST_START, RIGHT_DETECTION_POSITION_LANE_FIRST_START,
LEFT_DETECTION_POSITION_LANE_SECOND_START, RIGHT_DETECTION_POSITION_LANE_SECOND_START,
LEFT_DETECTION_POSITION_LANE_THIRD_START, RIGHT_DETECTION_POSITION_LANE_THIRD_START, SPEED_
DETECTION_POSITION_LANE_TOP, SPEED_DETECTION_POSITION_LANE_BOTTOM, LEFT_SPEED_DETECTION_
POSITION_LANE_FIRST, RIGHT_SPEED_DETECTION_POSITION_LANE_FIRST, LEFT_SPEED_DETECTION_
POSITION_LANE_SECOND, RIGHT_SPEED_DETECTION_POSITION_LANE_SECOND,
LEFT_SPEED_DETECTION_POSITION_LANE_THIRD, RIGHT_SPEED_DETECTION_POSITION_LANE_THIRD, LEAVE_
SPEED_DETECTION_POSITION_LANE_TOP, LEAVE_SPEED_DETECTION_POSITION_LANE_BOTTOM, LINE_CROSSING
_DETECTION_INTERVAL_IN_EACH_LANE, PIXEL_TO_REAL_LENGTH, PIXEL_HEIGHT_COMPENSATE, DETECTED_
LIGHT_COLOR, SPEED_LIMIT, LEFT_DETECTION_POSITION_LANE_FIRST_END)
    with detection_graph.as_default():
            with tf.compat.v1.Session(graph = detection_graph) as sess:
                #定义输入/输出张量用于 detection_graph
    image_tensor = detection_graph.get_tensor_by_name('image_tensor:0')
```

```python
        #每个框代表检测到特定对象图像的一部分
    detection_boxes = detection_graph.get_tensor_by_name('detection_boxes:0')
        #每个分数代表每个对象的置信度
        #分数与类标签一起显示在结果图像上
    detection_scores = detection_graph.get_tensor_by_name('detection_scores:0')
    detection_classes = detection_graph.get_tensor_by_name('detection_classes:0')
    num_detections = detection_graph.get_tensor_by_name('num_detections:0')
                #从输入视频中提取的所有帧
            while cap.isOpened():
                (ret, frame) = cap.read()
                    if not ret:
                    print ('end of the video file...')
                    break
                    input_frame = frame
                    #感兴趣的起始点(x,y)
                    #在特定区域检测车辆以提高性能,起始坐标和终点坐标
    traffic_light_area_img = frame[TRAFFIC_LIGHT_POS_TOP + OFFSET:TRAFFIC_LIGHT_POS_BOTTOM-
OFFSET,TRAFFIC_LIGHT_POS_LEFT + OFFSET:TRAFFIC_LIGHT_POS_RIGHT-OFFSET]
            #获取交通信号灯的颜色并将其传递给检测器
    predicted_color = color_recognition_api.color_recognition(traffic_light_area_img)
            vehicle_detection_api.set_current_light_color(predicted_color)
    interest_area = frame[INTEREST_AREA_Y_START:INTEREST_AREA_Y_END, INTEREST_AREA_X_START:
INTEREST_AREA_X_END]
                image_np_expanded = np.expand_dims(interest_area, axis = 0)
                    #实际检测
                        (boxes, scores, classes, num) = \
                    sess.run([detection_boxes, detection_scores,
                        detection_classes, num_detections],
                        feed_dict = {image_tensor: image_np_expanded})
                    #可视化检测结果
                (counter, csv_line) = \
                    vis_util.visualize_boxes_and_labels_on_image_array(
                    cap.get(1),
                    input_frame,
                    np.squeeze(boxes),
                    np.squeeze(classes).astype(np.int32),
                    np.squeeze(scores),
                    category_index,
                    use_normalized_coordinates = True,
                    line_thickness = 2,
                    skip_scores = True,
                    interest_area_xpos_start = INTEREST_AREA_X_START,
                    interest_area_ypos_start = INTEREST_AREA_Y_START,
                    interest_area_xpos_end = INTEREST_AREA_X_END,
                    interest_area_ypos_end = INTEREST_AREA_Y_END
                    )
                #在视频上画感兴趣区域 ROI
```

```
            draw_roi(input_frame,counter)
            cv2.namedWindow('vehicle detection',cv2.WINDOW_NORMAL)
        cv2.imshow('vehicle detection', input_frame)
            if cv2.waitKey(1) & 0xFF == ord('q'):
            break
        cap.release()
    cv2.destroyAllWindows()
```

11.3.4 模型生成

本部分包括超速检测、越线检测、配置预览和显示界面。

1. 超速检测

超速检测相关代码如下:

```
def search_overspeed_record(self):                  #搜索超速
    records = self.tree.get_children()
    for element in records:
        self.tree.delete(element)
    query = 'SELECT * FROM vehicles_info WHERE is_overspeed == 1'
    db_rows = self.run_query(query)
    for row in db_rows:
        self.tree.insert('',0, text = row[0], values = [row[1],row[2],row[3]])
    records = self.tree.get_children()
    for element in records:
        self.tree.bind("<Double-Button-1>", self.on_record_clicked)
```

2. 越线检测

越线检测相关代码如下:

```
def search_line_crossing_record(self):              #搜索越线
    records = self.tree.get_children()
    for element in records:
        self.tree.delete(element)
    query = 'SELECT * FROM vehicles_info WHERE is_line_crossing == 1'
    db_rows = self.run_query(query)
    for row in db_rows:
        self.tree.insert('',0, text = row[0], values = [row[1],row[2],row[3]])
    records = self.tree.get_children()
    for element in records:
        self.tree.bind("<Double-Button-1>", self.on_record_clicked)
```

3. 配置预览

配置预览相关代码如下:

```python
def roi_configuration_preview():                    # 配置预览
    is_preview_activated = is_roi_preview_activated.get()
    cap = cv2.VideoCapture(VIDEO_FILE_PATH)
    cv2.namedWindow('roi_preview',cv2.WINDOW_NORMAL)
    (ret, frame) = cap.read()
    while is_preview_activated == True:
        new_frame = frame.copy()
        draw_roi_preview(new_frame)
        cv2.imshow('roi_preview',new_frame)
        if cv2.waitKey(1) & 0xFF == ord('q'):
            is_roi_preview_activated.set('False')
            break
        time.sleep(0.5)
    cap.release()
    cv2.destroyAllWindows()
```

4. 显示界面

显示界面相关代码如下：

```
def draw_roi_preview(frame):                                                    # 画图显示
    input_frame = frame
    text_offset_y = 10
cv2.rectangle(input_frame,(interest_area_x_start.get(),interest_area_y_start.get()),
(interest_area_x_end.get(),interest_area_y_end.get()),(0,255,255),3)    # 矩形框
cv2.putText(input_frame,'region of interest',(interest_area_x_start.get(),interest_area_y_
start.get()-text_offset_y),1,cv2.FONT_HERSHEY_COMPLEX,(0, 255, 255),2)
cv2.rectangle(input_frame,(traffic_light_pos_left.get(),traffic_light_pos_top.get()),(traffic_
light_pos_right.get(),traffic_light_pos_bottom.get()),(0, 255, 0),2)       # 感兴趣区域
    cv2.putText(input_frame,'traffic light color detection
area',(traffic_light_pos_left.get(),traffic_light_pos_top.get()-text_offset_y),1,cv2.FONT_
HERSHEY_COMPLEX,(0, 255, 0),2)                                # 交通灯颜色检测区
        if(is_lane_first_available.get() == True):
            cv2.putText(input_frame,'line crossing detection area',(left_detection_position_
lane_first_start.get(),line_crossing_detection_pos_top.get()-text_offset_y),1,cv2.FONT_
HERSHEY_COMPLEX,(0, 240, 0),2)
cv2.line(input_frame,(left_detection_position_lane_first_start.get(),speed_detection_
position_lane_top.get()),(left_detection_position_lane_first_end.get(),speed_detection_
position_lane_top.get()+LINE_BOTTOM_HEIGHT),(0,240,0),4)              # 越线检测
cv2.line(input_frame,(right_detection_position_lane_first_start.get(),speed_detection_
position_lane_top.get()),(right_detection_position_lane_first_end.get(),speed_detection_
position_lane_top.get()+LINE_BOTTOM_HEIGHT),(0,255,0),4)              # 速度检测
cv2.putText(input_frame,'vehicle detection area',(left_speed_detection_position_lane_first.
get(),speed_detection_position_lane_top.get()-text_offset_y),1,cv2.FONT_HERSHEY_COMPLEX,
(18, 74, 115),2)
cv2.rectangle(input_frame,(left_speed_detection_position_lane_first.get(),speed_detection_
position_lane_top.get()),(right_speed_detection_position_lane_first.get(),speed_detection
```

```
_position_lane_bottom.get()),(18,74,115),3)
        cv2.putText(input_frame,'leave detection area',(left_speed_detection_position_lane_
first.get(),leave_speed_detection_position_lane_top.get()-text_offset_y),1,cv2.FONT_
HERSHEY_COMPLEX,(0,0,200),2)                                    #离开检测区
cv2.rectangle(input_frame,(left_speed_detection_position_lane_first.get(),leave_speed_
detection_position_lane_top.get()),(right_speed_detection_position_lane_first.get(),leave
_speed_detection_position_lane_bottom.get()),(0,0,200),2)       #矩形框结果
        if(is_lane_second_available.get() == True):
cv2.line(input_frame,(left_detection_position_lane_second_start.get(),speed_detection_
position_lane_top.get()),(left_detection_position_lane_second_end.get(),speed_detection_
position_lane_top.get()+LINE_BOTTOM_HEIGHT),(0,240,0),4)        #二次检测
cv2.line(input_frame,(right_detection_position_lane_second_start.get(),speed_detection_
position_lane_top.get()),(right_detection_position_lane_second_end.get(),speed_detection_
position_lane_top.get()+LINE_BOTTOM_HEIGHT),(0,255,0),4)
cv2.rectangle(input_frame,(left_speed_detection_position_lane_second.get(),speed_detection
_position_lane_top.get()),(right_speed_detection_position_lane_second.get(),speed_
detection_position_lane_bottom.get()),(18,74,115),3)
cv2.rectangle(input_frame,(left_speed_detection_position_lane_second.get(),leave_speed_
detection_position_lane_top.get()),(right_speed_detection_position_lane_second.get(),leave
_speed_detection_position_lane_bottom.get()),(0,0,200),2)       #矩形框检测结果
        if(is_lane_third_available.get() == True):
cv2.line(input_frame,(left_detection_position_lane_third_start.get(),speed_detection_
position_lane_top.get()),(left_detection_position_lane_third_end.get(),speed_detection_
position_lane_top.get()+LINE_BOTTOM_HEIGHT),(0,240,0),4)
cv2.line(input_frame,(right_detection_position_lane_third_start.get(),speed_detection_
position_lane_top.get()),(right_detection_position_lane_third_end.get(),speed_detection_
position_lane_top.get()+LINE_BOTTOM_HEIGHT),(0,255,0),4)        #三次检测
cv2.rectangle(input_frame,(left_speed_detection_position_lane_third.get(),speed_detection_
position_lane_top.get()),(right_speed_detection_position_lane_third.get(),speed_detection
_position_lane_bottom.get()),(18,74,115),3)
cv2.rectangle(input_frame,(left_speed_detection_position_lane_third.get(),leave_speed_
detection_position_lane_top.get()),(right_speed_detection_position_lane_third.get(),leave
_speed_detection_position_lane_bottom.get()),(0,0,200),2)       #矩形框检测结果
cv2.rectangle(input_frame,(line_crossing_detection_pos_left.get(),line_crossing_detection_
pos_top.get()
```

11.4 系统测试

程序运行效果如图11-3所示,导入参数及视频如图11-4所示,程序运行结果如图11-5所示,程序运行窗口可见记录如图11-6所示。

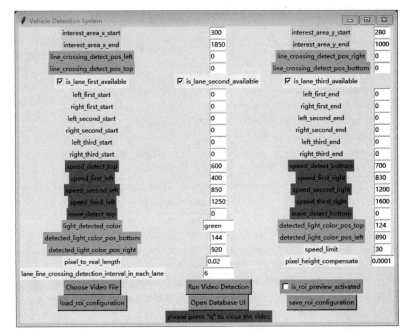

图 11-3　程序运行效果

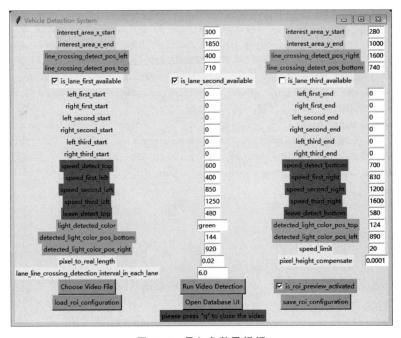

图 11-4　导入参数及视频

图 11-5 程序运行结果

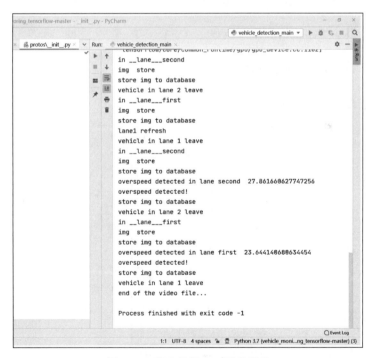

图 11-6 程序运行窗口可见记录

项目 12　验证码的生成与识别

PROJECT 12

本项目通过 VGG 卷积神经网络随机生成大量验证码图片用作训练集,对验证码的特征加以提取,训练出合适的模型,实现对验证码图片的精准识别。

12.1　总体设计

本部分包括系统整体结构和系统流程。

12.1.1　系统整体结构

系统整体结构如图 12-1 所示。

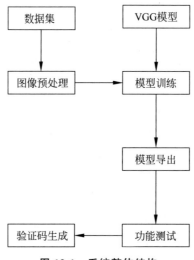

图 12-1　系统整体结构

12.1.2 系统流程

系统流程如图 12-2 所示。

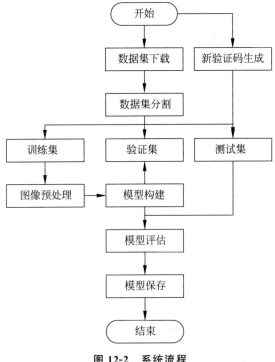

图 12-2 系统流程

12.2 运行环境

本部分包括 Python 环境、TensorFlow 环境和 VsCode 环境。

12.2.1 Python 环境

需要 Python 3.6 及以上配置,在 Windows 环境下推荐下载 Anaconda 完成 Python 所需的配置,下载地址为 https://www.anaconda.com/,也可以下载虚拟机在 Linux 环境下运行代码。

12.2.2 TensorFlow 环境

打开 Anaconda Prompt,输入清华仓库镜像,输入命令:

conda config -- add channels https://mirrors.tuna.tsinghua.edu.cn/anaconda/pkgs/free/

```
conda config - set show_channel_urls yes
```

创建 Python 3.6 的环境,名称为 TensorFlow,此时 Python 版本和后面 TensorFlow 的版本有匹配问题,此步选择 Python 3.6,输入命令:

```
conda create -n tensorflow python = 3.6
```

有需要确认的地方,都输入 y。

在 Anaconda Prompt 中激活 TensorFlow 环境,输入命令:

```
activate tensorflow
```

安装 CPU 版本的 TensorFlow,输入命令:

```
pip install tensorflow-gpu = = 1.15
```

安装完毕。

12.2.3 VsCode 环境

安装 VsCode,下载地址为 https://code.visualstudio.com/。

打开 VsCode,选择"文件"→New 命令新建文件,将新建的文件保存为 .py 格式,定义文件名称后即可开始编程。

导入 TensorFlow 的 OpenCV 库和 Numpy 库,以管理员身份运行 Anaconda Prompt,输入命令:

```
pip install opencv
```

有需要确认的地方,都输入 y。输入命令:

```
pip install numpy
```

有需要确认的地方,都输入 y。

此时,可能会报错提示找不到相关库,使用 conda 下载命令即可解决。配置完成后打开 VsCode,窗口左下角显示如图 12-3 所示。

图 12-3　VsCode 窗口左下角显示

选择 Python 3.6 TensorFlow 完成 VsCode 的环境配置。

12.3 模块实现

本项目包括 4 个模块：数据预处理、模型搭建、模型训练及保存、模型测试。下面分别给出各模块的功能介绍及相关代码。

12.3.1 数据预处理

本部分包括数据获取、数据集处理以及图像的预处理。

1. 数据获取

数据集来源于百度网盘，链接地址为 https://pan.baidu.com/s/1TbQWMdSNMWNlGchh_ouD7g，提取码为 gah2。其中包含 5000 张 JPG 格式的验证码图片及其对应的标签，图片包含 4 位数字（或字母），.csv 文件里面包含图片的序号及标签，如图 12-4 所示。

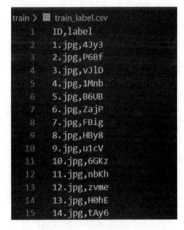

图 12-4 标签

2. 数据集处理

读取数据集，将其转换成二进制的.pkl 文件，相关代码如下：

```
def pre_process(self):
    train_data = pd.read_csv(self.train_path + 'train_label.csv')
                # 读取训练数据
    train_data = train_data.to_dict(orient = 'records')
    for i in range(len(train_data)):
        train_data[i]['ID'] = self.train_path + train_data[i]['ID']
    with open(self.train_save_path, 'wb') as f:
        pickle.dump(train_data[cfg.TEST_NUM:], f)
    with open(self.test_save_path, 'wb') as f:
        pickle.dump(train_data[:cfg.TEST_NUM], f)
```

3. 图像的预处理

为提高准确度,使图片便于识别,需要对扭曲模糊的图像做预处理,预处理前后的对比如图 12-5 所示。相关代码如下:

```python
def max_gray(image):                              # 灰度化
    b, g, r = cv2.split(image)
    index_1 = b > g
    result = np.where(index_1, b, g)
    index_2 = result > r
    result = np.where(index_2, result, r)
    return result
def normalize(a):
    amin, amax = a.min(), a.max()                 # 求最大/最小值
    result = (a-amin)/(amax-amin)                 # (矩阵元素-最小值)/(最大值-最小值)
    return result
def enhance(image, gamma = 1.6, norm = True):     # 图像增强
    fI = (image + 0.5)/256
    fI = fI + np.median(fI)
    dst = np.power(fI, gamma)
    if norm:
        dst = normalize(dst)
        # dst += np.clip(dst, cfg.CLIP_MIN, cfg.CLIP_MAX)
        # dst = normalize(dst)
    return dst
def image_process(image):
    image = en.enhance_value_max(image)
    image += ercode_dilate(image, 3)
    return image
def ercode_dilate(img, threshold):
    # 腐蚀参数,(threshold, threshold)为腐蚀矩阵大小
    kernel = np.ones((threshold, threshold), np.uint8)
    # 腐蚀图片
    img = cv2.dilate(img, kernel, iterations = 1)
    # 膨胀图片
    img = cv2.erode(img, kernel, iterations = 1)
    return img
if __name__ == "__main__":                        # 主函数
    import matplotlib.pyplot as plt
    for i in range(0, 10):
        image = cv2.imread('./test/{}.jpg'.format(i))
        dst = image_process(image)
        plt.imshow(image)
        plt.axis('off')
        plt.show()
```

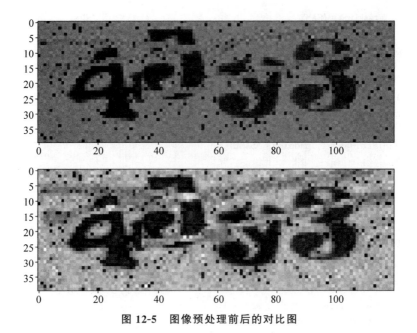

图 12-5 图像预处理前后的对比图

12.3.2 模型搭建

数据处理完成后需要定义模型结构并优化损失函数。

1. 定义模型结构

VGG-16 网络模型由 8 个卷积层和 3 个池化层构成。将其拆分成 4 份大小为 40×32 像素的图片,每 2 次卷积后都接 1 个池化层。最后的 2 个卷积层参数 padding=valid,起到降维的作用,从而得到 2×1×512 像素的特征图。通过 tf.contrib.layers.flatten() 函数将数据转换成一个向量,作为全连接层的预处理。在全连接层的最后,使用 tf.nn.dropout() 函数让神经元随机失活,起到正则化的作用,防止过拟合,其中 keep_pro 参数设置为 0.95。tf.nn.softmax() 作为激活函数,以一定的概率生成需要的预测值,相关代码如下:

```
def network(self, inputs):                      # 定义网络
    net = self.conv2d(
        inputs, filters = 64, k_size = 3,
        strides = 1, padding = 'SAME'
    )
    net = self.conv2d(                          # 二维卷积
        net, filters = 64, k_size = 3,
        strides = 1, padding = 'SAME'
    )
    net = self.max_pool2d(                      # 最大二维池化
        net, k = 3, strides = 2,
```

```python
        padding = 'SAME'
) # 参数为(20,15,3)
net = self.conv2d(
    net, filters = 128, k_size = 3,
    strides = 1, padding = 'SAME'
)
net = self.conv2d(                              # 二维卷积
    net, filters = 128, k_size = 3,
    strides = 1, padding = 'SAME'
)
net = self.max_pool2d(                          # 最大二维池化
    net, k = 3, strides = 2,
    padding = 'SAME'
) # 参数为(10,8,3)
net = tf.layers.batch_normalization(
    net, training = self.is_training
)
net = self.conv2d(                              # 二维卷积
    net, filters = 256, k_size = 1,
    strides = 1, padding = 'SAME'
)
net = self.conv2d(                              # 二维卷积
    net, filters = 256, k_size = 1,
    strides = 1, padding = 'SAME'
)
net = self.max_pool2d(                          # 最大二维池化
    net, k = 3, strides = 2,
    padding = 'SAME'
) # 参数为(5,4,3)
net = self.conv2d(                              # 二维卷积
    net, filters = 512, k_size = 3,
    strides = 1, padding = 'VALID'
)
net = self.conv2d(                              # 二维卷积
    net, filters = 512, k_size = 2,
    strides = 1, padding = 'VALID'
)
net = tf.contrib.layers.flatten(net)            # 扁平化
weights_1 = tf.Variable(tf.truncated_normal(
    [2 * 1 * 512, 128], stddev = 0.1, mean = 0))  # 加权
biases_1 = tf.Variable(                         # 偏置
    tf.random.normal(shape = [128])
)
net = tf.add(tf.matmul(net, weights_1), biases_1)
weights_3 = tf.Variable(tf.truncated_normal(
    [128, self.cls_num], stddev = 0.1, mean = 0))  # 加权
biases_3 = tf.Variable(                         # 偏置
```

```
                tf.random.normal(shape = [self.cls_num])
        )
        logits = tf.add(tf.matmul(net, weights_3), biases_3)
        if self.is_training:
            logits = tf.nn.dropout(logits, rate = 1-self.keep_rate)          #防止过拟合
        predictions = tf.nn.softmax(logits, axis = -1)
        return predictions, logits
    def conv2d(self, x, filters, k_size = 3,                    #定义二维卷积
strides = 1, padding = 'SAME', dilation = [1, 1]):
        return tf.layers.conv2d(
x, filters,
    kernel_size = [k_size, k_size], strides = [strides, strides],
dilation_rate = dilation,
padding = padding,
activation = tf.nn.relu,
use_bias = True
)
    def max_pool2d(self, x, k, strides, padding = 'SAME'):     #定义最大二维池化
        return tf.layers.max_pooling2d(inputs = x, pool_size = k,
strides = strides, padding = padding)
```

2. 优化损失函数

对于多类别的分类问题，使用交叉熵作为损失函数。

```
#定义损失函数
def loss_layer(self, y, y_hat):
        loss = tf.nn.softmax_cross_entropy_with_logits_v2(
            logits = y_hat, labels = y
        )
        return tf.reduce_mean(loss)
#y_hat 是上述模型得到的预测值 predictions,y = 64 是标签数量
```

12.3.3　模型训练及保存

搭建好模型框架后对模型进行训练，随机抓取一些数据作为 batch（批次）和验证集，相关代码如下：

```
def generate(self, batch_size, s = cfg.START_POINT, is_training = True):
        value = {}                                              #产生数据
        batch_images = []
        batch_labels = []
        batch_index = 0
        while batch_index < batch_size:
            #取一个 batch
            if is_training:
                image_path = self.train_dict[self.cursor]['ID']
```

```python
            label = self.train_dict[self.cursor]['label']
        else:
            image_path = self.test_dict[batch_index]['ID']
            label = self.test_dict[batch_index]['label']
        image = cv2.imread(image_path)
        tmp = randint(0, 2)
        image = self.moving(image)[tmp]
        label = self.transform_label(label)
        self.cursor += 1
        if self.cursor >= len(self.train_dict):
            # 如果已完成遍历,则打乱顺序并且游标归零
            np.random.shuffle(self.train_dict)
            self.cursor = 0
            self.epoch += 1
        batch_images.append(                            # 批次数据
            np.stack(
                [image[:, s[0]:s[0] + self.length],
                    image[:, s[1]:s[1] + self.length],
                    image[:, s[2]:s[2] + self.length],
                    image[:, s[3]:s[3] + self.length]]
            )
        )
        batch_labels.append(label)                      # 批次标签
        batch_index += 1
    value = {'images': np.stack(batch_images).reshape(
        (batch_size * 4, 40, self.length, 3)
    ),
        'labels': np.stack(batch_labels).reshape(
            (batch_size * 4, self.cls_num)
    )}
    value_list = []
    for i in range(batch_size * 4):
        tmp = image_process(value['images'][i])         # 处理的数据
        value_list.append(tmp)
    value_list = np.stack(value_list)
    value['images'] = np.expand_dims(value_list, axis=-1)
    return value
```

1. 模型训练

模型训练相关代码如下:

```python
def train_net(self):
    with tf.Session() as sess:
        sess.run(tf.compat.v1.global_variables_initializer())
        ckpt = tf.train.get_checkpoint_state(cfg.MODEL_PATH)
        if ckpt and ckpt.model_checkpoint_path:
```

```python
            #如果保存过模型,则在此基础上继续训练
            self.saver.restore(sess, ckpt.model_checkpoint_path)
            print('Model Reload Successfully!')
        for step in range(2000):                        #迭代 2000 次
            loss_list = []
            test_data = self.reader.generate(
                cfg.TEST_NUM, is_training = False)
                #TEST_NUM = 1000,训练集分离出来,用于验证健壮性
            test_dict = {
                self.x: test_data['images'],
                self.y: test_data['labels'],
                self.keep_rate: 1.0}
            for batch in range(cfg.BATCH_NUM):
            #每次训练 64 个 batch(可为 256 个)
                batch_data = self.reader.generate(self.batch_size, is_training = True)
            #每次抓取 32 个作为 batch(可为 128 个)
                feed_dict = {
                    self.x: batch_data['images'],
                    self.y: batch_data['labels'],
                    self.keep_rate: cfg.KEEP_RATE
                }
                loss = sess.run(                        #损失计算
                    [self.trian_step, self.loss],
                    feed_dict = feed_dict
                )
                print('batch:{}/{} loss:{} '.format(
                    batch, self.batch_num, loss
                ), end = '\r')
                loss_list.append(loss)
            loss_value = np.mean(
                np.array(loss_list)
            )
            pred = sess.run(self.y_hat, test_dict)
            acc = self.calculate_test_acc(test_data['labels'], pred)
            #准确度和损失函数写入.csv 文件
            f = open('准确度.csv','a',newline = '',encoding = 'utf-8')
            csv_writer = csv.writer(f)
            csv_writer.writerow([acc,loss_value])
            f.close()
            print('step:{}/{} loss:{} accuracy:{}'.format(
                step, self.steps, loss_value, acc
            ))
            self.saver.save(sess, cfg.MODEL_PATH + 'model.ckpt')
            if acc > cfg.STOP_ACC:
                print('Early Stop with acc:{}'.format(acc))
                break
```

其中，一个 batch 就是在一次前向/后向传播过程用到的训练样例数量。本项目中一次用 32 张图片进行训练，共训练 2000 张图片，如图 12-6 所示。

```
step:310/2000 loss:3.2041633129119873 accuracy:0.933
step:311/2000 loss:3.209059715270996 accuracy:0.943
step:312/2000 loss:3.2084131240844727 accuracy:0.944
step:313/2000 loss:3.2056941986083984 accuracy:0.947
step:314/2000 loss:3.2048895359039307 accuracy:0.942
step:315/2000 loss:3.2097220420837402 accuracy:0.934
step:316/2000 loss:3.2028770446777344 accuracy:0.934
step:317/2000 loss:3.2032899856567383 accuracy:0.941
step:318/2000 loss:3.205157518386841 accuracy:0.946
step:319/2000 loss:3.205052375793457 accuracy:0.948
step:320/2000 loss:3.2033209800720215 accuracy:0.946
step:321/2000 loss:3.2030587196350098 accuracy:0.945
step:322/2000 loss:3.211406707763672 accuracy:0.942
step:323/2000 loss:3.205319881439209 accuracy:0.939
step:324/2000 loss:3.2086009979248047 accuracy:0.94
step:325/2000 loss:3.2067928314208984 accuracy:0.946
step:326/2000 loss:3.2063393592834473 accuracy:0.938
step:327/2000 loss:3.2062149047851562 accuracy:0.948
step:328/2000 loss:3.20151948928833 accuracy:0.946
step:329/2000 loss:3.2055044174194336 accuracy:0.947
step:330/2000 loss:3.2095675468444824 accuracy:0.939
step:331/2000 loss:3.2047572135925293 accuracy:0.948
step:332/2000 loss:3.2089993953704834 accuracy:0.951
step:333/2000 loss:3.2029035091400146 accuracy:0.935
step:334/2000 loss:3.2096738815307617 accuracy:0.951
step:335/2000 loss:3.20542573928833 accuracy:0.956
```

图 12-6　训练结果

观察每次迭代后的损失函数与模型准确率，若保持稳定收敛，表明训练已经完成，继续训练则可能导致过拟合，为预防此现象，设置了门限 STOP_ACC＝0.955，若结果值超过它，则停止训练。另外，将每次训练的结果都保存为 .csv 文件，并且以图片的方式呈现，相关代码如下：

```python
import matplotlib.pyplot as plt            #加载画图模块
import numpy as np
import csv
import matplotlib as mpl
f = open('准确度.csv','r')
reader = csv.reader(f)                     #数据读取
accs = []
losses = []
nums = []
i = 0
for item in reader:
    accs.append(float(item[0]))
    losses.append(float(item[1]))
    i = i + 1
    nums.append(i)
f.close()
#绘制曲线
#解决中文显示问题
plt.rcParams['font.sans-serif'] = ['KaiTi']    #指定默认字体
```

```
plt.rcParams['axes.unicode_minus'] = False fig, ax1 = plt.subplots()
ax2 = ax1.twinx()
lns1 = ax1.plot(np.arange(i), losses, 'r',label = "Loss")
lns2 = ax2.plot(np.arange(i), accs, 'b', label = "Accuracy")
ax1.set_xlabel('训练轮次')
ax1.set_ylabel('训练损失值')
ax2.set_ylabel('训练准确率')
#合并图例
lns = lns1 + lns2
labels = ["损失", "准确率"]
# labels = [l.get_label() for l in lns]
plt.legend(lns, labels, loc = 7)
plt.show()
```

2. 模型保存

```
#保存模型在./mode/model.ckpt文件中
                self.saver.save(sess, cfg.MODEL_PATH + 'model.ckpt')
                if acc > cfg.STOP_ACC:
                    print('Early Stop with acc:{}'.format(acc))
                    break
```

模型被保存后,下次训练时可以继续使用。

12.3.4 模型测试

1. 新数据集的创建

下载 TTF 字体文件,再利用 PIL 库随机生成 4 位验证码。相关代码如下:

```
from PIL import ImageFont, ImageDraw, ImageFilter, Image    #导入各种模块
import random
import pandas as pd
import numpy as np
import os
import cv2
co = np.array(pd.read_csv('1.csv'))                          #读入数据
k = []
num = int(input('请输入您想要生成的验证码数目,保存在此目录/pic_data 中: '))
for i in co:
    k.append(tuple(i))
co = k
#字体路径
font_path = "RAVIE.TTF"
#位数
numbers = 1
#验证码大小
size = (120, 40)
```

```python
#背景颜色
bgcolor = (random.randint(0, 255), 0, 0)
#干扰线
draw_line = True
#干扰线数量范围
line_numbers = (1, 4)
def bgfontrange():
    global bgcolor
    bgcolor = random.choice(co)
source1 = [str(x) for x in range(0, 10)]  # 0~9
source2 = ["a", "b", "c", "d", "e", "f", "g", "h", "i", "j", "k", "l",
           "m", "n", "o", "p", "q", "r", "s", "t", "u", "v", "w", "x", "y","z"]
source3 = ["A", "B", "C", "D", "E", "F", "G", "H", "I", "J", "K", "L",
           "M", "N", "O", "P", "Q", "R", "S", "T", "U", "V", "W", "X", "Y","Z"]
source = []
source.extend(source3)
source.extend(source1)
source.extend(source2)
#生成4个随机数
def make_text():
    return "".join(random.sample(source, numbers))       # "",加上4个随机数
#随机绘线
def make_line(draw, width, height):
    begin = (random.randint(0, width), random.randint(0, height))
    end = (random.randint(0, width), random.randint(0, height))
    draw.line([begin, end], fill = (random.randint(170, 210),
random.randint(130, 170),
  random.randint(90, 130)),
    width = random.choice([2, 3]))                        #绘线
def create_points(draw, point_chance, width, height):
    chance = min(100, max(0, int(point_chance)))          #大小限制在[0,100]
    for w in range(width):
        for h in range(height):
            tmp = random.randint(0, 100)
            if tmp > 100 - chance:
                draw.point((w, h), fill = (random.randint(0, 200), 0, 0))
data = pd.DataFrame()
data1 = []
def rndChar():
    return chr(random.randint(65, 122))
def rndColorChar():
    return(random.randint(20, 110),
random.randint(20, 110), random.randint(20, 110))
#生成验证码
def make_codepng(index):
    width, height = size                                  #图片的宽度与高度
    #RGB是透明度
```

```python
    image = Image.new("RGB", (width, height), bgcolor)      #创建图片
    draw = ImageDraw.Draw(image)                            #绘图工具
    text = make_text()                                      #生成随机字符串
    font = ImageFont.truetype(font_path, 29)                #定义字体大小
    font_width, font_height = font.getsize(text)            #设置字体的宽度与高度
    s = ''
    for t in range(4):
        s1 = make_text()
        s = s + s1
        bias = random.randint(-5, 5)
        draw.text((25 * t + 10, (height-font_height)/2 + bias),
                  s1, font = font, fill = rndColorChar())
    text = s
    data1.append(s)
    if draw_line:                                           #绘线
        print("num : {}".format(index), end = '\r')
        num = random.randint(1, 3)                          #产生随机数
        for i in range(num):
            make_line(draw, width, height)
    create_points(draw, 6, width, height)
    filename = r"./pic_data/" + text + ".jpg"               #文件名
    #filename = text + '.jpg'
    with open(filename, "wb") as file:                      #打开文件
        image.save(file, format = 'JPEG')
if not os.path.exists("./pic_data"):
    os.makedirs("./pic_data")
for i in range(num):
    bgfontrange()                                           #随机生成背景颜色
    make_codepng(i)
```

运行时，会提示输入想要的验证码数量，保存在./pic_data文件夹中，如图 12-7 和图 12-8 所示。

图 12-7　输入提示

2. 模型测试

通过 generate() 函数将图像分成 4 个小矩阵，moving() 函数对图像进行随机偏移，保证测试图像的平移不变性。相关代码如下：

```python
def generate(im_path, length = cfg.WIDTH, s = cfg.START_POINT):
    #将 demo 下的每张图片数据切成 4 份，并返回 return 矩阵
    image = cv2.imread(im_path)
    result = np.stack(
```

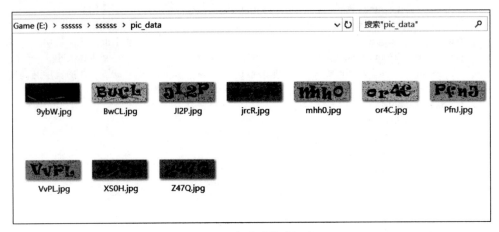

图 12-8 新生成的验证码

```
        [image[:, s[0]:s[0] + length], image[:, s[1]:s[1] + length],
         image[:, s[2]:s[2] + length], image[:, s[3]:s[3] + length]]
    )
    return result
def moving(image, bias = 2):                                          #移动处理
    size = image.shape
    image3 = image
    image4 = image
    image3 = np.concatenate(
        (image3[bias:, :], image3[:bias, :]), axis = 0)               #上
    image4 = np.concatenate(
(image4[size[0] - bias:, :], image4[:size[0] - bias, :]), axis = 0)   #下
return image, image3, image4
if __name__ == "__main__":                                            #主函数
    import os
    import matplotlib.pyplot as plt
    demo_path = './demo/'
    files = os.listdir(demo_path)                                     #加载文件
    im_paths = [os.path.join(demo_path, v)
                for v in files if v.endswith('.jpg')]
    net_obj = Net(is_training = False)
    with tf.Session() as sess:
        ckpt = tf.train.get_checkpoint_state(cfg.MODEL_PATH)
        if ckpt and ckpt.model_checkpoint_path:
            #如果保存过模型,则在此基础上继续训练
            net_obj.saver.restore(sess, ckpt.model_checkpoint_path)
            print('Model Reload Successfully!')
        else:
            raise Exception('no matching ckpt file')                  #异常处理
```

```python
    for path in im_paths:
        test_image = generate(path)                    #测试图像
        str_pred = ''
        for i in range(4):
            value = np.squeeze(test_image[i])
            value = image_process(value)
            value = np.expand_dims(value, axis = -1)
            new_values = moving(value)
            new_values = np.stack(new_values)
            predictions = sess.run(                    #预测
                net_obj.y_hat, feed_dict = {
                    net_obj.x: new_values,
                    net_obj.keep_rate: 1.0
                }
            )
            predictions = np.sum(predictions, axis = 0)
            str_pred += decode_label(predictions)      #结果输出
        print(path + ': ' + str_pred)
        raw = cv2.imread(path)
        plt.imshow(raw[:, :, [2, 1, 0]])
        plt.title('{}'.format(str_pred))
        plt.show()
```

12.4 系统测试

本部分包括训练准确率及测试效果。

12.4.1 训练准确率

模型的单字符准确率可达到95%甚至以上。随着训练次数的增多,模型在测试数据上的损失和准确率逐渐收敛,最终趋于稳定。为避免过拟合,在第368轮时停止训练,将每轮的训练结果保存于.csv文件中并绘制成折线图,如图12-9所示。验证集的准确率和损失函数如图12-10所示。

12.4.2 测试效果

利用上述新生成的数据集图片作为测试集,将图片放入./demo文件夹下,运行test.py,得到每张验证码图片及预测结果,在单字符95%准确率的情况下,能够识别验证码,如图12-11~图12-17所示。

351	0.945	3.2044914
352	0.938	3.2058272
353	0.944	3.20759
354	0.943	3.2040653
355	0.946	3.2043462
356	0.946	3.2059755
357	0.947	3.2081976
358	0.948	3.2027502
359	0.95	3.2087674
360	0.952	3.205964
361	0.955	3.2036595
362	0.956	3.2061305
363	0.939	3.204795
364	0.921	3.2040412
365	0.928	3.2101374
366	0.927	3.20462
367	0.932	3.2104223
368	0.95	3.202658

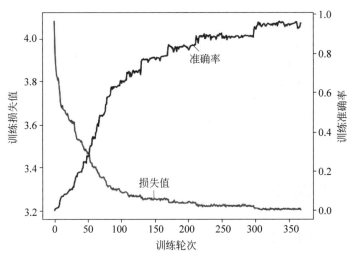

图 12-9 保存在.csv 文件中的训练结果

图 12-10 模型准确率和损失值

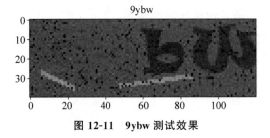

图 12-11 9ybw 测试效果

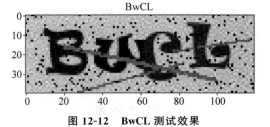

图 12-12 BwCL 测试效果

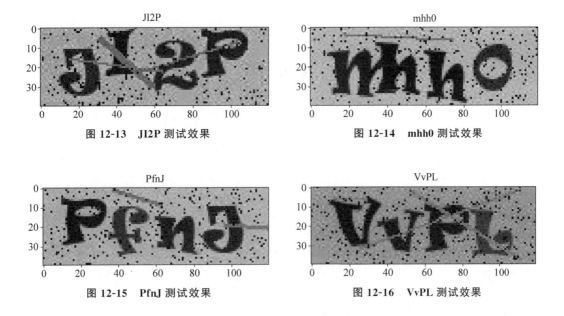

图 12-13　JI2P 测试效果

图 12-14　mhh0 测试效果

图 12-15　PfnJ 测试效果

图 12-16　VvPL 测试效果

图 12-17　模型测试结果

项目 13 基于 CNN 的交通标志识别

PROJECT 13

本项目针对出国自驾游特定场景,基于 Kaggle 上的交通标志数据集,采用 VGG 和 GoogLeNet 模型训练不同类型的卷积神经网络,通过修改网络的架构和参数,提高模型识别的准确率。

13.1 总体设计

本部分包括系统整体结构和系统流程。

13.1.1 系统整体结构

系统整体结构如图 13-1 所示。

图 13-1 系统整体结构

13.1.2 系统流程

系统流程如图 13-2 所示。

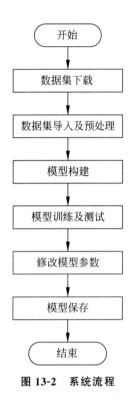

图 13-2 系统流程

13.2 运行环境

下载 Anaconda，下载地址为：https://www.anaconda.com/。
打开 Anaconda Prompt，用清华镜像安装 CPU 版本的 TensorFlow，输入命令：

pip install tensorflow == 1.14.0 -i https://pypi.tuna.tsinghua.edu.cn/simple

需要安装其他库，输入以下命令：

pip install numpy -i https://pypi.tuna.tsinghua.edu.cn/simple
pip install scikit-learn -i https://pypi.tuna.tsinghua.edu.cn/simple
pip install scikit-image -i https://pypi.tuna.tsinghua.edu.cn/simple
pip install imutils -i https://pypi.tuna.tsinghua.edu.cn/simple
pip install matplotlib -i https://pypi.tuna.tsinghua.edu.cn/simple

13.3 模块实现

本项目包括 3 个模块：数据预处理、模型构建、模型训练及保存。下面分别给出各模块的功能介绍及相关代码。

13.3.1 数据预处理

本项目使用德国交通标志识别基准数据集(GTSRB)，此数据集包含 50000 张在各种环境下拍摄的交通标志图像，下载地址为：https://www.kaggle.com/meowmeowmeowmeowmeow/gtsrb-german-traffic-sign/。数据集下载完成后，导入数据并进行预处理，相关代码如下：

```python
import matplotlib
from tensorflow.keras.preprocessing.image import ImageDataGenerator
from tensorflow.keras.utils import to_categorical
from tensorflow.keras.optimizers import Adam
from sklearn.metrics import classification_report
from skimage import transform
from skimage import exposure
from skimage import io
import matplotlib.pyplot as plt
import numpy as np
import random
import os
```

GTSRB 数据集已经划分为训练集和测试集，定义 load_split()函数导入训练集、测试集的图像数据和标签。因为属于同一类的图像相邻，需要打乱图像以保证训练效果。通过统计分析得到全部图像的分辨率，如图 13-3 所示，有极少数图像像素超过 100×100。为便于训练，将图像像素统一调整为 32×32。由于图像的对比度较低，调用 skimage 库的 equalize_adapthist()函数，使用自适应直方图均衡算法(CLAHE)增加图像的对比度。

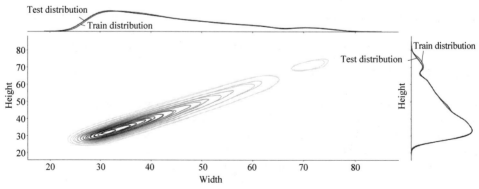

图 13-3　全部图像分辨率统计分析结果

load_split()函数的相关代码如下：

```python
def load_split(basePath, csvPath):
    # 初始化 data 和 labels 列表
    data = []
    labels = []
    # 加载存有训练集和测试集图像存储地址和标签的 csv 表格,去除空格,通过换行符识别各行
    # 并去除第一行标题行
    rows = open(csvPath).read().strip().split("\n")[1:]
    # 打乱 rows 的各行
    random.shuffle(rows)
    for (i, row) in enumerate(rows):
        # 每导入 1000 张图像后提示
        if i > 0 and i % 1000 == 0:
            print("[INFO] processed {} total images".format(i))
        # 取 csv 表格最后的两列:标签和存储地址
        (label, imagePath) = row.strip().split(",")[-2:]
        # 写出完整的图像存储地址
        imagePath = os.path.sep.join([basePath, imagePath])
        # 读取图像数据
        image = io.imread(imagePath)
        # 将图像统一调整为 32×32 像素
        image = transform.resize(image, (32, 32))
        # 增加图像的对比度
        image = exposure.equalize_adapthist(image, clip_limit=0.1)
        # 将当前图像的数据和标签添加到 data 和 labels 列表
        data.append(image)
        labels.append(int(label))
    data = np.array(data)
    labels = np.array(labels)
    return (data, labels)
```

导入图像各类别的具体名称,通过调用 load_split()函数获得训练集、测试集的图像数据和标签,将图像的数据范围从[0,225]调整为[0,1],图像标签采用 One-Hot 编码,相关代码如下：

```python
# 从 signnames.csv 表格中获取图像各类别的具体名称,该表格共两列,第二列是类别名称
labelNames = open("signnames.csv").read().strip().split("\n")[1:]
labelNames = [l.split(",")[1] for l in labelNames]
trainPath = os.path.sep.join(['gtsrb-german-traffic-sign', "Train.csv"])
testPath = os.path.sep.join(['gtsrb-german-traffic-sign', "Test.csv"])
print("[INFO] loading training and testing data...")
# 通过调用 load_split()函数获得训练集、测试集的图像数据和标签
(trainX, trainY) = load_split('gtsrb-german-traffic-sign', trainPath)
(testX, testY) = load_split('gtsrb-german-traffic-sign', testPath)
# 把 RGB 图像的数据范围从[0,225]调整为[0,1]
trainX = trainX.astype("float32") / 255.0
```

```
testX = testX.astype("float32") / 255.0
# One-Hot 编码图像的标签
numLabels = len(np.unique(trainY))
trainY = to_categorical(trainY, numLabels)
testY = to_categorical(testY, numLabels)
```

13.3.2 模型构建

本部分包括 VGG 模型和 GoogLeNet 模型简化版。

1. VGG 模型简化版

通过测试各种简化版模型,发现多减少网络的深度(卷积层、池化层、全连接层的层数),少减少网络的宽度(卷积层输出通道数),效果更好。由于本项目的图像尺寸较小,此版模型的卷积层输出通道数只减少为 VGG-11 的一半。输入图像经过 3 个卷积层、2 个最大池化层、1 个全连接层和 1 个 Softmax 层。卷积层的步幅为 1,通过填充使输出的宽和高与输入相同,前 2 个卷积层调整为 5×5,最后一个卷积层保持 3×3 不变,3 个卷积层的输出通道数依次为 32、64 和 64。2 个最大池化层分别位于第 2 和第 3 个卷积层后,池化窗口均为 2×2,步幅为 2,无填充,使输出的宽和高减半,每个最大池化层后接一个参数为 0.25 的 Dropout 层防止过拟合。最后是一个输出通道数为 256 的全连接层和 1 个 Softmax 层,全连接层后接 1 个参数为 0.5 的 Dropout 层防止过拟合。相关代码如下:

```
# 导入需要的软件包
from tensorflow.keras.models import Sequential
from tensorflow.keras.layers import BatchNormalization
from tensorflow.keras.layers import Conv2D
from tensorflow.keras.layers import MaxPooling2D
from tensorflow.keras.layers import Activation
from tensorflow.keras.layers import Flatten
from tensorflow.keras.layers import Dropout
from tensorflow.keras.layers import Dense
class VGGN:
    def build(width, height, depth, classes):
        # 使用 Keras 框架的 Sequential 模式编写代码
        model = Sequential()
        inputShape = (height, width, depth)
        chanDim = -1
    # 卷积核大小为 5×5,步幅为 1,输出通道数为 32,填充使得输出的宽和高与输入相同
        model.add(Conv2D(32, (5, 5), padding = "same", input_shape = inputShape))
    # ReLU 激活函数 + 批量归一化
        model.add(Activation("relu"))
        model.add(BatchNormalization(axis = chanDim))
    # 卷积核大小为 5×5,步幅为 1,输出通道数为 64,填充使得输出的宽和高与输入相同
        model.add(Conv2D(64, (5, 5), padding = "same"))
        model.add(Activation("relu"))
```

```python
        model.add(BatchNormalization(axis = chanDim))
# 池化窗口为 2×2,步幅为 2,不填充,输出的宽和高减半(变为 16×16)
        model.add(MaxPooling2D(pool_size = (2, 2)))
# 最大池化层后接一个参数为 0.25 的 Dropout 层防止过拟合
        model.add(Dropout(0.25))
# 卷积核大小为 3×3,步幅为 1,输出通道数 64,填充使得输出的宽和高与输入相同
        model.add(Conv2D(64, (3, 3), padding = "same"))
        model.add(Activation("relu"))
        model.add(BatchNormalization(axis = chanDim))
# 池化窗口为 2×2 的最大池化层,步幅为 2,不填充,输出的宽和高减半(变为 8×8)
        model.add(MaxPooling2D(pool_size = (2, 2)))
# 最大池化层后接一个参数为 0.25 的 Dropout 层防止过拟合
        model.add(Dropout(0.25))
        model.add(Flatten())
# 输出通道数为 256 的全连接层
        model.add(Dense(256))
        model.add(Activation("relu"))
        model.add(BatchNormalization())
# 全连接层后接一个参数为 0.5 的 Dropout 层防止过拟合
        model.add(Dropout(0.5))
# 最后是一个 Softmax 层,输出各类别的概率
        model.add(Dense(classes))
        model.add(Activation("softmax"))
        return model
```

2. GoogLeNet 简化版——MiniGoogLeNet

MiniGoogLeNet 由 Inception 模块、Downsample 模块和卷积模块组成,卷积模块包括卷积层、激活函数和批量归一化;Inception 模块由两个卷积核大小分别为 1×1 和 3×3 的卷积模块并联组成,这两个卷积模块都通过填充使输入/输出的高和宽相同,便于通道合并;Downsample 模块由一个卷积核大小为 3×3 的卷积模块和一个池化窗口为 3×3 的最大池化层并联组成,卷积模块和最大池化层不填充,步幅均为 2,使得输入经过后宽和高减半。MiniGoogLeNet 的输入图片经过一个卷积模块后输出通道数不同的 Inception 模块,最后是一个全局平均池化层和一个 Softmax 层,全局平均池化层将每个通道的高和宽变为 1,有效地减少了过拟合。相关代码如下:

```python
# 导入需要的软件包
from tensorflow.keras.layers import BatchNormalization
from tensorflow.keras.layers import Conv2D
from tensorflow.keras.layers import AveragePooling2D
from tensorflow.keras.layers import MaxPooling2D
from tensorflow.keras.layers import Activation
from tensorflow.keras.layers import Dropout
from tensorflow.keras.layers import Dense
from tensorflow.keras.layers import Flatten
```

```python
from tensorflow.keras.layers import Input
from tensorflow.keras.models import Model
from tensorflow.keras.layers import concatenate
from tensorflow.keras import backend as K
class MiniGoogLeNet:
    #定义卷积模块,x表示输入数据,K表示输出通道的数量,KX、KY表示卷积核的大小
    def conv_module(x,K,KX,KY,stride,chanDim,padding="same"):
        #卷积+激活函数+批量归一化,默认填充使得输出的宽和高不变
        x = Conv2D(K, (KX, KY), strides=stride, padding=padding)(x)
        x = Activation("relu")(x)
        x = BatchNormalization(axis=chanDim)(x)
        return x
    #定义Inception模块,x表示输入数据,numK1_1、numK3_3表示两个卷积模块的输出通道数量
    def inception_module(x,numK1_1,numK3_3,chanDim):
        #并联的两个卷积模块,卷积核大小分别为1×1和3×3
        conv1_1 = MiniGoogLeNet.conv_module(x,numK1_1,1,1,(1,1),chanDim)
        conv3_3 = MiniGoogLeNet.conv_module(x,numK3_3,3,3,(1,1),chanDim)
        x = concatenate([conv1_1,conv3_3],axis=chanDim)
        return x
    #定义Downsample模块,x表示输入数据,K表示卷积模块的输出通道数
    def downsample_module(x,K,chanDim):
        #并联的卷积模块和最大池化层,均使用3×3窗口,步幅为2,不填充,输出的宽和高减半
        conv3_3 = MiniGoogLeNet.conv_module(x,K,3,3,(2,2),chanDim,padding='valid')
        pool = MaxPooling2D((3,3),strides=(2,2))(x)
        x = concatenate([conv3_3,pool],axis=chanDim)
        return x
    #定义模型
    def build(width, height, depth, classes):
        inputShape = (height, width, depth)
        chanDim = -1
        #如果通道在前,将chanDim设为1
        if K.image_data_format() == "channels_first":
            inputShape = (depth, height, width)
            chanDim = 1
        #使用Keras的Model模式编写代码
        inputs = Input(shape=inputShape)
        #输入图片先经过一个卷积核大小为3×3,输出通道数为96的卷积模块
        x = MiniGoogLeNet.conv_module(inputs, 96, 3, 3, (1, 1),chanDim)
        #2个Inception模块(输出通道数分别为32+32、32+48)+1个Downsample模块
        x = MiniGoogLeNet.inception_module(x, 32, 32, chanDim)
        x = MiniGoogLeNet.inception_module(x, 32, 48, chanDim)
        x = MiniGoogLeNet.downsample_module(x, 80, chanDim)
        #4个Inception模块(输出通道数分别为112+48、96+64、80+80、48+96)+1个
          Downsample模块
        x = MiniGoogLeNet.inception_module(x, 112, 48, chanDim)
        x = MiniGoogLeNet.inception_module(x, 96, 64, chanDim)
        x = MiniGoogLeNet.inception_module(x, 80, 80, chanDim)
```

```
    x = MiniGoogLeNet.inception_module(x, 48, 96, chanDim)
    x = MiniGoogLeNet.downsample_module(x, 96, chanDim)
    #2个Inception模块+1个Downsample模块+1个全局平均池化层+1个Dropout层
    x = MiniGoogLeNet.inception_module(x, 176, 160, chanDim)
    x = MiniGoogLeNet.inception_module(x, 176, 160, chanDim)
    #输出7×7×(160+176)经过平均池化之后变成了1×1×376
    x = AveragePooling2D((7, 7))(x)
    x = Dropout(0.5)(x)
    #特征扁平化
    x = Flatten()(x)
    #Softmax层输出各类别的概率
    x = Dense(classes)(x)
    x = Activation("softmax")(x)
    #创建模型
    model = Model(inputs, x, name = "googlenet")
    return model
```

13.3.3 模型训练及保存

通过随机旋转等方法进行数据增强,选用 Adam 算法作为优化算法,随着迭代的次数增加降低学习速率,经过尝试,速率设为 0.001 时效果最好。调用之前的模型,以交叉熵为损失函数,使用 Keras 的 fit_generator()方法训练模型,最后评估并保存到磁盘,相关代码如下:

```
#设置初始学习速率、批量大小和迭代次数
INIT_LR = 1e-3
BS = 64
NUM_EPOCHS = 10
#使用随机旋转、缩放、水平/垂直移位、透视变换、剪切等方法进行数据增强(不用水平或垂直翻转)
aug = ImageDataGenerator(
    rotation_range = 10,
    zoom_range = 0.15,
    width_shift_range = 0.1,
    height_shift_range = 0.1,
    shear_range = 0.15,
    horizontal_flip = False,
    vertical_flip = False,
    fill_mode = "nearest")
print("[INFO] compiling model...")
#选用 Adam 算法作为优化算法,初始学习速率设为 0.001,随着迭代次数增加来降低学习速率
opt = Adam(lr = INIT_LR, decay = INIT_LR / (NUM_EPOCHS * 0.5))
#调用 MiniGoogLeNet,使用 VGG 网络时把 MiniGoogLeNet 更改为 VGGN
model = MiniGoogLeNet.build(width = 32, height = 32, depth = 3,
    classes = numLabels)
#编译模型,使用交叉熵作为损失函数
```

```python
model.compile(loss = "categorical_crossentropy", optimizer = opt,
    metrics = ["accuracy"])
# 使用 Keras 的 fit_generator()方法训练模型
print("[INFO] training network...")
H = model.fit_generator(
    aug.flow(trainX, trainY, batch_size = BS),
    validation_data = (testX, testY),
    steps_per_epoch = trainX.shape[0],
    epochs = NUM_EPOCHS,
    verbose = 1)
# 评估模型,打印分类报告
print("[INFO] evaluating network...")
predictions = model.predict(testX, batch_size = BS)
print(classification_report(testY.argmax(axis = 1),
    predictions.argmax(axis = 1), target_names = labelNames))
# 将模型存入磁盘
print("[INFO]serializing network to'{}'...".format('output/testmodel.pb'))
model.save('output/testmodel.pb')
# 绘制损失和准确率随迭代次数变化的曲线
N = np.arange(0, NUM_EPOCHS)
plt.style.use("ggplot")
plt.figure()
plt.plot(N, H.history["loss"], label = "train_loss")
plt.plot(N, H.history["val_loss"], label = "val_loss")
plt.plot(N, H.history["acc"], label = "train_acc")
plt.plot(N, H.history["val_acc"], label = "val_acc")
plt.title("Training Loss and Accuracy on Dataset")
plt.xlabel("Epoch #")
plt.ylabel("Loss/Accuracy")
plt.legend(loc = "lower left")
plt.savefig('output/test.png')
```

13.4 系统测试

本部分包括训练准确率及测试效果。

13.4.1 训练准确率

本部分包括 VGG 模型简化版和 MiniGoogLeNet 模型。

1. VGG 模型简化版

迭代 15 次后,准确率达到 98%,如图 13-4 所示。

2. MiniGoogLeNet 模型

迭代 10 次后,准确率达到 94%,如图 13-5 所示。

```
Epoch 11/15
612/612 [==============================] - 1243s 2s/step - loss: 0.0776 - acc: 0.9759 - val_loss: 0.0956 - val_acc: 0.9712
Epoch 12/15
612/612 [==============================] - 1091s 2s/step - loss: 0.0717 - acc: 0.9770 - val_loss: 0.0905 - val_acc: 0.9725
Epoch 13/15
612/612 [==============================] - 1207s 2s/step - loss: 0.0686 - acc: 0.9780 - val_loss: 0.0805 - val_acc: 0.9760
Epoch 14/15
612/612 [==============================] - 1229s 2s/step - loss: 0.0661 - acc: 0.9783 - val_loss: 0.1704 - val_acc: 0.9516
Epoch 15/15
612/612 [==============================] - 826s 1s/step - loss: 0.0612 - acc: 0.9808 - val_loss: 0.0684 - val_acc: 0.9805
[INFO] evaluating network...
                         precision    recall   f1-score   support
    Speed limit (20km/h)   1.00        1.00      1.00        60
    Speed limit (30km/h)   0.98        1.00      0.99       720
```

图 13-4　VGG 模型简化版训练结果

```
Epoch 6/10
612/612 [==============================] - 2880s 5s/step - loss: 0.0152 - acc: 0.9957 - val_loss: 0.1482 - val_acc: 0.9622
Epoch 7/10
612/612 [==============================] - 2881s 5s/step - loss: 0.0160 - acc: 0.9955 - val_loss: 0.1755 - val_acc: 0.9545
Epoch 8/10
612/612 [==============================] - 2871s 5s/step - loss: 0.0163 - acc: 0.9950 - val_loss: 0.1175 - val_acc: 0.9698
Epoch 9/10
612/612 [==============================] - 2875s 5s/step - loss: 0.0117 - acc: 0.9968 - val_loss: 0.1161 - val_acc: 0.9674
Epoch 10/10
612/612 [==============================] - 2866s 5s/step - loss: 0.0093 - acc: 0.9975 - val_loss: 0.2344 - val_acc: 0.9443
[INFO] evaluating network...
                         precision    recall   f1-score   support
    Speed limit (20km/h)   1.00        1.00      1.00        60
    Speed limit (30km/h)   0.95        0.99      0.97       720
```

图 13-5　MiniGoogLeNet 模型训练结果

13.4.2　测试效果

在互联网上找到 10 张德国交通标志图片,采用与训练阶段相同的方法对图片预处理后,利用之前保存的模型进行预测,相关代码如下:

```
# 导入需要的软件包
from tensorflow.keras.models import load_model
from skimage import transform
from skimage import exposure
from skimage import io
import numpy as np
import matplotlib.image as imgplt
import matplotlib.pyplot as plt
# 加载模型
model = load_model('output/testmodel.pb')
# 加载交通标志名称
labelNames = open("signnames1.csv").read().strip().split("\n")[1:]
labelNames = [l.split(",")[2] for l in labelNames]
labels = []
for i in range(1,11):
```

```python
        imagePath = 'C:/Users/Lenovo/Desktop/测试图片/' + str(i) + '.png'
        # 加载图片
        image = io.imread(imagePath)
        # 采用与训练阶段相同的方法对图片进行预处理
        image = transform.resize(image, (32, 32))
        image = exposure.equalize_adapthist(image, clip_limit = 0.1)
        image = image.astype("float32") / 255.0
        image = np.expand_dims(image, axis = 0)
        # 预测
        preds = model.predict(image)
        j = preds.argmax(axis = 1)[0]
        # 获得预测类别的具体名称
        label = labelNames[j]
        labels.append(label)
# 在 Jupyter 中正常显示中文
plt.rc('font', family = 'SimHei', size = 18)
fig = plt.figure(figsize = (20,9))
for i in range(1,11):
        imagePath = 'C:/Users/Lenovo/Desktop/测试图片/' + str(i) + '.png'
        x = imgplt.imread(imagePath)
        # 显示图片和预测的类别
        plt.subplot(2,5,i)
        plt.imshow(x)
        plt.title(labels[i-1])
```

模型预测效果如图 13-6 所示。

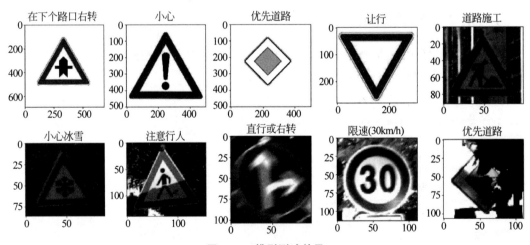

图 13-6　模型测试效果

项目 14　图像风格转移

PROJECT 14

本项目基于 TensorFlow 和 VGG-19 卷积神经网络,通过 Python 的 GUI 进行界面交互显示,实现图像的实时风格转移以及非实时的任意风格转移。

14.1　总体设计

本部分包括系统整体结构和系统流程。

14.1.1　系统整体结构

系统整体结构如图 14-1 所示。

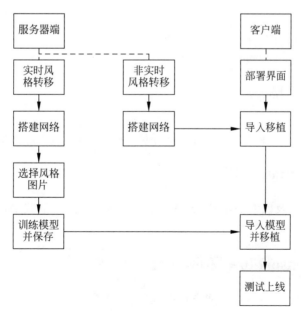

图 14-1　系统整体结构

14.1.2 系统流程

系统流程如图 14-2 所示。

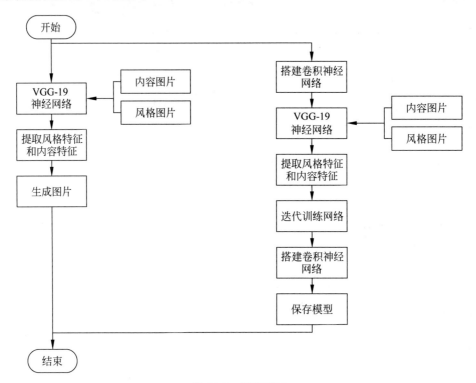

图 14-2　系统流程

14.2 运行环境

本部分包括 Python 环境、TensorFlow 环境、库安装和 VGG-19 网络下载。

14.2.1 Python 环境

需要 Python 3.6 及以上的配置，在 Windows 环境下推荐下载 Anaconda 完成 Python 所需的配置。

14.2.2 TensorFlow 环境

打开 Anaconda Prompt，输入清华仓库镜像，输入命令：

conda config -- add channels https://mirrors.tuna.tsinghua.edu.cn/anaconda/pkgs/free/

```
conda config-set show_channel_urls yes
```

创建 Python 3.5 的虚拟环境,名称为 TensorFlow,输入命令:

```
conda create -n tensorflow python=3.5
```

激活 TensorFlow 环境,输入命令:

```
activate tensorflow
```

安装 CPU 版本的 TensorFlow(版本适用为 TensorFlow=1.9.0),输入命令:

```
conda install tensorflow==1.9.0
```

推荐使用 GPU,其对于矩阵运算有更高的速度,输入命令:

```
conda install tensorflow-gpu==1.9.0
```

14.2.3 库安装

主要使用 Scipy 库和 Numpy 库。
Scipy 库安装 1.2.0 及以下版本,输入命令:

```
conda install scip==1.2.0
```

Numpy 库安装,输入命令:

```
conda install numpy
```

其他库没有版本要求,可以自行安装。

14.2.4 VGG-19 网络下载

VGG-19 网络下载地址为:http://www.vlfeat.org/matconvnet/models/beta16/imagenet-vgg-verydeep-19.mat,需要训练 VGG-19 网络进行特征提取。

14.3 模块实现

本项目包括 3 个模块:实时风格转移、非实时风格转移和交互界面设计。下面分别给出各模块的功能介绍及相关代码。

14.3.1 实时风格转移

实时风格转移需要提前训练好转换网络,在转换时,将图片输入网络中即可得到转换图片。缺点是风格固定,如果需要其他风格,需要重新对网络进行训练;优点是转换具有实时性,一张图片只需 5s 即可完成风格转移。

实时风格转移的内部流程主要需要两张神经网络:一张是需要进行训练的转换网络,另一张是 VGG-19 网络。将内容图片和风格图片输入 VGG-19 网络,分别提取内容图片的特征以及风格特征,计算出内容损失以及风格损失,加权求出损失函数,通过减小损失函数对转换网络进行训练,多次迭代得到训练好的转换网络,如图 14-3 所示。

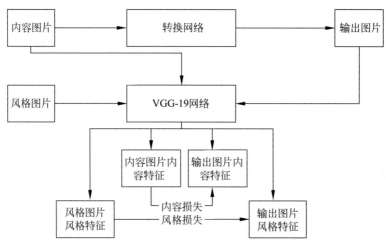

图 14-3　内部流程图

1. 定义模型结构

搭建转换网络,转换网络由 3 个卷积层、5 个残差层和 3 个反卷积层构成,如图 14-4 所示。

相关代码如下:

```
#定义卷积层
def _conv_layer(net,num_filters,filter_size,strides,relu = True):
    weight_init = _conv_init_vars(net,num_filters,filter_size)
    strides_shape = [1,strides,strides,1]
    net = tf.nn.conv2d(net,weight_init,strides_shape,padding = 'SAME')
    net = _instance_norm(net)
    if relu:
        net = tf.nn.relu(net)
    return net
#转置卷积层
def _conv_tranpose_layer(net, num_filters, filter_size, strides):
    weights_init = _conv_init_vars(net, num_filters, filter_size, transpose = True)
    #创建权重变量
    batch_size, rows, cols, in_channels = [i.value for i in net.get_shape()]
    new_rows, new_cols = int(rows * strides), int(cols * strides)
    #计算输出矩阵的大小
    new_shape = [batch_size, new_rows, new_cols, num_filters]          #新矩阵的大小
```

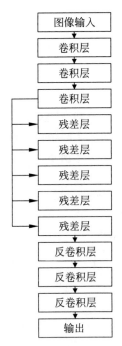

图 14-4 转换网络结构

```
    tf_shape = tf.stack(new_shape)
    strides_shape = [1,strides,strides,1]                      #步长
    net = tf.nn.conv2d_transpose(net, weights_init, tf_shape, strides_shape, padding = 'SAME')
                                                                #转置卷积
    net = _instance_norm(net)
    return tf.nn.relu(net)                                      #进行激活函数的运算
#残差层
def _residual_block(net, filter_size = 3):
    tmp = _conv_layer(net, 128, filter_size, 1)
    return net + _conv_layer(tmp, 128, filter_size, 1, relu = False)
#创建网络
def net(image):
    image = image / 255.0
    conv1 = _conv_layer(image, 32, 9, 1)
    conv2 = _conv_layer(conv1, 64, 3, 2)
    conv3 = _conv_layer(conv2, 128, 3, 2)
    resid1 = _residual_block(conv3, 3)
    resid2 = _residual_block(resid1, 3)
    resid3 = _residual_block(resid2, 3)
    resid4 = _residual_block(resid3, 3)
    resid5 = _residual_block(resid4, 3)
    conv_t1 = _conv_tranpose_layer(resid5, 64, 3, 2)
    conv_t2 = _conv_tranpose_layer(conv_t1, 32, 3, 2)
```

```
conv_t3 = _conv_layer(conv_t2, 3, 9, 1, relu = False)
preds = tf.nn.tanh(conv_t3)
output = image + preds
return tf.nn.tanh(output) * 127.5 + 255. / 2
```

2. VGG-19 网络预处理

实时风格转移模型的损失函数需要用到 VGG-19 网络,主要作用是特征提取。下载好的 VGG-19 网络是权重参数。因此,对 VGG-19 网络需要进行一些预处理,自行搭建 VGG-19 网络,填入参数后网络方可正常使用,如图 14-5 所示。

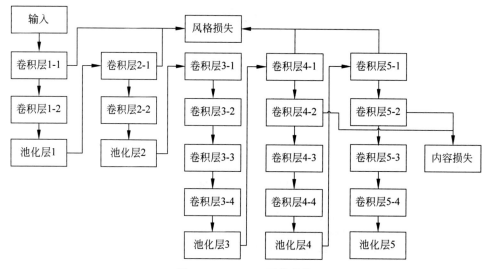

图 14-5　VGG-19 网络结构

预处理相关代码如下:

```
class VGG:
    LAYERS = (
        'conv1_1','relu1_1','conv1_2','relu1_2','pool1',
        'conv2_1','relu2_1','conv2_2','relu2_2','pool2',
        'conv3_1','relu3_1','conv3_2','relu3_2','conv3_3',
        'relu3_3','conv3_4','relu3_4','pool3',
        'conv4_1','relu4_1','conv4_2','relu4_2','conv4_3',
        'relu4_3','conv4_4','relu4_4','pool4',
        'conv5_1','relu5_1','conv5_2','relu5_2','conv5_3',
        'relu5_3','conv5_4','relu5_4'
    )
    #读取 VGG 网络的权重数据
    def __init__(self,data_path):
        self.data_path = data_path
        self.data = scipy.io.loadmat(data_path)
        self.weights = self.data['layers'][0]
```

```python
# 对图片进行一般化处理
def preprocess(self, image):
    return image - 125
def unprocess(self, image):
    return image + 125
def _is_convolutional_layer(self, name):
    return name[:4] == 'conv'
def _is_relu_layer(self, name):
    return name[:4] == 'relu'
def _is_pooling_layer(self, name):
    return name[:4] == 'pool'
def _conv_layer_from(self, input, weights, bias):
    conv = tf.nn.conv2d(input, tf.constant(weights), strides = (1, 1, 1, 1),
                        padding = 'SAME')
    return tf.nn.bias_add(conv, bias)
def _pooling_layer_from(self, input):
    return tf.nn.max_pool(input, ksize = (1, 2, 2, 1), strides = (1, 2, 2, 1),
                          padding = 'SAME')
def net(self, input_image):
    net = {}
    current_layer = input_image
    for i, name in enumerate(self.LAYERS):
        if (self._is_convolutional_layer(name)):
            # 将 VGG 网络中的权重与偏置量全部取出
            # 只有卷积层会提取权重
            kernels = self.weights[i][0][0][2][0][0]
            bias = self.weights[i][0][0][2][0][1]
            bias = bias.reshape(-1)                    # 将 bias 转换为一行
            current_layer = self._conv_layer_from(current_layer, kernels, bias)
                                                        # 搭建一层神经网络
        elif self._is_relu_layer(name):
            current_layer = tf.nn.relu(current_layer)
        elif self._is_pooling_layer(name):
            current_layer = self._pooling_layer_from(current_layer)
        net[name] = current_layer  # 将每一层的输出保存起来,方便以后提取计算损失函数
    assert len(net) == len(self.LAYERS)
    # 将 net 返回,net 是一个字典,每一层的名字为键,对应的值为这一层的输出
    return net
```

3. 损失函数

损失函数主要由内容损失函数和风格损失函数加权得到。
内容损失函数为

$$L_i = \frac{1}{2 \times M \times N} \sum_{ij} (\boldsymbol{X}_{ij} - \boldsymbol{P}_{ij})^2 \tag{14-1}$$

其中,$\boldsymbol{X}$ 为噪声图片特征矩阵,$\boldsymbol{P}$ 为内容图片特征矩阵,M 是 $\boldsymbol{P}$ 的长×宽,N 是通道数。

风格损失函数为

$$L_i = \frac{1}{4 \times M^2 \times N^2} \sum_{ij} (G_{ij} - A_{ij})^2 \qquad (14\text{-}2)$$

其中，M 是特征矩阵的长×宽，N 是特征矩阵的信道数，G 是生成图片特征的 Gram 矩阵，A 是风格图片特征的 GRAM 矩阵。

其中，Gram 矩阵为

$$G_{ij}^l = \sum_k F_{ik}^l F_{jk}^l \qquad (14\text{-}3)$$

损失函数的实现代码如下：

```
#内容损失
def content_loss(self,content_input_batch,content_layer,content_weight):
    #输入内容图片,创建一个VGG网络用来提取损失函数
    content_loss_net = self.vgg.net(self.vgg.preprocess(content_input_batch))
    return content_weight * (2 * tf.nn.l2_loss(
        content_loss_net[content_layer]-self.transform_loss_net[content_layer]) /
                (_tensor_size(content_loss_net[content_layer])))
#风格损失
def style_loss(self, style_image, style_layers, style_weight):
    style_image_placeholder = tf.placeholder('float', shape = style_image.shape)
    style_loss_net = self.vgg.net(style_image_placeholder)
    with tf.Session() as sess:
        style_loss = 0
        style_preprocessed = self.vgg.preprocess(style_image)
        for layer in style_layers:
            style_image_gram = self._calculate_style_gram_matrix_for(style_loss_net, style_image_placeholder, layer, style_preprocessed)
            input_image_gram = self._calculate_input_gram_matrix_for(self.transform_loss_net, layer)
            style_loss += (2 * tf.nn.l2_loss(input_image_gram - style_image_gram) / style_image_gram.size)
        return style_weight * (style_loss)
#计算风格图片的 Gram 矩阵
def _calculate_style_gram_matrix_for(self, network, image, layer, style_image):image_feature
 = network[layer].eval(feed_dict = {image: style_image})
    image_feature = np.reshape(image_feature, (-1, image_feature.shape[3]))
    return np.matmul(image_feature.T, image_feature) / image_feature.size
#计算输入图片的 Gram 矩阵
def _calculate_input_gram_matrix_for(self,network,layer):image_feature = network[layer]
    batch_size, height, width, number = map(lambda i: i.value, image_feature.get_shape())
    size = height * width * number
    image_feature = tf.reshape(image_feature, (batch_size, height * width, number))
    return tf.matmul(tf.transpose(image_feature,perm = [0,2,1]),image_feature) / size
```

4. 图片的预处理

图片的预处理需要加载图片、保存图片，并从文件夹中提取图片。考虑迭代训练，需要

一张风格图片,但是需要几万张内容图片来训练网络。将内容图片放到一个文件夹中,从中提取。提取出来的图片不能直接输入网络,需要对图片格式进行转换。

```python
def load_image(image_path, img_size = None):
    assert exists(image_path),"image {} does not exist".format(image_path)
    img = scipy.misc.imread(image_path)
    if (len(img.shape) != 3) or (img.shape[2] != 3):
        img = np.dstack((img, img, img))
    #如果是单通道图片,则叠到一起
    if (img_size is not None):
        img = scipy.misc.imresize(img, img_size)
    img = scipy.misc.imresize(img,[480,640,3])
    img = img.astype("float32")
    return np.array(img)
#保存图片
def save_image(img, path):
    scipy.misc.imsave(path, np.clip(img, 0, 255).astype(np.uint8))
#从文件夹取图片
def get_files(img_dir):
    files = list_files(img_dir)
    picture = []
    for i in files:
        a = os.path.join(img_dir, i)
        picture.append(a)
    return picture
#为文件名建立队列
def list_files(in_path):
    files = []
    for (dirpath, dirnames, filenames) in os.walk(in_path):
        files.extend(filenames)
        break
    return files
```

5. 模型训练及保存

提取文件夹中的图片进行训练。每训练 50 次,保存一次模型,转换好的图片也需进行保存。

```python
def train():                                    #训练定义
    model = models.Model(settings.CONTENT_IMAGE, settings.STYLE_IMAGE1, settings.STYLE_IMAGE2)
    with tf.Session()as sess:
        sess.run(tf.global_variables_initializer())
        cost = loss(sess, model)                #损失
        optimizer = tf.train.AdamOptimizer(1.0).minimize(cost)   #优化器
        sess.run(tf.global_variables_initializer())
        sess.run(tf.assign(model.net['input'], model.random_img))
```

```python
for step in range(int(settings.TRAIN_STEPS)):
    sess.run(optimizer)
    if step % 50 == 0:
        print('step{}is down'.format(step))
        settings.SHOW = 'STEP ' + str(step) + 'is down'
        img = sess.run(model.net['input'])
        img += settings.IMAGE_MEAN_VALUE
        img = img[0]
        img = np.clip(img, 0, 255).astype(np.uint8)
        scipy.misc.imsave('{}-{}.jpg'.format(settings.OUTPUT_IMAGE, step), img)
        settings.OUTPUT_PATH = settings.OUTPUT_IMAGE + '-' + str(step) + '.jpg'
img = sess.run(model.net['input'])
img += settings.IMAGE_MEAN_VALUE
img = img[0]
img = np.clip(img, 0, 255).astype(np.uint8)
scipy.misc.imsave('{}.jpg'.format(settings.OUTPUT_IMAGE), img)
```

14.3.2 非实时风格转移

本部分包括系统流程、模型结构和损失函数计算。

1. 系统流程

非实时的风格转移需要提供内容图片和风格图片,用于生成转换后图片。优点是能够自由挑选风格,缺点是图片需要现场迭代。因此,生成一张图片需要花费很长时间,而且模型没有保存,下次生成时还需要花费相同的时间。在非实时的风格转移中,首先搭建一张VGG-19 网络,生成一张随机噪声图片作为转换后的图片;其次,通过 VGG-19 网络进行内容特征以及风格特征提取并进行加权作为损失函数;再次,使用损失函数训练模型;最后,随机的噪声图片会被训练为一张转换后的图片。模型结构如图 14-6 所示。

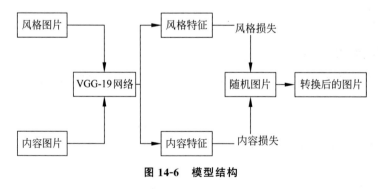

图 14-6 模型结构

2. 模型结构

本模块需要搭建一张 VGG-19 网络,代码如下:

```python
def vggnet(self):  # 创建 VGG 网络
    vgg = scipy.io.loadmat(settings.VGG_MODEL_PATH)
    vgg_layers = vgg['layers'][0]
    net = {}
    net['input'] = tf.Variable(np.zeros([1, settings.IMAGE_HEIGHT, settings.IMAGE_WIDTH,
3]), dtype=tf.float32)
    net['conv1_1'] = self.conv_relu(net['input'], self.get_wb(vgg_layers, 0))
    net['conv1_2'] = self.conv_relu(net['conv1_1'], self.get_wb(vgg_layers, 2))
    net['pool1'] = self.pool(net['conv1_2'])
    net['conv2_1'] = self.conv_relu(net['pool1'], self.get_wb(vgg_layers, 5))
    net['conv2_2'] = self.conv_relu(net['conv2_1'], self.get_wb(vgg_layers, 7))
    net['pool2'] = self.pool(net['conv2_2'])
    net['conv3_1'] = self.conv_relu(net['pool2'], self.get_wb(vgg_layers, 10))
    net['conv3_2'] = self.conv_relu(net['conv3_1'], self.get_wb(vgg_layers, 12))  # 卷积
    net['conv3_3'] = self.conv_relu(net['conv3_2'], self.get_wb(vgg_layers, 14))
    net['conv3_4'] = self.conv_relu(net['conv3_3'], self.get_wb(vgg_layers, 16))
    net['pool3'] = self.pool(net['conv3_4'])  # 池化
    net['conv4_1'] = self.conv_relu(net['pool3'], self.get_wb(vgg_layers, 19))
    net['conv4_2'] = self.conv_relu(net['conv4_1'], self.get_wb(vgg_layers, 21))
    net['conv4_3'] = self.conv_relu(net['conv4_2'], self.get_wb(vgg_layers, 23))
    net['conv4_4'] = self.conv_relu(net['conv4_3'], self.get_wb(vgg_layers, 25))
    net['pool4'] = self.pool(net['conv4_4'])
    net['conv5_1'] = self.conv_relu(net['pool4'], self.get_wb(vgg_layers, 28))
    net['conv5_2'] = self.conv_relu(net['conv5_1'], self.get_wb(vgg_layers, 30))
    net['conv5_3'] = self.conv_relu(net['conv5_2'], self.get_wb(vgg_layers, 32))
    net['conv5_4'] = self.conv_relu(net['conv5_3'], self.get_wb(vgg_layers, 34))
    net['pool5'] = self.pool(net['conv5_4'])
    return net
# 定义卷积层
def conv_relu(self, input, wb):
    conv = tf.nn.conv2d(input, wb[0], strides=[1, 1, 1, 1], padding='SAME')
    relu = tf.nn.relu(conv + wb[1])
    return relu
# 定义池化层
def pool(self, input):
    return tf.nn.max_pool(input, ksize=[1, 2, 2, 1], strides=[1, 2, 2, 1], padding='SAME')
# 获取网络的参数
def get_wb(self, layers, i):
    w = tf.constant(layers[i][0][0][2][0][0])
    bias = layers[i][0][0][2][0][1]
    b = tf.constant(np.reshape(bias, (bias.size)))
    return w, b
# 获取随机图片
def get_random_img(self):
    noise_image = np.random.uniform(-20, 20, [1, settings.IMAGE_HEIGHT, settings.IMAGE_
WIDTH, 3])
    random_img = noise_image * settings.NOISE + self.content * (1 - settings.NOISE)
```

```
        return random_img
```

3. 计算损失函数

相关代码如下:

```python
def loss(sess, model):                                           #定义损失
    content_layers = settings.CONTENT_LOSS_LAYERS
    sess.run(tf.assign(model.net['input'], model.content))
    content_loss = 0.0
    for layer_name, weight in content_layers:
        p = sess.run(model.net[layer_name])
        x = model.net[layer_name]
        M = p.shape[1] * p.shape[2]
        N = p.shape[3]
        content_loss += (1.0 / (2 * M * N)) * tf.reduce_sum(tf.pow(p - x, 2)) * weight
    content_loss /= len(content_layers)                          #内容损失
    style_layers = settings.STYLE_LOSS_LAYERS
    sess.run(tf.assign(model.net['input'], model.style1))
    style1_loss = 0.0
    for layer_name, weight in style_layers:
        a = sess.run(model.net[layer_name])
        x = model.net[layer_name]
        M = a.shape[1] * a.shape[2]
        N = a.shape[3]
        A = gram(a, M, N)
        G = gram(x, M, N)
        style1_loss += (1.0 / (4 * M * M * N * N)) * tf.reduce_sum(tf.pow(G - A, 2)) * weight
    style1_loss /= len(style_layers)                             #风格损失
    sess.run(tf.assign(model.net['input'], model.style2))
    style2_loss = 0.0
    for layer_name, weight in style_layers:
        a = sess.run(model.net[layer_name])
        x = model.net[layer_name]
        M = a.shape[1] * a.shape[2]
        N = a.shape[3]
        A = gram(a, M, N)
        G = gram(x, M, N)
        style2_loss += (1.0 / (4 * M * M * N * N)) * tf.reduce_sum(tf.pow(G - A, 2)) * weight
    style2_loss /= len(style_layers)
    loss = settings.ALPHA * content_loss + settings.BETA_1 * style1_loss + settings.BETA_2 * style2_loss                                        #总损失
    return loss
def gram(x, size, deep):
    x = tf.reshape(x, (size, deep))
```

```
    g = tf.matmul(tf.transpose(x), x)
    return g
```

14.3.3 交互界面设计

交互界面使用 Python 内置的 GUI 进行搭建,主要分两部分:一是实时风格转移;二是非实时风格转移。用户可自行选择需要转换的图片,如图 14-7 所示。

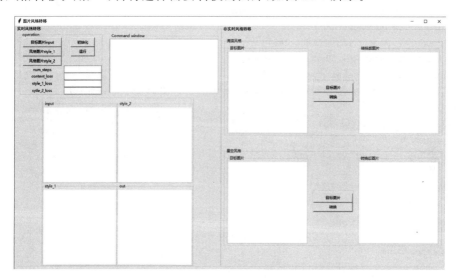

图 14-7 用户界面展示

```
def wtf():
    test.main('C:\Users\pipizhu\Desktop\result\hailang','fuck',settings.hailang)
    message = '转换完成'
    text_inset(message)
    show_image_shit(settings.SAVE_PATH + '\' + 'fuck.jpg')
def xingkong():
    test.main('C:\Users\pipizhu\Desktop\result\xingkong','shit',settings.xingkong)
    message = '转换完成'
    text_inset(message)
    show_image_xinkong(settings.SAVE_PATH + '\' + 'shit.jpg')
def shit():
    train.train()
    message = '运行完成'
    text_inset(message)
    show_image(settings.OUTPUT_PATH)
def show_image_shit(File):
    image = Image.open(File)
    image = image.resize((260, 260))
    filename = ImageTk.PhotoImage(image)
```

```python
        canvas_show_output_hailang.image = filename              #保留图像参考
        canvas_show_output_hailang.create_image(0, 0, anchor = 'nw', image = filename)
def show_image_xinkong(File):
        image = Image.open(File)
        image = image.resize((260, 260))
        filename = ImageTk.PhotoImage(image)
        canvas_show_output_xinkong.image = filename              #保留图像参考
        canvas_show_output_xinkong.create_image(0, 0, anchor = 'nw', image = filename)
def show_image(File):
        image = Image.open(File)
        image = image.resize((240, 240))
        filename = ImageTk.PhotoImage(image)
        canvas_show_output.image = filename                      #保留图像参考
        canvas_show_output.create_image(0, 0, anchor = 'nw', image = filename)
a = 1.0
def text_inset(content):
        global a
        txtMsgList.insert(a, content)
        b = int(a)
        txtMsgList.insert(str(b) + '.end', '\n')
        a = a + 1
def printcoords_content():
        File = filedialog.askopenfilename(parent = Frame_trans, initialdir = "C:/", title = 'Choose an image.')
        image = Image.open(File)
        image = image.resize((240, 240))
        filename = ImageTk.PhotoImage(image)
        canvas_show_input_content.image = filename               #保留图像参考
        canvas_show_input_content.create_image(0, 0, anchor = 'nw', image = filename)
        settings.CONTENT_IMAGE = File
        message = "内容图片路径 = " + File
        text_inset((message))
def printcoords_style_1():
        File = filedialog.askopenfilename(parent = Frame_trans, initialdir = "C:/", title = 'Choose an image.')
        image = Image.open(File)
        image = image.resize((240, 240))
        filename = ImageTk.PhotoImage(image)
        canvas_show_input_style_1.image = filename               #保留图像参考
        canvas_show_input_style_1.create_image(0, 0, anchor = 'nw', image = filename)
        settings.STYLE_IMAGE1 = File
        message = "风格 1 图片路径 = " + File
        text_inset((message))
def printcoords_style_2():
        File = filedialog.askopenfilename(parent = Frame_trans, initialdir = "C:/", title = 'Choose an image.')
        image = Image.open(File)
```

```python
            image = image.resize((240, 240))
            filename = ImageTk.PhotoImage(image)
            canvas_show_input_style_2.image = filename        #保留图像参考
            canvas_show_input_style_2.create_image(0,0,anchor = 'nw',image = filename)
            settings.STYLE_IMAGE2 = File
            message = "风格2图片路径 = " + File
            text_inset((message))
    def show():
        if entry1.get():
            settings.TRAIN_STEPS = entry1.get()
        if entry2.get():
            settings.ALPHA = entry2.get()
        if entry3.get():
            settings.BETA_1 = entry3.get()
        if entry4.get():
            settings.BETA_2 = entry4.get()
        message1 = "num_steps = " + settings.TRAIN_STEPS
        message2 = "content_loss" + settings.ALPHA
        message3 = "style_1_loss" + settings.BETA_1
        message4 = "style_2_loss" + settings.BETA_2
        settings.TRAIN_STEPS = float(settings.TRAIN_STEPS)
        settings.ALPHA = float(settings.ALPHA)
        settings.BETA_1 = float(settings.BETA_1)
        settings.BETA_2 = float(settings.BETA_2)
        text_inset(message1)
        text_inset(message2)
        text_inset(message3)
        text_inset(message4)
    #左上角的配置窗口
    Frame_trans = ttk.LabelFrame(win,text = '实时风格转移')
    Frame_fast = ttk.LabelFrame(win,text = '快速风格转移')
    Frame_trans.grid(column = 0,row = 0,sticky = 'ns')
    Frame_fast.grid(column = 1,row = 0,sticky = 'ns')
    #空行
    '''
    label_t1 = ttk.Label(Frame_setting)
    label_t1.grid(column = 0, row = 2, sticky = 'N')
    label_t2 = ttk.Label(Frame_setting)
    label_t2.grid(column = 0, row = 2, sticky = 'N')
    '''
    #命令行窗口
    Frame_command = ttk.LabelFrame(Frame_trans, text = 'Command window')
    Frame_command.grid(column = 1, row = 0,padx = 8, pady = 4)
    Frame_command_text = Frame(Frame_command,width = 400, height = 13, bg = 'white')
    Frame_command_text.grid(column = 0, row = 0, sticky = 'N')
    txtMsgList = Text(Frame_command_text,width = 50 ,height = 13)
    txtMsgList.tag_config('greencolor', foreground = '#008C00')#创建tag
```

```python
txtMsgList.grid()
#输入图片
Frame_picture_input = ttk.LabelFrame(Frame_trans)
Frame_picture_input.grid(column = 0, row = 1,columnspan = 2,padx = 15,)
picture_input_content = ttk.LabelFrame(Frame_picture_input, text = 'input')
picture_input_content.grid(column = 0, row = 0)
picture_input_style_1 = ttk.LabelFrame(Frame_picture_input, text = 'style_1')
picture_input_style_1.grid(column = 0, row = 1)
picture_input_style_2 = ttk.LabelFrame(Frame_picture_input, text = 'style_2')
picture_input_style_2.grid(column = 1, row = 0)
picture_show = ttk.LabelFrame(Frame_picture_input, text = 'out')
picture_show.grid(column = 1, row = 1)
canvas_show_input_content = tk.Canvas(picture_input_content, width = 240, height = 240,
bg = "white")                                           #显示输入内容
canvas_show_input_content.grid(column = 0, row = 0, sticky = 'N')
canvas_show_input_style_1 = tk.Canvas(picture_input_style_1, width = 240, height = 240,
bg = "white")                                           #显示输入风格1
canvas_show_input_style_1.grid(column = 0, row = 0, sticky = 'N')
canvas_show_input_style_2 = tk.Canvas(picture_input_style_2, width = 240, height = 240,
bg = "white")                                           #显示输入风格2
canvas_show_input_style_2.grid(column = 1, row = 0, sticky = 'N')
canvas_show_output = tk.Canvas(picture_show, width = 240, height = 240, bg = "white")
                                                        #显示输出
canvas_show_output.grid(column = 1, row = 1, sticky = 'N')
def printcoords_content_hailang():                      #定义海浪风格
    File = filedialog.askopenfilename(parent = Frame_trans, initialdir = "C:/",title =
'Choose an image.')
    image = Image.open(File)
    image = image.resize((260, 260))
    filename = ImageTk.PhotoImage(image)
    canvas_show_input_hailang.image = filename          #保留图像参考
    canvas_show_input_hailang.create_image(0,0,anchor = 'nw',image = filename)
    settings.hailang = File
    message = "原图片路径 = " + File
    text_inset((message))
def printcoords_content_xinkong():                      #定义星空风格
    File = filedialog.askopenfilename(parent = Frame_trans, initialdir = "C:/",title =
'Choose an image.')
    image = Image.open(File)
    image = image.resize((260, 260))
    filename = ImageTk.PhotoImage(image)
    canvas_show_input_xinkong.image = filename          #保留图像参考
    canvas_show_input_xinkong.create_image(0,0,anchor = 'nw',image = filename)
    settings.xingkong = File
    message = "原图片路径 = " + File
    text_inset((message))
#非实时风格转移的配置窗口
```

```python
#海浪风格
Frame_hailang = ttk.LabelFrame(Frame_fast,text = '海浪风格')
Frame_hailang.grid(column = 0,row = 0,padx = 8,pady = 20)
picture_input_hailang = ttk.LabelFrame(Frame_hailang,text = '目标图片')
picture_input_hailang.grid(column = 0,row = 0,padx = 8,pady = 4)
picture_output_hailang = ttk.LabelFrame(Frame_hailang, text = '转换后图片')
picture_output_hailang.grid(column = 2,row = 0, padx = 8, pady = 4)
canvas_show_input_hailang = tk.Canvas(picture_input_hailang, width = 260, height = 260,
    bg = "white")
canvas_show_input_hailang.grid(column = 0, row = 0, sticky = 'N')
canvas_show_output_hailang = tk.Canvas(picture_output_hailang, width = 260, height = 260,
    bg = "white")                                          #输出海浪风格
canvas_show_output_hailang.grid(column = 0, row = 0, sticky = 'N')
Frame_hailang_button = ttk.LabelFrame(Frame_hailang)
Frame_hailang_button.grid(column = 1,row = 0,padx = 8,pady = 4)
button_input_hailang = tk.Button(Frame_hailang_button,text = ' 目标图片 ',width = 17,command
    = printcoords_content_hailang)                        #海浪输入按钮
button_input_hailang.grid(column = 0, row = 0, sticky = 'N')
button_trans_hailang = tk.Button(Frame_hailang_button,text = ' 转换 ',width = 17,command = wtf)
button_trans_hailang.grid(column = 0, row = 1, sticky = 'N')
#星空风格
Frame_xinkong = ttk.LabelFrame(Frame_fast,text = '星空风格')
Frame_xinkong.grid(column = 0,row = 1,padx = 8,pady = 20)
picture_input_xinkong = ttk.LabelFrame(Frame_xinkong,text = '目标图片') #输入图片
picture_input_xinkong.grid(column = 0,row = 0,padx = 8,pady = 4)
picture_output_xinkong = ttk.LabelFrame(Frame_xinkong, text = '转换后图片')
picture_output_xinkong.grid(column = 2,row = 0, padx = 8, pady = 4)
canvas_show_input_xinkong = tk.Canvas(picture_input_xinkong, width = 260, height = 260,
    bg = "white")                                          #显示输入图片
canvas_show_input_xinkong.grid(column = 0, row = 0, sticky = 'N')
canvas_show_output_xinkong = tk.Canvas(picture_output_xinkong, width = 260, height = 260,
    bg = "white")                                          #输出星空风格图片
canvas_show_output_xinkong.grid(column = 0, row = 0, sticky = 'N')
Frame_xinkong_button = ttk.LabelFrame(Frame_xinkong)
Frame_xinkong_button.grid(column = 1,row = 0,padx = 8,pady = 4)
button_input_xinkong = tk.Button(Frame_xinkong_button,text = ' 目标图片 ',width = 17,command
    = printcoords_content_xinkong)                        #输入星空风格按钮
button_input_xinkong.grid(column = 0, row = 0, sticky = 'N')
button_trans_xinkon = tk.Button(Frame_xinkong_button,text = ' 转换 ',width = 17,command = xingkong)
button_trans_xinkon.grid(column = 0, row = 1, sticky = 'N')
Frame_setting = ttk.LabelFrame(Frame_trans,text = 'operation')
Frame_setting.grid(column = 0,row = 0,padx = 15,sticky = 'N')
button_input_content = tk.Button(Frame_setting,text = '目标图片 input ',width = 17,command =
    printcoords_content)                                  #输入内容按钮
button_input_content.grid(column = 0, row = 0, padx = 8,sticky = 'N')
button_input_style_1 = tk.Button(Frame_setting, text = ' 风格图片 style_1 ',width = 17 ,
    command = printcoords_style_1)                        #输入风格 1 按钮
```

```python
button_input_style_1.grid(column = 0, row = 1, padx = 8, sticky = 'N')
button_input_style_2 = tk.Button(Frame_setting, text = ' 风格图片 style_2 ', width = 17,
    command = printcoords_style_2)                          # 输入风格2按钮
button_input_style_2.grid(column = 0, row = 2, padx = 8, sticky = 'N')
button_output = tk.Button(Frame_setting, text = '初始化', width = 10, command = show)
button_output.grid(column = 1, row = 0, sticky = 'N')
button_show = tk.Button(Frame_setting, text = '运行', width = 10, command = shit)
button_show.grid(column = 1, row = 1, sticky = 'N')
label1 = ttk.Label(Frame_setting, text = "num_steps")        # 标签1
label1.grid(column = 0, row = 3, sticky = 'S')
var_1 = StringVar()
entry1 = ttk.Entry(Frame_setting, textvariable = var_1, width = 17)
entry1.grid(column = 1, row = 3, sticky = 'N')
settings.TRAIN_STEPS = entry1.get()
label2 = ttk.Label(Frame_setting, text = "content_loss")     # 标签2
label2.grid(column = 0, row = 4, sticky = 'N')
var_2 = StringVar()
entry2 = ttk.Entry(Frame_setting, textvariable = var_2, width = 17)
entry2.grid(column = 1, row = 4, sticky = 'N')
label3 = ttk.Label(Frame_setting, text = "style_1_loss")     # 标签3
label3.grid(column = 0, row = 5, sticky = 'N')
var_3 = StringVar()
entry3 = ttk.Entry(Frame_setting, textvariable = var_3, width = 17)
entry3.grid(column = 1, row = 5, sticky = 'N')
label4 = ttk.Label(Frame_setting, text = "sytle_2_loss")     # 标签4
label4.grid(column = 0, row = 6, sticky = 'N')
var_4 = StringVar()
entry4 = ttk.Entry(Frame_setting, textvariable = var_4, width = 17)
entry4.grid(column = 1, row = 6, sticky = 'N')
win.mainloop()
```

14.4 系统测试

本部分包括非实时风格转移测试和实时风格转移测试。

14.4.1 非实时风格转移测试

直接通过网络将图片进行转换并保存,观察效果,如图14-8～图14-10所示。

14.4.2 实时风格转移测试

将实时风格转移的转换网络对一个风格图进行2000次的迭代训练,输入一张内容图片,观察效果,如图14-11～图14-13所示。

图 14-8 非实时风格内容图

图 14-9 非实时风格图

图 14-10 转换后的图

图 14-11 实时风格内容图

图 14-12 实时风格图

图 14-13 网络训练 2000 次后进行的转换测试图

项目 15 口罩识别系统

PROJECT 15

本项目通过使用佩戴不同类型口罩的图片作为训练集,进行特征筛选和提取,选择 YOLO V3 目标检测算法对模型进行训练,实现判定人员是否佩戴口罩的功能。

15.1 总体设计

本部分包括系统整体结构和系统流程。

15.1.1 系统整体结构

系统整体结构如图 15-1 所示。

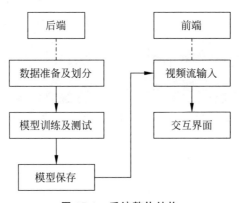

图 15-1 系统整体结构

15.1.2 系统流程图

系统流程如图 15-2 所示。

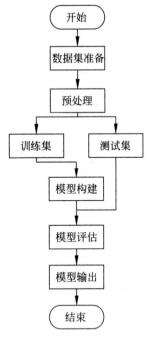

图 15-2 系统流程

15.2 运行环境

需要 Python 3.7 及以上配置，使用 PyCharm 作为开发平台，下载地址为：https://www.jetbrains.com/pycharm/download/。也可以下载 Anaconda 完成 Python 环境的配置，下载地址为：https://www.anaconda.com/。

安装 OpenCV 之前需要先安装 Numpy、Matplotlib 等。打开 Anaconda Prompt，安装不低于 3.3 版本的 OpenCV，输入命令：

```
pip install opencv-python
```

打开 Anaconda Prompt，安装 5.14.2 版本的 PyQt，输入命令：

```
pip install pyqt5
```

打开 Anaconda Prompt，安装 1.0.2 版本的 PyTorch，输入命令：

```
pip install pytorch
```

15.3 模块实现

本部分包括 4 个模块：数据预处理、模型训练及保存、页面显示和视频流输入、模型生成。下面分别给出各模块的功能介绍及相关代码。

15.3.1 数据预处理

从 MAFA 数据集中搜集口罩佩戴图片。它本身并不是针对口罩监测而制作的，而是对面部遮挡制作的数据集，该数据集由 30811 张无遮挡的人脸和 35806 张有遮挡的人脸图片组成。

创建数据集的相关代码如下：

```python
#导入相应数据包
from torchvision.datasets import ImageFolder
from torchvision import transforms as tfs
from torch.utils.data import DataLoader
```

对数据进行预处理，调整图片的大小，进行水平翻转做数据增强使之标准化，相关代码如下：

```python
#对数据集进行预处理
data_tf = tfs.Compose([
    tfs.Resize((256, 384)),
    tfs.RandomHorizontalFlip(),
    tfs.ToTensor(),
    tfs.Normalize([0.5, 0.5, 0.5], [0.5, 0.5, 0.5])
])
```

使用 PyTorch 的 ImageFolder 模块读取数据，使用 DataLoader 模块进行数据加载。具体步骤和技术详情请参考：https://pytorch.org/docs/stable/torchvision/datasets.html # imagefolder 和 https://pytorch.org/docs/stable/data.html # torch.utils.data.DataLoader。相关代码如下：

```python
# ImageFolder 读取数据
train_set = ImageFolder('./mask_dataset/train/', transform = data_tf)
test_set = ImageFolder('./mask_dataset/test/', transform = data_tf)
# DataLoader 进行数据加载
train_data = DataLoader(train_set, batch_size = 4, shuffle = True)
test_data = DataLoader(test_set, batch_size = 2, shuffle = True)
```

15.3.2 模型训练及保存

完成数据集的准备和读取之后开始训练模型。

1. 模型定义

定义的模型网络结构中包含 4 个大小为 3×3 的卷积层，在每 1 个卷积层后连接 1 个批量标准化层（Batch Normalization Layer）、1 个 ReLU 激活函数层以及 1 个 2×2 的最大池化层，最后是 3 个全连接层，相关代码如下：

```python
class MaskCNN(nn.Module):
    #定义模型结构
    def __init__(self):
        super().__init__()
                #定义4个卷积层及池化层
        self.conv1 = nn.Sequential(
                nn.Conv2d(3, 64, 3),
                nn.BatchNorm2d(64),
                nn.ReLU(True)
                )
        self.pooling1 = nn.MaxPool2d(2, 2)
            self.conv2 = nn.Sequential(
                nn.Conv2d(64, 128, 3),
                nn.BatchNorm2d(128),
                nn.ReLU(True)
                )
        self.pooling2 = nn.MaxPool2d(2, 2)
            self.conv3 = nn.Sequential(
                nn.Conv2d(128, 64, 3),
                nn.BatchNorm2d(64),
                nn.ReLU(True)
                )
        self.pooling3 = nn.MaxPool2d(2, 2)
          self.conv4 = nn.Sequential(
                nn.Conv2d(64, 3, 3),
                nn.BatchNorm2d(3),
                nn.ReLU(True))
        self.pooling4 = nn.MaxPool2d(2, 2)
                #定义全连接层
        self.fc1 = nn.Sequential(
                nn.Linear(924, 500),
                nn.ReLU(True))
            self.fc2 = nn.Sequential(
                nn.Linear(500, 100),
                nn.ReLU(True))
            self.fc3 = nn.Sequential(
                nn.Linear(100, 1))
    #定义数据传送方向
    def forward(self, x):
        #依次经过卷积层和池化层
        x = self.pooling1(self.conv1(x))
```

```python
        x = self.pooling2(self.conv2(x))
        x = self.pooling3(self.conv3(x))
        x = self.pooling4(self.conv4(x))
        # 将数据拉平为一维数据
        x = x.view(x.shape[0], -1)
        # 经过全连接层
        x = self.fc1(x)
        x = self.fc2(x)
        x = self.fc3(x)
        x = x.float()
        return x
```

2. 模型训练

确定模型结构之后进行训练。由于在构建数据集时,将口罩类别分为普通口罩和 N95 口罩,因此这是一个二分类问题,选择二元交叉熵作为损失函数,并在 PyTorch 中实现,相关代码如下:

```python
criterion = nn.BCEWithLogitsLoss()                    # 定义二元交叉熵损失函数
```

确定优化器,采用梯度下降方法 Adam 算法优化模型参数,加入 L2 正则化防止模型出现过拟合,PyTorch 中体现在优化器的参数 weight_decay 上,设置的正则化系数为 0.0001,学习率设置为 0.001。相关代码如下:

```python
optimizer = torch.optim.Adam(net.parameters(), lr = 0.001, weight_decay = 1e-4)
                                                       # 使用 Adam 作为优化器
```

确定损失函数和优化器之后,将网络实例化,开始训练网络。为了便于复用,将训练过程封装为一个函数,名为 train,输入参数依次为网络、训练集、测试集、迭代次数、优化器和损失函数。相关代码如下:

```python
def train(net, train_data, valid_data, num_epochs, optimizer, criterion):
    if torch.cuda.is_available():
        # 如果 GPU 可用,将网络加载进 GPU
        net = net.cuda()
    # 获取当前时间作为起点
    prev_time = datetime.now()
    # 开始迭代训练
    for epoch in range(num_epochs):
        train_loss = 0
        train_acc = 0
        net = net.train()
        # 加载训练数据
        for im, label in train_data:
            if torch.cuda.is_available():
                im = im.cuda()
                label = torch.unsqueeze(label, 1) # (bs, 3, h, w)
```

```python
            label = label.float().cuda()  # (bs, h, w)
        else:
            im = im
            label = torch.unsqueeze(label, 1)
            label = label.float()
# 前向传播
        output = net(im)
        loss = criterion(output, label)
# 反向传播
        optimizer.zero_grad()
        loss.backward()
        optimizer.step()
# 计算损失和准确率
        train_loss += loss.data
        train_acc += get_acc(output, label)
# 获取训练一个 epoch 结束的时间
    cur_time = datetime.now()
# 计算训练一个 epoch 用时
    h, remainder = divmod((cur_time - prev_time).seconds, 3600)
    m, s = divmod(remainder, 60)
    time_str = "Time %02d:%02d:%02d" % (h, m, s)
# 测试训练集准确度
    if valid_data is not None:                        # 获取模型参数
        valid_loss = 0
        valid_acc = 0
        net = net.eval()
        for im, label in valid_data:
            if torch.cuda.is_available():
                im = im.cuda()
                label = torch.unsqueeze(label, 1)  # (bs, 3, h, w)
                label = label.float().cuda()
            else:
                im = im
                label = torch.unsqueeze(label, 1)
                label = label.float()
            output = net(im)
            loss = criterion(output, label)           # 损失
            valid_loss += loss.data
            valid_acc += get_acc(output, label)       # 准确度
        epoch_str = (
            "Epoch %d. Train Loss: %f, Train Acc: %f, Valid Loss: %f, Valid Acc: %f, "
            % (epoch, train_loss / len(train_data),
               train_acc / len(train_data), valid_loss / len(valid_data),
               valid_acc / len(valid_data)))
    else:
        epoch_str = ("Epoch %d. Train Loss: %f, Train Acc: %f, " %
                     (epoch, train_loss / len(train_data),
```

```
                            train_acc / len(train_data)))
            prev_time = cur_time
            print(epoch_str + time_str)
```

训练输出结果如图 15-3 所示。

```
epoch: 0, Train Loss: 1.307801, Train Acc: 0.673874, Eval Loss: 0.492519, Eval Acc: 0.868826
epoch: 1, Train Loss: 0.382268, Train Acc: 0.893573, Eval Loss: 0.325435, Eval Acc: 0.906844
epoch: 2, Train Loss: 0.298539, Train Acc: 0.914062, Eval Loss: 0.265722, Eval Acc: 0.923619
epoch: 3, Train Loss: 0.247401, Train Acc: 0.928871, Eval Loss: 0.219947, Eval Acc: 0.937483
epoch: 4, Train Loss: 0.206424, Train Acc: 0.939666, Eval Loss: 0.182583, Eval Acc: 0.947517
epoch: 5, Train Loss: 0.173956, Train Acc: 0.949544, Eval Loss: 0.153908, Eval Acc: 0.955624
epoch: 6, Train Loss: 0.148721, Train Acc: 0.956890, Eval Loss: 0.133055, Eval Acc: 0.961704
epoch: 7, Train Loss: 0.130514, Train Acc: 0.961887, Eval Loss: 0.116623, Eval Acc: 0.966318
epoch: 8, Train Loss: 0.116745, Train Acc: 0.965902, Eval Loss: 0.103472, Eval Acc: 0.970133
epoch: 9, Train Loss: 0.105424, Train Acc: 0.968733, Eval Loss: 0.092620, Eval Acc: 0.973270
epoch: 10, Train Loss: 0.095491, Train Acc: 0.971932, Eval Loss: 0.083275, Eval Acc: 0.976002
epoch: 11, Train Loss: 0.086400, Train Acc: 0.974913, Eval Loss: 0.075457, Eval Acc: 0.978434
epoch: 12, Train Loss: 0.078967, Train Acc: 0.976479, Eval Loss: 0.068364, Eval Acc: 0.980899
epoch: 13, Train Loss: 0.073167, Train Acc: 0.978545, Eval Loss: 0.062272, Eval Acc: 0.982554
epoch: 14, Train Loss: 0.067566, Train Acc: 0.980011, Eval Loss: 0.056764, Eval Acc: 0.984153
epoch: 15, Train Loss: 0.062223, Train Acc: 0.981293, Eval Loss: 0.051842, Eval Acc: 0.985935
epoch: 16, Train Loss: 0.056789, Train Acc: 0.983725, Eval Loss: 0.047402, Eval Acc: 0.987151
epoch: 17, Train Loss: 0.051871, Train Acc: 0.985291, Eval Loss: 0.043244, Eval Acc: 0.988167
epoch: 18, Train Loss: 0.048869, Train Acc: 0.985624, Eval Loss: 0.039567, Eval Acc: 0.989517
epoch: 19, Train Loss: 0.045717, Train Acc: 0.986824, Eval Loss: 0.036309, Eval Acc: 0.990633
```

图 15-3　训练结果

3. 模型保存

在 PyTorch 中,有两种保存方式:保存完整的模型和只保存模型的参数。PyTorch 官方建议保存模型参数,相关代码如下:

```
torch.save(net.state_dict(), 'params.pkl')
```

15.3.3　页面显示和视频流输入

模型训练完成后,需要使用 PyQt 和 OpenCV 进行页面显示、视频流输入及读取。

1. 页面显示

定义整体界面布局以及两个按钮,作用分别是打开相机进行相关任务和关闭相机。截取视频画面通过定时器完成,定时器设置计时 30ms,每 30ms 从摄像头中取一帧页面进行显示。相关代码如下:

```
def set_ui(self):
    self.__layout_main = QtWidgets.QHBoxLayout()        #总布局
    self.__layout_fun_button = QtWidgets.QVBoxLayout()  #按键布局
    self.__layout_data_show = QtWidgets.QVBoxLayout()   #数据(视频)显示布局
    self.button_open_camera = QtWidgets.QPushButton('打开相机')
    self.button_close = QtWidgets.QPushButton('退出')
    self.button_open_camera.setMinimumHeight(50)
```

```python
                self.button_close.setMinimumHeight(50)
                self.button = QtWidgets.QPushButton("click")
                self.button_close.move(10, 100)                    # 移动按键
                # 信息显示
                self.label_show_camera = QtWidgets.QLabel()        # 定义显示视频的 Label
            self.label_show_camera.setFixedSize(641,481)           # 给显示视频的 Label 设置大小
                # 把按键加入布局中
                self.__layout_fun_button.addWidget(self.button_open_camera)
                self.__layout_fun_button.addWidget(self.button_close)
                self.__layout_fun_button.addWidget(self.button)
                self.__layout_main.addLayout(self.__layout_fun_button)
                self.__layout_main.addWidget(self.label_show_camera)
                self.setLayout(self.__layout_main)
                # 初始化所有槽函数
        def slot_init(self):
                self.button_open_camera.clicked.connect(
                self.button_open_camera_clicked)
                self.timer_camera.timeout.connect(self.show_camera)
                self.button_close.clicked.connect(self.close)
                # 槽函数之一
        def button_open_camera_clicked(self):
                if self.timer_camera.isActive() == False:
                        flag = self.cap.open(self.CAM_NUM)
                        if flag == False:   # flag 表示 open()成不成功
                                msg = QtWidgets.QMessageBox.warning(self, 'warning', "请检查相机与计算机
是否连接正确", buttons = QtWidgets.QMessageBox.Ok)
                        else:
                                self.timer_camera.start(30)
                                # 定时器设置计时 30ms,每 30ms 从摄像头中取一帧显示
                                self.button_open_camera.setText('关闭相机')
                else:
                        self.timer_camera.stop()                   # 关闭定时器
                        self.cap.release()                         # 释放视频流
                        self.label_show_camera.clear()             # 清空视频显示区域
                        self.button_open_camera.setText('打开相机')
class SecondWindow(QtWidgets.QWidget):
    def __init__(self, parent = None):
        super().__init__(parent)
    def set_but(self):
        self.btn.setText("click")
if __name__ == '__main__':                                         # 主函数
    app = QtWidgets.QApplication(sys.argv)
    ui = Ui_MainWindow()                                           # 实例化 Ui_MainWindow
    ui.show()
    sys.exit(app.exec_())
```

2. 视频流输入及读取

通过设置定时器每 30ms 从视频中取一帧显示，并且定义视频大小、色彩转换，把读取到的视频数据变成 QImage 形式。相关代码如下：

```python
class Ui_MainWindow(QtWidgets.QWidget):                  # 界面窗口类
    def __init__(self, parent = None):
        super().__init__(parent)                         # 父类的构造函数
        self.timer_camera = QtCore.QTimer()              # 定义定时器,用于控制显示视频的帧率
        self.cap = cv2.VideoCapture()                    # 视频流
        self.CAM_NUM = 0
        self.set_ui()                                    # 初始化程序界面
        self.slot_init()                                 # 初始化槽函数
    def show_camera(self):
        flag, self.image = self.cap.read()               # 从视频流中读取数据
        show = cv2.resize(self.image, (640, 480))
        show = cv2.cvtColor(show, cv2.COLOR_BGR2RGB)     # 视频色彩转换回 RGB 颜色
        showImage = QtGui.QImage(show.data, show.shape[1], show.shape[0],
                 QtGui.QImage.Format_RGB888)             # 把读取到的视频数据变成 QImage 形式
        self.label_show_camera.setPixmap(QtGui.QPixmap.fromImage(showImage))
```

15.3.4 模型生成

把训练好的模型放入 project/model 文件夹内。在 pytorch_inference 脚本中通过 PyTorch 导入模型及代码：

```python
model = load_pytorch_model('models/face_mask_detection.pth')
```

在 pytorch_inference 脚本中调用模型进行检测，调用代码：

```python
y_bboxes, y_scores, = model.forward(torch.tensor(img_arr).float())
```

直接调用模型中的 forward 方法完成一次前向传播过程，得到预测的 bunding_box 和预测准确率值。相关代码如下：

```python
#定义一个检测按钮
self.detect = QtWidgets.QPushButton('检测')
#将按钮加到界面布局中
self.__layout_fun_button.addWidget(self.detect)
#定义单击检测按钮后的行为,调用 Detection 函数
self.detect.clicked.connect(self.Detection)
#定义 Detection 函数
def Detection(self):
    #获取图片类别:佩戴口罩或未佩戴口罩
    class_info = self.get_class()
    #根据图片类别调用不同的提醒弹窗函数
    if class_info == 0:
```

```python
            self.open_second_ui()
        else:
            self.open_third_ui()
# 定义获取图片类别函数 getClass
def getClass(self):
    flag, self.image = self.cap.read()                    # 从视频流中读取数据
    show = cv2.resize(self.image, (640, 480))             # 帧的大小重新设置为 640×480
    show = cv2.cvtColor(show, cv2.COLOR_BGR2RGB)          # 视频色彩转换回 RGB 颜色
    output_info = output_info = inference(show,
                                          iou_thresh = 0.5,
                                          target_shape = (260, 260),
                                          draw_result = True,
                                          show_result = False)
    return output_info[0][0]
# 定义两个提醒弹窗函数
class SecondWindow(QtWidgets.QWidget):
    def __init__(self):
        super(SecondWindow, self).__init__()
        self.resize(400, 300)
        self.setWindowTitle("允许进入")
        self.label = QtWidgets.QLabel(self)
        self.label.setGeometry(QtCore.QRect(100, 80, 181, 91))
        self.label.setText("您佩戴的口罩符合标准,允许进入")
class thirdWindow(QtWidgets.QWidget):
    def __init__(self):
        super(thirdWindow, self).__init__()
        self.resize(400, 300)
        self.setWindowTitle("禁止进入")
        self.label = QtWidgets.QLabel(self)
        self.label.setGeometry(QtCore.QRect(100, 80, 200, 91))
        self.label.setText("您未佩戴口罩,禁止进入")
# inference 函数
def inference(image,
              conf_thresh = 0.5,
              iou_thresh = 0.4,
              target_shape = (160, 160),
              draw_result = True,
              show_result = True
              ):
    output_info = []
    height, width, _ = image.shape
    image_resized = cv2.resize(image, target_shape)
    image_np = image_resized / 255.0  # 归一化到 0~1
    image_exp = np.expand_dims(image_np, axis = 0)
    image_transposed = image_exp.transpose((0, 3, 1, 2))
    y_bboxes_output, y_cls_output = pytorch_inference(model, image_transposed)
    # 删除批次尺寸,因为批次始终为 1
```

```python
        y_bboxes = decode_bbox(anchors_exp, y_bboxes_output)[0]
        y_cls = y_cls_output[0]
        # 为了加快速度，请执行单 NMS，而不是多 NMS
        bbox_max_scores = np.max(y_cls, axis=1)
        bbox_max_score_classes = np.argmax(y_cls, axis=1)
        # keep_idx 是 NMS 之后的活动边界框
        keep_idxs = single_class_non_max_suppression(y_bboxes,
                                                    bbox_max_scores,
                                                    conf_thresh=conf_thresh,
                                                    iou_thresh=iou_thresh,
                                                    )
    for idx in keep_idxs:
        conf = float(bbox_max_scores[idx])
        class_id = bbox_max_score_classes[idx]
        bbox = y_bboxes[idx]
        # 裁剪坐标，避免该值超出图像边界
        xmin = max(0, int(bbox[0] * width))
        ymin = max(0, int(bbox[1] * height))
        xmax = min(int(bbox[2] * width), width)
        ymax = min(int(bbox[3] * height), height)
        if draw_result:
            if class_id == 0:
                color = (0, 255, 0)
            else:
                color = (255, 0, 0)
            cv2.rectangle(image, (xmin, ymin), (xmax, ymax), color, 2)     # 矩形框
            cv2.putText(image, id2class[class_id], (xmin + 2, ymin - 2),
                        cv2.FONT_HERSHEY_SIMPLEX, 0.8, color)                # 输出文本
        output_info.append([class_id, conf, xmin, ymin, xmax, ymax])
    # if show_result:
    #     Image.fromarray(image).show()
    return output_info
```

15.4 系统测试

本部分包括训练准确率及测试效果。

15.4.1 训练准确率

训练时的准确率和损失情况如图 15-4 所示。随着训练次数的增加，模型在数据上的损失值逐渐降低并趋于收敛，而准确率则逐渐增加，最终达到 98.6%，意味着模型训练取得了比较好的效果。

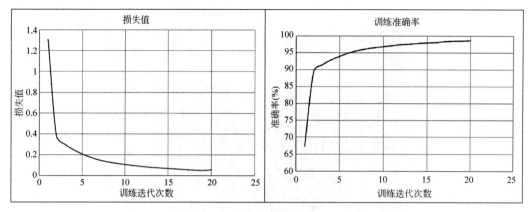

图 15-4　训练准确率和损失情况

15.4.2　测试效果

将程序脚本及模型下载后,通过两种方式运行:一是用 PyCharm 打开程序项目,运行 main.py 脚本;二是在命令行中进入项目文件夹,通过 python main.py 命令运行。初始界面如图 15-5 所示。

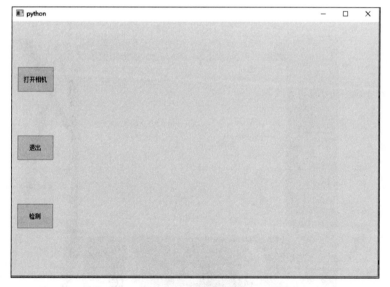

图 15-5　程序初始界面

单击"打开相机"按钮,程序将调用便携式计算机的摄像头,并实时显示画面,在画面中出现人脸检测框,如图 15-6 所示。

图 15-6 打开相机界面

开启相机,单击"检测"按钮,对是否佩戴口罩进行检测,并通过弹窗分别给出对应提示,如图 15-7 所示。

图 15-7 检测提示界面

佩戴口罩、未佩戴口罩两个示例如图15-8所示。

图15-8　测试示例

项目 16 垃圾分类微信小程序

PROJECT 16

本项目基于 TensorFlow 和 VGG-16 卷积神经网络训练垃圾分类模型,通过服务器实现分类模型移植到移动端,并在微信小程序中进行应用。

16.1 总体设计

本部分包括系统整体结构和系统流程。

16.1.1 系统整体结构

系统整体结构如图 16-1 所示。

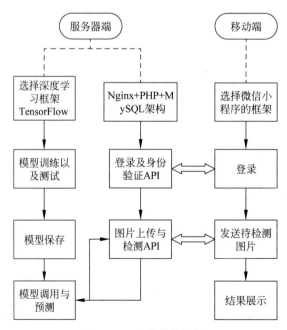

图 16-1 系统整体结构

16.1.2 系统流程

系统流程如图 16-2 所示，小程序后台系统结构如图 16-3 所示。

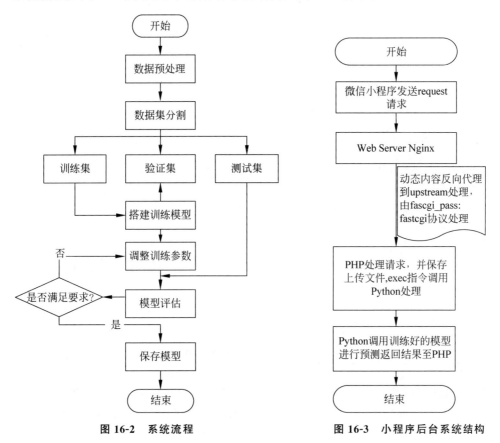

图 16-2　系统流程　　　　　　　　图 16-3　小程序后台系统结构

16.2　运行环境

本部分包括 Python 环境、TensorFlow 环境、微信小程序及后台服务器等环境。

16.2.1　Python 环境

建议安装 Anconda 34.2.0 版本，Windows 版本下的 TensorFlow 暂时不支持 Python 2.7，需要安装 Python 3.7。

16.2.2　TensorFlow 环境

由于 Keras 默认 TensorFlow 为后端，本项目采用 TensorFlow 作为 Keras 后端。打开 Anaconda Prompt。

（1）在 C:\Users\Lenovo\Anaconda3\envs 新建环境目录，例如，DeepLearning。

（2）在命令行窗口输入以下命令激活 DeepLearning：

```
C:\Users\Lenovo\Anaconda3 > activate DeepLearning
```

（3）输入以下命令安装 tensorflowgpu：

```
conda install tensorflow-gpu
```

Anaconda 会自动显示匹配所需的安装包，输入 y，安装完成。

（4）安装 tensorflow-gpu 后，在新建的 DeepLearning 环境中安装基础库。

（5）在 CMD 命令行或者 Powershell 中安装 Keras，输入以下命令：

```
pip install keras
```

验证 Keras 是否安装成功，在命令行中输入 Python 命令变成命令行环境后输入：

```
import keras
```

未报错，则 Keras 成功安装。

16.2.3　微信小程序及后台服务器环境

（1）下载微信开发者工具 v1.02.2003112，扫描二维码登录后进行程序编写。

（2）服务器配置为 2 核 CPU、4GB 内存、CentOS 7.6 64 位操作系统、1Mbps 带宽。

（3）配置出入安全组，开放端口有 22、80、443、3306、5000。

（4）后台需要安装 Nginx、编译工具和库文件，首先运行：

```
yum -y install make zlib zlib-devel gcc-c++ libtool openssl openssl-devel
```

下载 Nginx 源文件：

```
wget http://nginx.org/download/nginx-1.6.2.tar.gz
```

解压并编译安装。

（5）安装 PHP，相关代码如下：

```
rpm -Uvh https://dl.fedoraproject.org/pub/epel/epel-release-latest-7.noarch.rpm
rpm -Uvh https://mirror.webtatic.com/yum/el7/webtatic-release.rpm
yum install php72w php72w-cli php72w-common php72w-devel php72w-embedded php72w-fpm php72w-gd php72w-mbstring php72w-mysqlnd php72w-opcache php72w-pdo php72w-xml
```

（6）安装 Python 环境。

16.3 模块实现

本项目包括4个模块:数据预处理、创建模型与编译、模型训练及保存、模型生成。下面分别给出各模块的功能介绍及相关代码。

16.3.1 数据预处理

数据下载地址为:https://pan.baidu.com/s/1VhJnSRKlpUOHRp1P4MZ0pQ,提取码:8dsd。该数据集包含10种垃圾:纸箱、玻璃、金属、纸、塑料、其他废品、LED灯泡、瓜子、白菜、纽扣电池,每种垃圾数据集大约包含300~400张图片,压缩后的尺寸为512×384。筛选出具有明显特征的(主要是单一物体、单一背景)垃圾图片,数据集信息如表16-1所示。

表 16-1 数据集信息

序号	中文名	英文名	数据集大小
1	玻璃	glass	497张图片
2	纸	paper	590张图片
3	硬纸板	cardboard	400张图片
4	塑料	plastic	479张图片
5	金属	metal	407张图片
6	一般垃圾	trash	134张图片
7	纽扣电池	button battery	171张图片
8	瓜子	melon seed	228张图片
9	白菜	Chinese cabbage	258张图片
10	LED灯泡	LED lamp	240张图片

首先,对图片进行缩放和翻转,并压缩为(150,150);其次,把10%的数据作为测试集,训练数据进行分批,每批16张图片。

相关代码如下:

```
def processing_data(data_path):
    #数据处理,data_path:数据集路径,返回处理后的训练集数据、测试集数据
    train_data = ImageDataGenerator(
        #对图片的每个像素值均乘上放缩因子,把像素值放缩到0和1之间有利于模型的收敛
        rescale=1. / 225,
        #浮点数,剪切强度(逆时针方向的剪切变换角度)
        shear_range=0.1,
        #随机缩放的幅度,若为浮点数,则相当于[lower,upper] = [1 - zoom_range, 1 + zoom_range]
        zoom_range=0.1,
        #浮点数,图片宽度的某个比例,数据提升时图片水平偏移的幅度
```

```
                    width_shift_range = 0.1,
                    #浮点数,图片高度的某个比例,数据提升时图片竖直偏移的幅度
                    height_shift_range = 0.1,
                    #布尔值,进行随机水平翻转
                    horizontal_flip = True,
                    #布尔值,进行随机垂直翻转
                    vertical_flip = True,
                    #在0和1之间浮动,用作验证集的训练数据的比值
                    validation_split = 0.1
        )
                    #生成测试集,可以参考生成训练集的方法
        validation_data = ImageDataGenerator(
                    rescale = 1. / 255,
                    validation_split = 0.1)
        train_generator = train_data.flow_from_directory(
                    #提供的路径下面需要有子目录
                    data_path,
                    #整数元组(height,width),默认为(256,256),所有图像调整为该尺寸
                    target_size = (150, 150),
                    #一批数据的大小
                    batch_size = 16,
                    #categorical, binary, sparse, input 或 None 之一
                    #默认为 categorical,返回 One-Hot 编码标签
                    class_mode = 'categorical',
                    #数据子集 ("training" 或 "validation")
                    subset = 'training',
                    seed = 0)
        validation_generator = validation_data.flow_from_directory(
                    data_path,
                    target_size = (150, 150),
                    batch_size = 16,
                    class_mode = 'categorical',
                    subset = 'validation',
                    seed = 0)
        return train_generator, validation_generator
```

16.3.2　创建模型与编译

　　数据加载进模型后定义模型结构,并优化损失函数。直接调用 VGG-16 模型作为卷积神经网络,包括 13 个卷积层、3 个全连接层、5 个池化层,后接全连接层,神经元数目为 256,连接 ReLU 激活函数,再接全连接层,神经元个数为 6,得到 6 维的特征向量,用于 6 个垃圾的分类训练,输入 Softmax 层进行分类,得到分类结果的概率输出。

　　由于批量梯度下降法在更新每一个参数时需要所有的训练样本,训练过程会随着样本数量的加大而变得异常缓慢。

构建优化器，optimizer＝SGD(lr＝1e-3,momentum＝0.9)，采用随机梯度下降学习算法。对数据分批训练，一批 16 张图片，分 20 轮训练。相关代码如下：

```
def model(train_generator, validation_generator, save_model_path):
    vgg16_model = VGG16(weights = 'imagenet', include_top = False, input_shape = (150,150,3))
    # 训练模型定义
    top_model = Sequential()
    top_model.add(Flatten(input_shape = vgg16_model.output_shape[1:]))
    top_model.add(Dense(256, activation = 'relu'))
    top_model.add(Dropout(0.5))
    top_model.add(Dense(10, activation = 'softmax'))
    model = Sequential()
    model.add(vgg16_model)
    model.add(top_model)
    # 编译模型，采用 compile()函数：https://keras.io/models/model/#compile
    model.compile(
            # 优化器主要有 Adam、SGD、rmsprop 等方式
            optimizer = SGD(lr = 1e-3, momentum = 0.9),
            # 损失函数，多分类采用 categorical_crossentropy
            loss = 'categorical_crossentropy',
            # 除了损失函数值之外的特定指标,分类问题一般都是准确率
            metrics = ['accuracy'])
    model.fit_generator(
            # 一个生成器或 Sequence 对象的实例
            generator = train_generator,
            # epochs：整数,数据的迭代总轮数
            epochs = 20,
            # 一个 epoch 包含的步数,应该等于数据集的样本数量除以批量大小
            steps_per_epoch = 2704 // 16,
            # 验证集
            validation_data = validation_generator,
            # 在验证集上,一个 epoch 包含的步数,应该等于数据集的样本数量除以批量大小
            validation_steps = 272 //16,
            )
    model.save(save_model_path)
    return model
```

16.3.3　模型训练及保存

在定义模型架构并编译之后，通过训练集训练模型，使模型识别图片。这里，使用训练集和测试集拟合并保存模型。

```
def evaluate_mode(validation_generator, save_model_path):
    # 加载模型
    model = load_model('results/knn.h5')
    # 获取验证集的损失率和准确率
```

```python
loss, accuracy = model.evaluate_generator(validation_generator)
print("\nLoss: %.2f, Accuracy: %.2f%%" % (loss, accuracy * 100))
```

通过观察训练集和测试集的损失函数、准确率的大小来评估模型的训练程度,进行模型训练的进一步决策。一般来说,训练集和测试集的损失函数(或准确率)不变且基本相等时为模型训练的最佳状态。

```python
#把图片转换为Numpy数组
#导入图片
img = ('test.jpg')
pil_im = Image.open(img)
img = Image.open(img)
"img.show()"
#对图片预处理
img = img.resize((150, 150))
img = image.img_to_array(img)
x = np.expand_dims(img, axis = 0)
#导入模型
model_path = 'results/knn.h5'
model = load_model(model_path)
#模型预测
y = model.predict(x)
```

16.3.4 模型生成

本部分包括后台服务器和微信小程序前端两部分。

1. 后台服务器

后台服务器相关代码如下:

```
Nginx 部分配置:
    http{
    server{
        listen 443 ssl;                            //因为采用https协议,所以需要预定义443端口
        server_name www.xxxxx.com;                 //此处填写域名
        ssl_certificate /etc/nginx/certs/3558126.pem;      //填写SSL证书位置
        ssl_certificate_key /etc/nginx/certs/3558126.key;
        ssl_session_timeout 5m;
        ssl_ciphers ECDHE-RSA-AES128-GCM-SHA256:ECDHE:ECDH:AES:HIGH:!NULL:!aNULL:!MD5:!ADH:!RC4;
        ssl_protocols TLSv1 TLSv1.1 TLSv1.2;
        ssl_prefer_server_ciphers on;
        location ~ * \.php$ {
            root /;                                //设置根目录
            fastcgi_pass 127.0.0.1:9000;           //设置fastcgi服务器地址
            fastcgi_index index.php;
```

```
                fastcgi_param SCRIPT_FILENAME /$fastcgi_script_name;
                include fastcgi_params;                //设置传递参数
        }
        location / {
            proxy_pass http://localhost:7777;
            proxy_http_version 1.1;
            proxy_read_timeout 600s;
            proxy_send_timeout 12s;
                root /wepro/;
                }
}
```

图片接口 API 相关代码如下:

```php
<?php
header('Content-Type:text/html; charset = utf-8');        //定义 php 头,定义文本类型
$allowedExts = array("gif","jpeg","jpg","png");
$temp = explode(".", $_FILES["file"]["name"]);            //以点为分割符号分割文件名
$extension = end($temp);                                   //获取文件后缀
$form_data = $_FILES['file']['tmp_name'];
if ((($_FILES["file"]["type"] == "image/jpeg")            //检验文件格式和大小
    ||($_FILES["file"]["type"] == "image/jpg")
    ||($_FILES["file"]["type"] == "image/pjpeg")
    ||($_FILES["file"]["type"] == "image/png"))
    &&($_FILES["file"]["size"]< 20 * 1024 * 1024)
    ) {
    $img_url = "/wepro/imgsto/". $_FILES["file"]["name"];
    move_uploaded_file($_FILES["file"]["tmp_name"],//存储文件至相应路径
$img_url);
    chmod($img_url, 777);                                  //使图片为可读可写权限
    $tmp = "/wepro/tmp/tmp.txt";
    $locale = 'de_DE.UTF-8';                               //设置编码格式
    setlocale(LC_ALL, $locale);
    putenv('LC_ALL = '. $locale);
    # echo exec('locale charmap');
    if (($myfile = fopen($tmp, 'w + ')) === false) {
        echo "fail create file";
        exit();
    }                                                      //创建识别结果的文本文件
    fclose($myfile);
    chmod($tmp, 0777);
    exec("./py/trah_env/bin/python3 hello.py $img_url >> /wepro/tmp/tmp.txt", $callback,
$status);                                                  //运行调用模型的 Python 文件
# print_r($callback);
$timeflag = true;                                          //设置判断是否完成识别
while ($timeflag) {
    sleep(1);                                              //设置休眠时间,防止占用大量服务器资源
```

```php
                if (filesize($tmp)!= 0) {                    //判断tmp文件的大小
                    if (($result = fopen($tmp, 'r')) === false) {
                        echo "fail open";
                        exit();
                    }
                    if (($content = fread($result, filesize($tmp))) === false) {
                        echo "fail read";
                    }
                    echo $content;                            //返回识别结果
                    fclose($result);
                    $timeflag = false;                        //设置识别完成,结束循环
#                   echo file_get_contents($tmp);}
                }
        }
    }
    unlink($tmp);                                             //删除识别结果文件
?>
```

2. 微信小程序前端

微信小程序前端相关代码如下：

```javascript
//index.js
//获取应用实例
const app = getApp()
Page({
  data: {
    progress_txt: '正在识别中...',               //进度文本
    judge: true,                                  //判断是否完成识别
    count: 0,                                     //设置计数器初始为0
    countTimer: null,                             //设置定时器初始为null
    userInfo: {},
    hasUserInfo: false,
    canIUse: wx.canIUse('button.open-type.getUserInfo'),
    tempImagePath: []
  },
  drawProgressbg: function() {
    //使用wx.createContext获取绘图上下文context
    var ctx = wx.createCanvasContext('canvasProgressbg', this)
    ctx.setLineWidth(4);                          //设置圆环宽度
    ctx.setStrokeStyle('#20183b');                //设置圆环颜色
    ctx.setLineCap('round');                      //设置圆环端点形状
    ctx.beginPath();                              //开始一个新的路径
    ctx.arc(110, 110, 100, 0, 2 * Math.PI, false);
    //设置圆心(110,110),半径为100的路径到当前路径
    ctx.stroke();                                 //对当前路径进行描边
    ctx.draw();
```

```
  },
  drawCircle: function(step) {
    var context = wx.createCanvasContext('canvasProgress', this);
    //设置渐变
    var gradient = context.createLinearGradient(200, 100, 100, 200);
    gradient.addColorStop("0", "#2661DD");
    gradient.addColorStop("0.5", "#40ED94");
    gradient.addColorStop("1.0", "#5956CC");
    context.setLineWidth(10);
    context.setStrokeStyle(gradient);
    context.setLineCap('round')
    context.beginPath();
    //参数 step 为绘制的圆环周长
    context.arc(110, 110, 100, -Math.PI / 2, step * Math.PI - Math.PI / 2, false);
    context.stroke();
    context.draw()
  },
  countInterval: function() {
    //设置倒计时,定时器每 100ms 执行一次,计数器 count+1,耗时 6s 绘一圈
    this.countTimer = setInterval(() => {
      if (this.data.count <= 400) {
        //绘制彩色圆环进度条
        this.drawCircle(this.data.count / (400 / 2))
        this.data.count++;
      } else {
        this.setData({
          progress_txt: "识别成功",
          judge: true
        });
        clearInterval(this.countTimer);
      }
    }, 100)
  },
  onLoad: function() {
    if (app.globalData.userInfo) {                    //设置用户信息
      this.setData({
        userInfo: app.globalData.userInfo,
        hasUserInfo: true
      })
    } else if (this.data.canIUse) {
      //由于 getUserInfo 是网络请求,可能会在页面 onLoad 之后才返回
      //所以此处加入 callback 以防止这种情况
      app.userInfoReadyCallback = res => {
        this.setData({
          userInfo: res.userInfo,
          hasUserInfo: true
        })
```

```javascript
      }
    } else {
      //在没有 open-type = getUserInfo 版本的兼容处理
      wx.getUserInfo({
        success: res => {
          app.globalData.userInfo = res.userInfo
          this.setData({
            userInfo: res.userInfo,
            hasUserInfo: true
          })
        }
      })
    }
  },
  choose_image() {                                    //图片处理函数
    var that = this
    wx.chooseImage({
      count: 1,
      sizeType: ['original', 'compressed'],           //选择图片质量
      sourceType: ['album', 'camera'],                //选择拍照或者从图库上传
      success: (res) => {
        //tempFilePath 可以作为 img 标签的 src 属性显示图片
        const tempFilePaths = res.tempFilePaths
        wx.setStorageSync('img', tempFilePaths)       //缓存图片地址方便后续调用
        this.setData({
          tempImagePath: res.tempFilePaths,
          judge: false
        })
        that.drawProgressbg();                        //开始运行进度条进程
        that.countInterval()
        wx.uploadFile({                               //上传图片
          filePath: this.data.tempImagePath[0],
          name: 'file',
          url: 'https://www.xxxxxx.com/wepro/timg_upload.php',
                                                      //上传文件接口地址
          formatData: {
            'method': 'addImg'
          },
          success(res) {
            that.setData({
              judge: false
            })
            var tmp = res.data
            wx.redirectTo({
              url: '/pages/result/result?class = ' + tmp    //跳转至结果页面
            })
          },
```

```
          fail(e) {
            console.log(e)
          }
        })
      }
    })
  },
  error(e) {
    console.log(e.detail)
  },
}
})
// result.js
var message = require('../../utils/trash.js');
Page({
  //页面的初始数据
  data: {
    judge: false,
    class: "",
    result: "",
    info: "",
    img: ""
  },
  compara: function(e) {                          //对比识别结果 判断垃圾属于哪个分类
    var a = this.data.class
    if (a == e) {
      return true
    } else {
      return false
    }
  },
  onShow: function() {
    var tmp = this.data.class                     //获取页面传递的值
    var dic = message.dict                        //设置分类词典
    var that = this;
    var judge = that.data.class
    for (var index in dic) {                      //检测函数
      var trash_class = index
      var tmp = dic[`${index}`]
      var flag = true
      for (var c in tmp['class']) {
        console.log(tmp['class'][c])
        if (judge == tmp['class'][c]) {           //判断此垃圾属于4种中的哪一种
          console.log(trash_class)                //停止循环,设置基本页面信息
          that.setData({
            result: index,
            info: tmp['info']
```

```
            })
            return
          }
        }
      }
    },
    onLoad: function(options) {
      var tmp = wx.getStorageSync('img')[0]        //获取识别图片的信息
      var tmp2 = options.class.replace(/[\r][\n]/g, "").trim();
                                                   //处理传值的信息,去除空格换行
      this.setData({                               //保存基本信息
        class: tmp2,
        img: tmp
      })
    },
})
```

16.4 系统测试

本部分包括训练准确率、测试效果及模型应用。

16.4.1 训练准确率

训练准确率如图 16-4 所示。

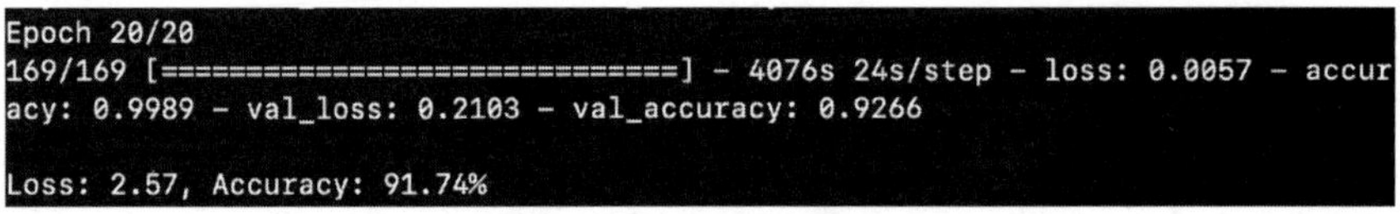

```
Epoch 20/20
169/169 [==============================] - 4076s 24s/step - loss: 0.0057 - accur
acy: 0.9989 - val_loss: 0.2103 - val_accuracy: 0.9266
Loss: 2.57, Accuracy: 91.74%
```

图 16-4　训练准确率

16.4.2 测试效果

将数据代入模型进行测试,分类的标签与原始数据进行对比,如图 16-5 所示。

16.4.3 模型应用

本部分包括程序下载运行、应用使用说明和测试结果示例。

1. 程序下载运行

微信搜索"基于图像识别的你是什么垃圾"小程序,单击即可进入。

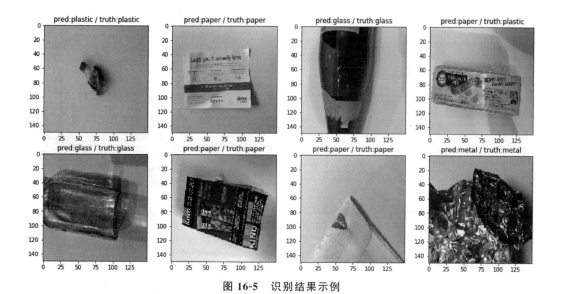

图 16-5　识别结果示例

2．应用使用说明

打开 App，初始界面如图 16-6 所示。

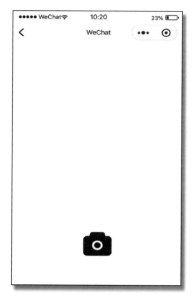

图 16-6　应用初始界面

界面采用简约的设计，只有一个按钮。单击按钮后可以上传图片进行垃圾分类识别，如图 16-7 所示。

上传图片后出现进度条对用户进行反馈，识别完成后跳转至结果页面。

3. 测试结果示例

在结果页面单击加号按钮可以返回首页,如图 16-8 所示。

图 16-7　预测结果进度界面　　　　　图 16-8　测试结果示例

项目 17 基于 OpenCV 的人脸识别程序

PROJECT 17

本项目使用 OpenCV 进行数据采集,在 TensorFlow 环境中建立并训练模型,最终实现人脸识别功能。

17.1 总体设计

本部分主要包括系统整体结构和系统流程。

17.1.1 系统整体结构

系统整体结构如图 17-1 所示。

图 17-1 系统整体结构

17.1.2 系统流程

系统流程如图 17-2 所示,OpenCV 工作流程如图 17-3 所示。

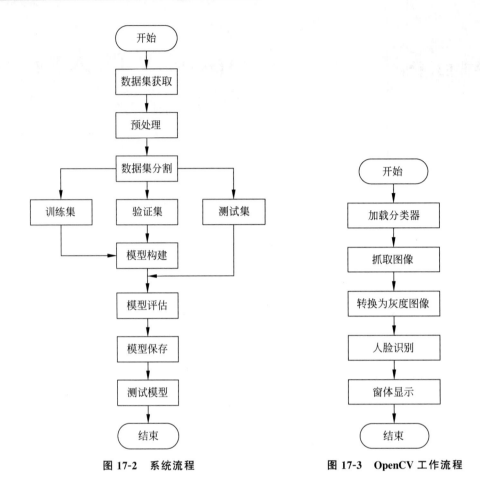

图 17-2　系统流程　　　　　图 17-3　OpenCV 工作流程

17.2　运行环境

本部分包括 Python 环境和 TensorFlow 环境。

17.2.1　Python 环境

需要 Python 3.6 及以上配置,在 Windows 环境下推荐下载 Anaconda 完成 Python 所需的配置,下载地址为:https://www.anaconda.com/。

17.2.2　TensorFlow 环境

打开 Anaconda Prompt,输入清华仓库镜像,输入命令:

```
conda config -- add channels https://mirrors.tuna.tsinghua.edu.cn/anaconda/pkgs/free/
conda config -set show_channel_urls yes
```

创建 Python 3.5 的环境,名称为 TensorFlow,此时 Python 版本和后面 TensorFlow 的版本有匹配问题,此步选择 Python 3.x,输入命令:

```
conda create -n tensorflow python = 3.5
```

有需要确认的地方,都输入 y。

在 Anaconda Prompt 中激活 TensorFlow 环境,输入命令:

```
activate tensorflow
```

安装 CPU 版本的 TensorFlow,输入命令:

```
pip install -upgrade -- ignore-installed tensorflow
```

安装完毕。

17.3 模块实现

本项目包括 3 个模块:数据预处理、模型构建、模型训练。下面分别给出各模块的功能介绍及相关代码。

17.3.1 数据预处理

Haar 人脸分类器是能够识别出人脸 OpenCV 提供的众多分类器的一种,通过 detectMultiScale()函数调用,相关代码如下:

```
faceRects = classfier.detectMultiScale(grey,scaleFactor = 1.2,minNeighbors = 3,minSize = (32,32))
for faceRect in faceRects:          #单独框出每张人脸
            x, y, w, h = faceRect
            #将当前帧保存为图片
            img_name = '%s/%d.jpg'%(path_name, num)
            image = frame[y-10: y + h + 10, x-10: x + w + 10]
            cv2.imwrite(img_name, image)
            #画出矩形框
            cv2.rectangle(frame,(x-10,y-10),(x + w + 10,y + h + 10),color, 2)
```

调用 OpenCV 中的 haarcascade_frontalface_alt2.xml 分类器,实现人脸检测,如果检测到人脸,就会把当前帧保存为图片,并在摄像头拍到的视频中框出人脸。利用这个函数实现人脸数据的采集并最终完成对比,如图 17-4 所示。

将采集的图片进行人工筛选,删除不是人脸的照片,并统一将图片设置成 64×64 的正方形图片,在不是正方形图片中短的一边添加黑色像素,相关代码如下:

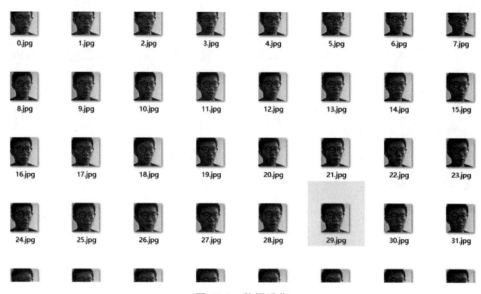

图 17-4　数据采集

```
def resize_image(image, height = IMAGE_SIZE, width = IMAGE_SIZE):
    top, bottom, left, right = (0, 0, 0, 0)
    #获取图片尺寸
    h, w, _ = image.shape
    #对于长宽不相等的图片,找到最长的一边
    longest_edge = max(h, w)
    #计算短边需要增加多少像素宽度使其与长边等长
    if h < longest_edge:
        dh = longest_edge - h
        top = dh // 2
        bottom = dh - top
    elif w < longest_edge:
        dw = longest_edge - w
        left = dw // 2
        right = dw - left
    else:
        pass
    BLACK = [0, 0, 0]
    constant = cv2.copyMakeBorder(image, top, bottom, left, right, cv2.BORDER_CONSTANT, value = BLACK)
    return cv2.resize(constant, (height, width))
```

处理图片后,对每组数据设置标签,这里将图像设置为0,其他设置为1,相关代码如下:

```
labels = np.array([0 if label.endswith('songyanyi') else 1 for label in labels])
```

17.3.2　模型构建

数据加载进模型后,划分数据集和定义模型结构。

1. 划分数据集

按照交叉验证的原则将数据集划分成三部分：训练集、验证集、测试集。交叉验证属于机器学习中常用的精度测试方法，目的是提升模型的可靠性和稳定性。步骤如下：大部分数据用于模型训练，小部分数据用于对训练后的模型验证，验证结果与验证集真实值（即标签值）比较并计算出差平方和，此项工作重复进行，直至所有验证结果与真实值相同，交叉验证结束，模型交付使用。

导入 sklearn 库的交叉验证模块，利用函数 train_test_split() 划分训练集和验证集，具体代码如下：

```
train_images, valid_images, train_labels, valid_labels = train_test_split(images, labels, test_size = 0.3, random_state = random.randint(0, 100))
    _, test_images, _, test_labels = train_test_split(images, labels, test_size = 0.5, random_state = random.randint(0, 100))
```

```
(1960, 64, 64, 3)
1372 train samples
588 valid samples
980 test samples
```

图 17-5 划分的数据集

划分数据集如图 17-5 所示。

2. 定义模型结构

定义 4 个卷积层，每个卷积层后连接 1 个激活层，每 2 个卷积层连接 1 个最大池化层。同样，最大池化层和全连接层之后，在模型中引入 Dropout 进行正则化，用以消除模型的过拟合问题。模型中共有 4 个卷积层、5 个激活函数层、2 个池化层、3 个 Dropout 层、2 个全连接层、1 个 Flatten 层和 1 个分类层。相关代码如下：

```
self.model.add(Convolution2D(32, 3, 3, border_mode = 'same',
                    input_shape = dataset.input_shape))        #二维卷积层
        self.model.add(Activation('relu'))                      #激活函数层
        self.model.add(Convolution2D(32, 3, 3))                 #二维卷积层
        self.model.add(Activation('relu'))                      #激活函数层
            self.model.add(MaxPooling2D(pool_size = (2, 2)))    #池化层
        self.model.add(Dropout(0.25))                           #Dropout 层
        self.model.add(Convolution2D(64,3,3, border_mode = 'same'))  #二维卷积层
        self.model.add(Activation('relu'))                      #激活函数层
        self.model.add(Convolution2D(64, 3, 3))                 #二维卷积层
        self.model.add(Activation('relu'))                      #激活函数层
        self.model.add(MaxPooling2D(pool_size = (2, 2)))        #池化层
        self.model.add(Dropout(0.25))                           #Dropout 层
        self.model.add(Flatten())                               #Flatten 层
        self.model.add(Dense(512))                              #Dense 层，又被称作全连接层
        self.model.add(Activation('relu'))                      #激活函数层
        self.model.add(Dropout(0.5))                            #Dropout 层
        self.model.add(Dense(nb_classes))                       #Dense 层
        self.model.add(Activation('softmax'))                   #分类层
```

17.3.3 模型训练

在定义模型架构和编译之后，通过训练集训练模型，使模型可以识别人脸。

采用 SGD+momentum 的优化器进行训练，SGD 是著名的随机下降算法，但存在下降方向完全依赖于当前训练样本的问题，因此一般配合 momentum 动量值使用。训练中损失模型选择 categorical_crossentropy，使用 accuracy 参数值作为模型评价指标。相关代码如下：

```
sgd = SGD(lr = 0.01, decay = 1e-6,
          momentum = 0.9, nesterov = True)          #采用 SGD+momentum 的优化器进行训练
    self.model.compile(loss = 'categorical_crossentropy',
                       optimizer = sgd,
                       metrics = ['accuracy'])
```

使用实时数据提升训练规模，增加训练量，通过 model.fit_generator() 进行模型训练并保存，相关代码如下：

```
self.model.fit_generator(datagen.flow(dataset.train_images, dataset.train_labels, batch_size = batch_size), samples_per_epoch = dataset.train_images.shape[0], nb_epoch = nb_epoch, validation_data = (dataset.valid_images,dataset.valid_labels))
    MODEL_PATH = './songyanyi.face.model.h5'
    def save_model(self, file_path = MODEL_PATH):
        self.model.save(file_path)
```

其中，nb_epoch 参数指定模型需要训练多少轮次，batch_size 的作用是指定每次迭代训练样本的数量。

17.4 系统测试

计算机端模型训练效果如图 17-6 所示。

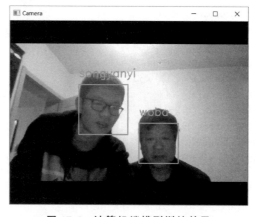

图 17-6　计算机端模型训练效果

项目 18　基于 CGAN 的线稿自动上色

PROJECT 18

本项目基于 GAN 的分支——条件生成对抗网络设计线稿自动上色模型,实现为画师提供预期上色参考及黑白漫画上色的功能。

18.1　总体设计

本部分包括系统整体结构和系统流程。

18.1.1　系统整体结构

系统整体结构如图 18-1 所示。

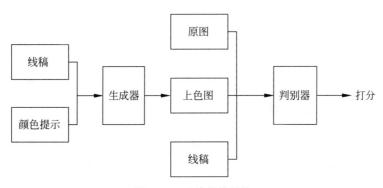

图 18-1　系统整体结构

18.1.2　系统流程

系统流程如图 18-2 所示。

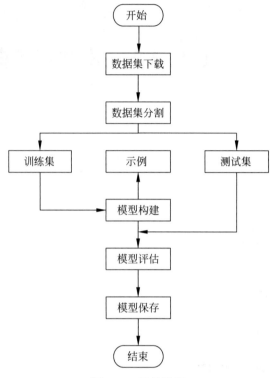

图 18-2 系统流程

18.2 运行环境

本部分包括 Python 环境和 TensorFlow 环境。

18.2.1 Python 环境

需要 Python 3.6 及以上配置,在 Windows 环境下推荐下载 Anaconda 完成 Python 所需配置,下载地址为:https://www.anaconda.com/,也可以下载虚拟机在 Linux 环境下运行代码。

18.2.2 TensorFlow 环境

打开 Anaconda Prompt,创建 Python 3.6 的环境,名称为 TensorFlow,此时 Python 版本和后面 TensorFlow 的版本有匹配问题,此步选择 Python 3.6,输入命令:

```
conda create -n tensorflow python = 3.6
```

有需要确认的地方,都输入 y。
在 Anaconda Prompt 中激活 TensorFlow 环境,输入命令:

```
activate tensorflow
```

TensorFlow GPU 版本要与计算机的显卡配置相匹配,安装 GPU 版本的 TensorFlow 1.12.0,输入命令:

```
conda install tensorflow-gpu == 1.12.0
```

对应版本的 CUDA 和 cuDNN 安装好,无须单独安装。
安装 OpenCV 3.4.2,输入命令:

```
conda install opencv == 3.4.2
```

安装完毕。

18.3 模块实现

本项目包括 4 个模块:数据预处理、模型构建、模型训练及保存和模型生成。下面分别给出各模块的功能介绍及相关代码。

18.3.1 数据预处理

在地址 https://safebooru.donmai.us/中爬取 8200 张漫画作为原始数据集,并在此基础上提取颜色、线稿方案。

1. 数据集爬取

为方便模型训练,将爬取的图片统一裁剪为 512×512 的大小。保持原图片比例,将原图片按长宽的比例放大后进行裁剪,相关代码如下:

```python
if height > width:                                          # 如果高度大于宽度
    scalefactor = (maxsize * 1.0) / width
    res = cv2.resize(image, (int(width * scalefactor), int(height * scalefactor)), interpolation = cv2.INTER_CUBIC)
    cropped = res[0:maxsize, 0:maxsize]
if width > height:                                          # 如果宽度大于高度
    scalefactor = (maxsize * 1.0) / height
    res = cv2.resize(image, (int(width * scalefactor), int(height * scalefactor)), interpolation = cv2.INTER_CUBIC)
    center_x = int(round(width * scalefactor * 0.5))
    print(center_x)
    cropped = res[0:maxsize, center_x - maxsize/2:center_x + maxsize/2]
```

2. 颜色方案提取

为增强网络泛化能力,需尽量减少颜色提示的细节,将原图随机去除色块后进行均值模糊,如图 18-3 所示。

```
#颜色方案提取(模拟凌乱的颜色提示)
def imgprocess(self, cimg, sampling = False):
    if sampling:                                              #采样处理
        cimg = cimg * 0.3 + np.ones_like(cimg) * 0.7 * 255
    else:                                                     #不采样处理
        for i in range(30):
            randx = randint(0,205)
            randy = randint(0,205)
            cimg[randx:randx + 50, randy:randy + 50] = 255
    return cv2.blur(cimg,(100,100))                           #均值模糊
```

图 18-3 颜色方案效果

在图 18-3 中,左边 4 幅图为原图,右边 4 幅图为提取的颜色方案。

3. 线稿提取

方案一:

调用 OpenCV 库,通过转灰度图、二值化、归一化以及维度扩充获取线稿,如图 18-4 所示,代码如下:

```
#线稿提取,转灰度图、二值化、归一化
base_edge = np.array([cv2.adaptiveThreshold(cv2.cvtColor(ba, cv2.COLOR_BGR2GRAY), 255,
cv2.ADAPTIVE_THRESH_MEAN_C, cv2.THRESH_BINARY, blockSize = 9, C = 2) for ba in base]) / 255.0
#维度扩充
base_edge = np.expand_dims(base_edge, 3)
```

图 18-4　方案一线稿效果

方案二：

方案一所得到的线稿效果不理想，噪声大，所以对方案进行了改进。从 GitHub 下载训练好的线稿提取模型，链接地址为：https://github.com/lllyasviel/sketchKeras。需配置好函数接口，正确载入模型。如图 18-5 所示，与方案一相比，方案二更接近真实线稿。

图 18-5　方案二线稿效果

18.3.2　模型构建

本部分包括定义模型结构及优化损失函数。

1. 定义模型结构

本部分包括生成器和判别器的功能介绍及相关代码。

1) 生成器

生成器类似 U-Net 的网络结构，由 5 层编码器和 5 层解码器构成，如图 18-6 所示。

此处使用的卷积滤波器大小是 5×5，步长为 2，对于每一层解码器，特征矩阵的高度和宽度减半。再通过 5 层编码器进行反卷积操作，每一层特征矩阵的宽度、高度加倍，同时将特征维度减半。最后还原成 256×256 的特征矩阵和三个颜色通道的特征维度。

此外，将编码器和解码器部分的对应层相连接，便于后面的网络直接调用，这样的特征融合形式有利于网络参数更新。

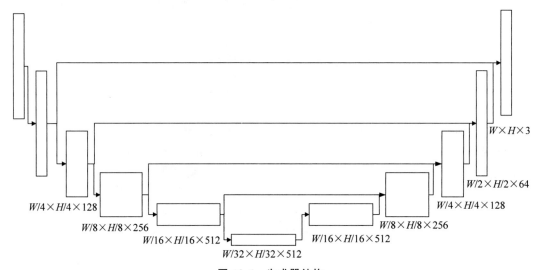

图 18-6　生成器结构

```python
#生成器
def generator(self, img_in):
    s = self.output_size
    s2, s4, s8, s16, s32, s64, s128 = s // 2, s // 4, s // 8, s // 16, s // 32, s // 64, s // 128
    # image = 256 * 256 * input_c_dim; self.gf_dim = 64
    #编码器部分
    e1 = conv2d(img_in, self.gf_dim, name = 'g_e1_conv') # e1 = 128 * 128 * self.gf_dim
    e2 = bn(conv2d(lrelu(e1), self.gf_dim * 2, name = 'g_e2_conv'))
    #e2 = 64 * 64 * self.gf_dim * 2
    e3 = bn(conv2d(lrelu(e2), self.gf_dim * 4, name = 'g_e3_conv'))
    #e3 = 32 * 32 * self.gf_dim * 4
    e4 = bn(conv2d(lrelu(e3), self.gf_dim * 8, name = 'g_e4_conv'))
    #e4 = 16 * 16 * self.gf_dim * 8
    e5 = bn(conv2d(lrelu(e4), self.gf_dim * 8, name = 'g_e5_conv'))
```

```
    #e5 = 8 * 8 * self.gf_dim * 8
    #解码器部分
        self.d4, self.d4_w, self.d4_b = deconv2d(tf.nn.relu(e5), [self.batch_size, s16, s16,
self.gf_dim * 8], name = 'g_d4', with_w = True)
        d4 = bn(self.d4)
        d4 = tf.concat([d4, e4], 3)
        #d4 = 16 * 16 * self.gf_dim * 8 * 2
        self.d5, self.d5_w, self.d5_b = deconv2d(tf.nn.relu(d4), [self.batch_size, s8, s8,
self.gf_dim * 4], name = 'g_d5', with_w = True)
        d5 = bn(self.d5)
        d5 = tf.concat([d5, e3], 3)
        #d5 = 32 * 32 * self.gf_dim * 4 * 2
        self.d6, self.d6_w, self.d6_b = deconv2d(tf.nn.relu(d5), [self.batch_size, s4, s4,
self.gf_dim * 2], name = 'g_d6', with_w = True)
        d6 = bn(self.d6)
        d6 = tf.concat([d6, e2], 3)
        #d6 = 64 * 64 * self.gf_dim * 2 * 2
        self.d7, self.d7_w, self.d7_b = deconv2d(tf.nn.relu(d6), [self.batch_size, s2, s2,
self.gf_dim], name = 'g_d7', with_w = True)
        d7 = bn(self.d7)
        d7 = tf.concat([d7, e1], 3)
        #d7 = 128 * 128 * self.gf_dim * 1 * 2
        self.d8, self.d8_w, self.d8_b = deconv2d(tf.nn.relu(d7), [self.batch_size, s, s, self.
output_colors], name = 'g_d8', with_w = True)
        #d8 = 256 * 256 * output_c_dim
        return tf.nn.tanh(self.d8)
```

2）判别器

判别器的网络结构为 5 层，如图 18-7 所示，前 4 层是卷积层，与生成器的编码器部分类似，每层操作后，特征矩阵的宽度和高度减半、维度加倍。最后一层是全连接层，将特征矩阵映射到最终概率，输出 0～1 的值，直观的解释就是给生成器生成的图片打分，值越接近 1 越好。

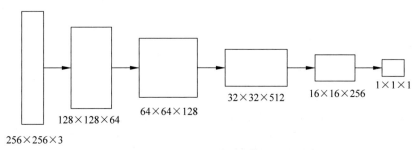

图 18-7 判别器结构

```python
# 判别器
def discriminator(self, image, y = None, reuse = False):
    # image = 256x256x(input_c_dim + output_c_dim)
    if reuse:
        tf.get_variable_scope().reuse_variables()
    else:
        assert tf.get_variable_scope().reuse == False
    # self.df_dim = 64
    h0 = lrelu(conv2d(image, self.df_dim, name = 'd_h0_conv'))
    # h0 = 128x128xself.df_dim
    h1 = lrelu(self.d_bn1(conv2d(h0, self.df_dim * 2, name = 'd_h1_conv')))
    # h1 = 64x64xself.df_dim * 2
    h2 = lrelu(self.d_bn2(conv2d(h1, self.df_dim * 4, name = 'd_h2_conv')))  # h2 = 32x32xself.df_dim * 4
    h3 = lrelu(self.d_bn3(conv2d(h2, self.df_dim * 8, d_h = 1, d_w = 1, name = 'd_h3_conv')))
    # h3 = 16x16xself.df_dim * 8
    h4 = linear(tf.reshape(h3, [self.batch_size, -1]), 1, 'd_h3_lin')
    return tf.nn.sigmoid(h4), h4
```

2. 优化损失函数

d_loss_real 是真实图片输入判别器中的结果和预期为 1 的结果之间的交叉熵。d_loss_fake 是生成器生成的图片输入判别器中的结果和预期为 0 的结果之间的交叉熵。判别器的损失函数 d_loss 是 d_loss_fake 和 d_loss_real 之和。

```python
# 定义判别器的损失函数
self.d_loss_real = tf.reduce_mean(tf.nn.sigmoid_cross_entropy_with_logits(logits = disc_true_logits, labels = tf.ones_like(disc_true_logits)))
self.d_loss_fake = tf.reduce_mean(tf.nn.sigmoid_cross_entropy_with_logits(logits = disc_fake_logits, labels = tf.zeros_like(disc_fake_logits)))
self.d_loss = self.d_loss_real + self.d_loss_fake
```

生成器的损失函数 g_loss 是生成器生成的图片输入判别器中的结果和预期为 1 的结果之间的交叉熵。

```python
# 定义生成器的损失函数
self.g_loss = tf.reduce_mean(tf.nn.sigmoid_cross_entropy_with_logits(logits = disc_fake_logits, labels = tf.ones_like(disc_fake_logits))) + self.l1_scaling * tf.reduce_mean(tf.abs(self.real_images - self.generated_images))
# Adam 是常用的梯度下降方法优化模型参数
# 分别从每个模型中收集变量,让它们可以被分开训练
t_vars = tf.trainable_variables()
self.d_vars = [var for var in t_vars if 'd_' in var.name]
self.g_vars = [var for var in t_vars if 'g_' in var.name]
# 定义优化器
with tf.variable_scope('d_optim', reuse = tf.AUTO_REUSE):
    self.d_optim = tf.train.AdamOptimizer(0.0002, beta1 = 0.5).minimize(self.d_loss, var_list =
```

```
        self.d_vars, )
        with tf.variable_scope('g_optim', reuse = tf.AUTO_REUSE):
            self.g_optim = tf.train.AdamOptimizer(0.0002, beta1 = 0.5).minimize(self.g_loss, var_list =
        self.g_vars, )
```

18.3.3 模型训练及保存

本部分包括模型训练及模型保存。

1. 模型训练

在定义模型架构和编译之后,通过训练集训练模型,使模型基于颜色提示对线稿进行上色。由于内存限制,设置批量大小为 4。

```
#模型训练
def train(self):
    #载入模型
    self.loadmodel()
    #读取 imgs 文件夹里的 jpg 格式图片
    data = glob(os.path.join("imgs", "*.jpg"))
    print(data[0])
    #选取 4 张图片合为一张,以便展示训练效果
    base = np.array([get_image(sample_file) for sample_file in data[0:self.batch_size]])
    base_normalized = base/255.0
    #获取线稿,此时获得的线稿图片为灰度图,因此维度为 1
    base_edge = np.array([threshold_demo(show_active_img_and_save_denoise_filter(ba))for
ba in data[0:self.batch_size]]) / 255.0
    # base_edge = np.array([transform(show_active_img_and_save_denoise_filter(ba)) for ba
in data[0:self.batch_size]]) / 255.0
    #维度扩充
    base_edge = np.expand_dims(base_edge, 3)
    #调用自定义的 imgprocess()函数获得颜色提示点
    base_colors = np.array([self.imgprocess(ba) for ba in base]) / 255.0
    #将示例图片的原图、线稿、颜色方案保存到 results 文件夹中
    ims("results/base.png",merge_color(base_normalized, [self.batch_size_sqrt, self.batch_
size_sqrt]))
    ims("results/base_line.jpg",merge(base_edge, [self.batch_size_sqrt, self.batch_size_
sqrt]))
    ims("results/base_colors.jpg",merge_color(base_colors, [self.batch_size_sqrt, self.
batch_size_sqrt]))
    #获得 imgs 文件夹中图片长度,确定图片个数
    datalen = len(data)
    #开始训练
    for e in range(1000):
        for i in range(datalen // self.batch_size):
            #一次训练 4 张图片
            batch_files = data[i * self.batch_size:(i + 1) * self.batch_size]
```

```python
        batch = np.array([get_image(batch_file) for batch_file in batch_files])
        batch_normalized = batch/255.0
        '''
        #线稿提取方案一(OpenCV边缘检测算法)
        bas_edge = np.array([cv2.adaptiveThreshold(cv2.cvtColor(show_active_img_and_
save_denoise_filter(ba), cv2.COLOR_BGR2GRAY), 255,cv2.ADAPTIVE_THRESH_MEAN_C, cv2.THRESH_
BINARY, blockSize = 9, C = 2)for ba in batch_files]) / 255.0
        '''
        #线稿提取方案二(调用已训练好的线稿提取模型)
        batch_edge = np.array([transform(show_active_img_and_save_denoise_filter(ba))
for ba in batch_files]) / 255.0
        batch_edge = np.expand_dims(batch_edge, 3)
        #获取颜色提示
        batch_colors = np.array([self.imgprocess(ba) for ba in batch]) / 255.0
        #数据输入判别器和生成器
        d_loss, _ = self.sess.run([self.d_loss, self.d_optim], feed_dict = {self.real_
images: batch_normalized, self.line_images: batch_edge, self.color_images: batch_colors})
        g_loss, _ = self.sess.run([self.g_loss, self.g_optim], feed_dict = {self.real_
images: batch_normalized, self.line_images: batch_edge, self.color_images: batch_colors})
        #打印生成器、判别器损失函数
        print("%d: [%d / %d] d_loss %f, g_loss %f" % (e, i, (datalen//self.batch_
size), d_loss, g_loss))
        #每200次迭代后保存一次训练效果图至results文件夹
        if i % 200 == 0:
            recreation = self.sess.run(self.generated_images, feed_dict = {self.real_
images: base_normalized, self.line_images: base_edge, self.color_images: base_colors})
            ims("results/" + str(e * 100000 + i) + ".jpg", merge_color(recreation,
[self.batch_size_sqrt, self.batch_size_sqrt]))
        #每500次迭代保存一次模型至checkpoint文件夹
        if i % 500 == 0:
            self.save("./checkpoint", e * 100000 + i)
```

其中,一个 batch 就是在一次前向/后向传播过程用到的训练样例数量,一次用 4 张图片进行训练。epoch 是训练的循环轮数,每个 epoch 迭代一个完整训练集,每个 epoch 训练 2000 个 batch,共 8000 张图片,如图 18-8 和图 18-9 所示。

图 18-8 为训练了 40 个 epoch 之后的损失函数。图 18-9 中左边 4 张图片是训练了 10 个 epoch 之后的效果图,右边 4 张图片是训练 40 个 epoch 之后的效果图。

通过观察训练集损失函数以及训练结果图片来评估模型的训练程度,进行模型训练的进一步决策。一般来说,训练集损失函数越小,训练效果越好。

2. 模型保存

调用 TensorFlow 的 tf.train.Saver()进行模型保存。

```python
#模型保存
def save(self, checkpoint_dir, step):
```

```
40: [1018 / 2000] d_loss 0.339064, g_loss 8.101345
40: [1019 / 2000] d_loss 0.624775, g_loss 10.888014
40: [1020 / 2000] d_loss 0.717385, g_loss 12.346249
40: [1021 / 2000] d_loss 0.605913, g_loss 10.994989
40: [1022 / 2000] d_loss 0.471260, g_loss 13.503658
40: [1023 / 2000] d_loss 1.083526, g_loss 10.527475
40: [1024 / 2000] d_loss 0.432427, g_loss 9.023579
40: [1025 / 2000] d_loss 0.304633, g_loss 11.593304
40: [1026 / 2000] d_loss 0.323134, g_loss 12.362459
40: [1027 / 2000] d_loss 0.199150, g_loss 12.508513
40: [1028 / 2000] d_loss 0.480603, g_loss 10.004042
40: [1029 / 2000] d_loss 0.298838, g_loss 11.532845
40: [1030 / 2000] d_loss 1.361192, g_loss 11.111380
40: [1031 / 2000] d_loss 0.905775, g_loss 11.249300
40: [1032 / 2000] d_loss 0.280949, g_loss 12.259300
40: [1033 / 2000] d_loss 0.140020, g_loss 11.299433
40: [1034 / 2000] d_loss 1.409916, g_loss 8.980406
40: [1035 / 2000] d_loss 2.663800, g_loss 14.890905
40: [1036 / 2000] d_loss 0.736856, g_loss 9.901164
40: [1037 / 2000] d_loss 0.906262, g_loss 15.893844
40: [1038 / 2000] d_loss 0.404512, g_loss 14.859211
40: [1039 / 2000] d_loss 0.321734, g_loss 7.904107
40: [1040 / 2000] d_loss 0.852163, g_loss 11.337540
40: [1041 / 2000] d_loss 0.806573, g_loss 10.955511
40: [1042 / 2000] d_loss 0.496660, g_loss 11.666111
40: [1043 / 2000] d_loss 0.423448, g_loss 13.127735
40: [1044 / 2000] d_loss 1.024247, g_loss 7.266569
40: [1045 / 2000] d_loss 2.340706, g_loss 12.936205
40: [1046 / 2000] d_loss 1.189296, g_loss 9.553530
40: [1047 / 2000] d_loss 0.300690, g_loss 10.838240
40: [1048 / 2000] d_loss 0.781089, g_loss 11.920691
40: [1049 / 2000] d_loss 0.513205, g_loss 11.717654
40: [1050 / 2000] d_loss 0.421007, g_loss 10.589517
```

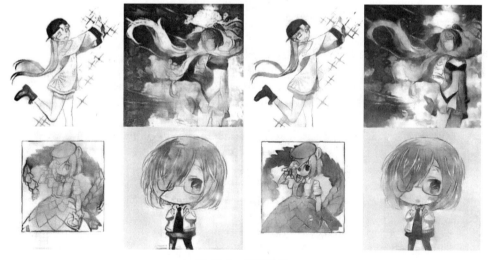

图 18-8 训练集损失函数

图 18-9 训练效果

```
model_name = "model"
model_dir = "tr"
checkpoint_dir = os.path.join(checkpoint_dir, model_dir)
if not os.path.exists(checkpoint_dir):
```

```
                os.makedirs(checkpoint_dir)
    self.saver.save(self.sess,
                    os.path.join(checkpoint_dir, model_name),
                    global_step = step)
```

程序会生成并保存 4 种类型的文件,如图 18-10 所示。

(1) checkpoint:文本文件,记录了模型文件的路径信息列表;

(2) model.data-00000-of-00001:网络权重信息;

(3) model-10000.index:模型中的变量参数(权重)信息;

(4) model-10000.meta:模型的计算图结构信息(模型的网络结构)。

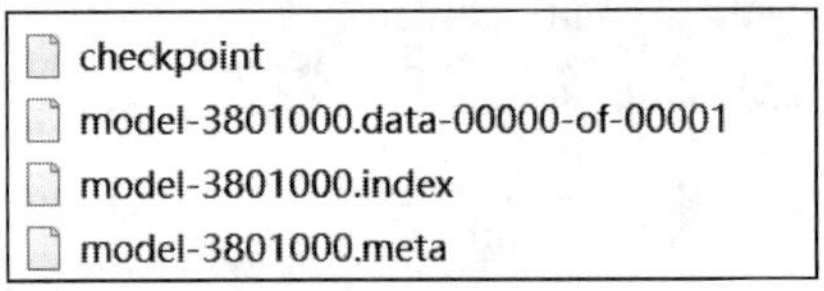

> checkpoint
> model-3801000.data-00000-of-00001
> model-3801000.index
> model-3801000.meta

图 18-10　模型保存信息

18.3.4　模型应用

本部分实现对输入线稿的上色过程。

```
#读取需要上色的线稿图片以及对应的颜色提示
data_line = glob(os.path.join("true"," * _line.jpg"))
data_color = glob(os.path.join("true"," * _color.jpg"))
#开始对线稿进行上色
for i in range(min(100,datalen // self.batch_size)):
    batch_files_line = data_0[i * self.batch_size:(i + 1) * self.batch_size]
    batch_files_color = data_1[i * self.batch_size:(i + 1) * self.batch_size]
    batch_line = np.array([cv2.resize(imread(batch_file), (512,512)) for batch_file in batch_files_line])                                                    #缩放图像
    batch_color = np.array([cv2.resize(imread(batch_file), (512,512)) for batch_file in batch_files_color])
    batch_normalized = batch_0/255.0
    #cv2.adaptiveThreshold()为图像自适应阈值二值化函数
    #base_edge = np.array([ cv2.adaptiveThreshold ( cv2. cvtColor ( ba, cv2. COLOR _ BGR2GRAY), 255,
    #cv2.ADAPTIVE_THRESH_MEAN_C, cv2.THRESH_BINARY, blockSize = 9, C = 2)
    #for ba in batch_0]) / 255.0
    batch_edge = np.array([transform(show_active_img_and_save_denoise_filter(ba)) for ba in batch_files_line]) / 255.0
    batch_edge = np.expand_dims(batch_edge, 3)                                          #边缘处理
    #batch_colors = np.array([self.imageblur(ba,True) for ba in batch_color]) / 255.0
    batch_colors = np.array([self.imgprocess(ba, True) for ba in batch_color]) / 255.0 #颜色处理
    recreation = self.sess.run(self.generated_images, feed_dict = {self.real_images: batch_
```

normalized, self.line_images: batch_edge, self.color_images: batch_colors})
        ims("results/sample_" + str(i) + ".jpg",merge_color(recreation, [self.batch_size_sqrt, self.batch_size_sqrt]))

18.4　系统测试

本部分包括训练效果、测试效果及模型使用说明。

18.4.1　训练效果

如图 18-11 所示，左边是第 0 个 epoch，右边是第 3 个 epoch，随着 epoch 次数的增多，d_loss 和 g_loss 逐渐收敛，说明训练效果越来越理想。

图 18-11　模型损失函数

如图 18-12 所示，左边 4 张图片是原图，右边 4 张图片是训练 40 个 epoch 的结果图。

18.4.2　测试效果

将测试集图片输入模型进行测试和分析，如图 18-13 所示。
图 18-13 中，自左到右分别为：线稿、颜色方案、上色结果。
模型实现基于颜色提示的线稿上色功能，且该上色功能并不是单纯的颜色填充，而是能体现出一定的明暗变化与色彩过渡。

图 18-12 训练结果

图 18-13 模型测试效果

18.4.3　模型使用说明

本项目的运行环境为 Python 3.6，tensorflow-gpu == 1.12.0，OpenCV 3.4.2，Keras 2.2.4。根据自己设备配置相应版本的环境，建议配置 GPU 版本的 TensorFlow。

在项目文件夹下创建名为 imgs 的文件夹，运行 download_imgs.py，以爬取漫画数据集，爬取的数据会保存到 imgs 文件夹中，也可以使用已有的漫画数据集，但注意将图片像素统一为 512×512，格式为 jpg。

在命令行下运行 python main.py train 和 python main.py sample。

项目 19 基于 ACGAN 的动漫头像生成

PROJECT 19

本项目基于 ACGAN 模型,通过网络爬取动漫图片进行模型训练,实现特定标签动漫头像的生成,以满足特定场景的应用。

19.1 总体设计

本部分包括系统整体结构和系统流程。

19.1.1 系统整体结构

系统整体结构如图 19-1 所示。

图 19-1 系统整体结构

19.1.2 系统流程

系统流程如图 19-2 所示,ACGAN 原理如图 19-3 所示。

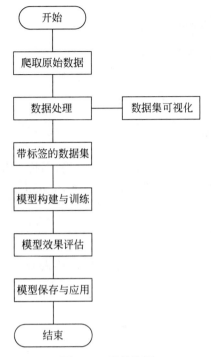

图 19-2　系统流程

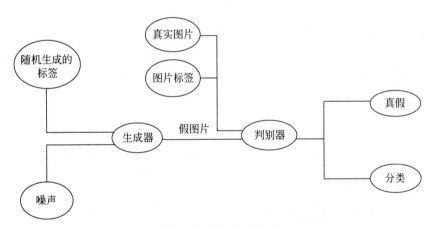

图 19-3　ACGAN 原理

19.2　运行环境

本部分包括 Python 环境、TensorFlow 环境、OpenCV 环境和 Illustration2Vec。

19.2.1　Python 环境

需要 Python 3.7 配置，在 Windows 环境下推荐下载 Anaconda 完成 Python 所需的配置，下载地址为：https://www.anaconda.com/，也可以下载虚拟机在 Linux 环境下运行代码。

19.2.2　TensorFlow 环境

打开 Anaconda Prompt，创建 Python 3.7 的环境，名称为 TensorFlow，此时 Python 版本和后面 TensorFlow 的版本有匹配问题，此步选择 Python 3.x，输入命令：

```
conda create -n tensorflow python = 3.7
```

有需要确认的地方，都输入 y。
在 Anaconda Prompt 中激活 TensorFlow 环境，输入命令：

```
activate tensorflow
```

安装 GPU 版本的 TensorFlow，输入命令：

```
pip install -i https://pypi.tuna.tsinghua.edu.cn/simple tensorflow-gpu == 1.14.0
```

安装完毕。

19.2.3　OpenCV 环境

安装 OpenCV 库，输入命令：

```
pip install opencv-python
```

安装完毕后，下载 cascade file 并将其放入项目目录，输入命令：

```
wget https://raw.githubusercontent.com/nagadomi/lbpcascade_animeface/master/lbpcascade_animeface.xml
```

19.2.4　Illustration2Vec

在 https://github.com/rezoo/illustration2vec/releases 仓库中下载名为 i2v 的文件夹，以 Python 文件(Illustration2Vec 的包)放在项目目录下。

19.3　模块实现

本项目包括 4 个模块：数据获取、数据处理、模型构建、模型训练及保存。下面分别给出各模块的功能介绍及相关代码。

19.3.1 数据获取

考虑生成式对抗网络需要对大量样本进行迭代训练,选择从网站上爬取大量图片作为合适的数据集(爬虫)。数据来源于 konachan.net,总共爬取约 3 万张图片,相关代码如下:

```
#采用request+beautiful库
start = 1
end = 8000
for i in range(2000, end + 1):
    url = 'http://konachan.net/post?page = % d&tags = ' % i
    html = requests.get(url).text
    soup = BeautifulSoup(html, 'html.parser')
    for img in soup.find_all('img', class_ = "preview"):
        target_url = '' + img['src']
        filename = os.path.join('imgs', target_url.split('/')[-1])
        download(target_url, filename)
    print('% d / % d' % (i, end))
```

数据会保存在相应的文件夹中,如图 19-4 所示。

图 19-4 爬取图片保存文件夹

19.3.2 数据处理

本部分包括截取头像、提取标签、提取特征向量及降维可视化。

1. 截取头像

使用 OpenCV 下 objdetect 模块中用作目标检测级联分类器的一个类（CascadeClassifier），开源的 lbpcascade_animeface.xml 用于动漫人脸检测的级联分类器，进行图片中二次元人脸的检测和截取。

```
images = glob.glob(''E:/images/imgs/ * .jpg'')
#加载 xml 级联分类器
cascade = cv2.CascadeClassifier('lbpcascade_animeface.xml')
image = cv2.imread(img_path)
#将 BGR 格式转换成灰度图片
gray = cv2.cvtColor(image, cv2.COLOR_BGR2GRAY)
#直方图均衡化，增强对比度，便于检测
gray = cv2.equalizeHist(gray)
#多尺度检测，返回所有人脸的矩形位置
faces = cascade.detectMultiScale(gray, scaleFactor = 1.1, minNeighbors = 5, minSize = (48, 48))
for (x, y, w, h) in faces:
#截取人脸
faces = cascade.detectMultiScale(gray, scaleFactor = 1.1, minNeighbors = 5, minSize = (64, 64))
#将截取后的人脸缩放为 128 × 128
face = cv2.resize(face, (128, 128))
#保存
cv2.imwrite(os.path.join(OUTPUT_DIR, '%d.jpg' % num), face)
```

截取的头像如图 19-5 所示。

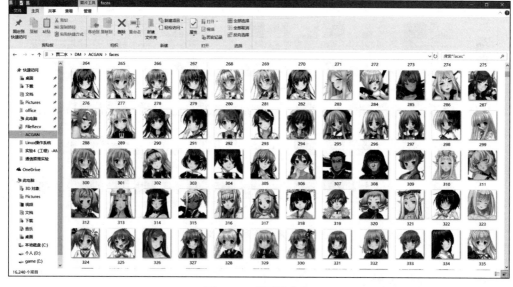

图 19-5　截取的头像

2. 提取标签

人物特征的分类需要对截取的头像进行标签提取。Illustration2Vec 是基于卷积神经网络的任务标签开源库,用于生成人物特征的标签。Illustration2Vec 可以指定一个阈值,并提取概率高于阈值的标签;或者指定一些标签,并返回对应标签的概率。

```
import i2v
from imageio import imread
# 加载已有的 Caffe 分类模型
illust2vec = i2v.make_i2v_with_chainer('illust2vec_tag_ver200.caffemodel', 'tag_list.json')
img = imread('imgs/us.jpg')
# 返回具有一对标签及其置信度的字典
tags = illust2vec.estimate_plausible_tags([img], threshold = 0.5)
print(tags)
tags = illust2vec.estimate_specific_tags([img], ['blue eyes', 'red hair'])
print(tags)
```

3. 提取特征向量及降维可视化

调用第三方的库函数提取语义特征向量,随机选取 2000 个头像,使用 TSNE 进行降维,最后可视化。

```
# 加载模型
illust2vec = i2v.make_i2v_with_chainer("illust2vec_ver200.caffemodel")
# 提取语义特征向量
vector = illust2vec.extract_feature([image])[0]
# 将提取的特征向量保存
vec_all.append(vector)
# 随机选取 2000 个头像
data_index = np.arange(img_all.shape[0])
np.random.shuffle(data_index)
data_index = data_index[:2000]
# 降维
tsne = TSNE(perplexity = 30, n_components = 2, init = 'pca', n_iter = 5000)
two_d_vectors = tsne.fit_transform(vec_all[data_index, :])
puzzles = np.ones((4800, 4800, 3))
xmin = np.min(two_d_vectors[:, 0])
xmax = np.max(two_d_vectors[:, 0])
ymin = np.min(two_d_vectors[:, 1])
ymax = np.max(two_d_vectors[:, 1])
# 可视化,输出.png 格式的图片
for i, vector in enumerate(two_d_vectors):
    x, y = two_d_vectors[i, :]
    x = int((x - xmin) / (xmax - xmin) * (6400 - 128) + 64)
    y = int((y - ymin) / (ymax - ymin) * (6400 - 128) + 64)
    puzzles[y - 64: y + 64, x - 64: x + 64, :] = img_all[data_index[i]]
imsave('降维可视化.png', puzzles)
```

降维可视化结果如图 19-6 所示。

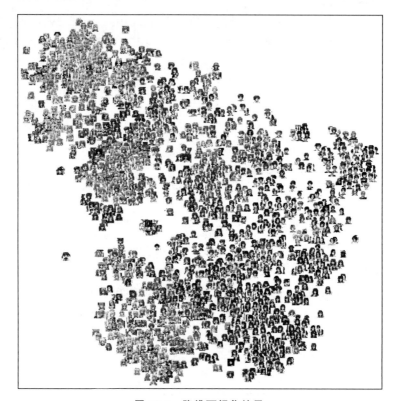

图 19-6　降维可视化结果

19.3.3　模型构建

数据加载进模型后,需要定义常量、参数、辅助函数、模型结构并优化损失函数。

1. 定义常量、参数、辅助函数

本部分相关代码如下:

```
#定义常量
batch_size = 64
z_dim = 128
WIDTH = 128
HEIGHT = 128
LABEL = 34
LAMBDA = 0.05
BETA = 3
#创建训练样本文件夹
OUTPUT_DIR = 'samples'
if not os.path.exists(OUTPUT_DIR):
```

```
    os.mkdir(OUTPUT_DIR)
# 定义 placeholder(占位符,在 TensorFlow 中类似于函数参数,在执行时赋具体值)
X = tf.placeholder(dtype = tf.float32, shape = [batch_size, HEIGHT, WIDTH, 3], name = 'X')
X_perturb = tf.placeholder(dtype = tf.float32, shape = [batch_size, HEIGHT, WIDTH, 3], name = 'X_perturb')
Y = tf.placeholder(dtype = tf.float32, shape = [batch_size, LABEL], name = 'Y')
noise = tf.placeholder(dtype = tf.float32, shape = [batch_size, z_dim], name = 'noise')
noise_y = tf.placeholder(dtype = tf.float32, shape = [batch_size, LABEL], name = 'noise_y')
is_training = tf.placeholder(dtype = tf.bool, name = 'is_training')
# 定义学习率
# 定义迭代步数变量
global_step = tf.Variable(0, trainable = False)
add_global = global_step.assign_add(1)
# 定义初始学习率为 0.0002
initial_learning_rate = 0.0002
# 学习率衰减函数,表示每经过 20000 次的迭代,学习率变为原来的 0.5
learning_rate = tf.train.exponential_decay(initial_learning_rate, global_step = global_step, decay_steps = 20000, decay_rate = 0.5)
# 定义一些辅助函数
# 定义激活函数 Leaky ReLU
def lrelu(x, leak = 0.2):
    return tf.maximum(x, leak * x)
# 交叉熵损失函数,对于给定 logits 计算 sigmoid 的交叉熵;衡量分类任务中的概率误差,可以用它
来计算多标签分类任务,即一幅图片可以同时具有多个类别标签
def sigmoid_cross_entropy_with_logits(x, y):
    return tf.nn.sigmoid_cross_entropy_with_logits(logits = x, labels = y)
# 定义二维卷积层
def conv2d(inputs, kernel_size, filters, strides, padding = 'same', use_bias = True):
    return tf.layers.conv2d(inputs = inputs, kernel_size = kernel_size, filters = filters, strides = strides, padding = padding, use_bias = use_bias)
# 定义 BN 层,Batch Normalization 通过减少内部协变量加速神经网络的训练
def batch_norm(inputs, is_training = is_training, decay = 0.9):
    return tf.contrib.layers.batch_norm(inputs, is_training = is_training, decay = decay)
```

2. 定义模型结构

本部分包括判别器和生成器的相关代码。

1) 判别器

判别器使用 10 个残差块,输出端有两支,分别完成判别和分类任务。

```
# 定义残差块,filters 是维数
def d_block(inputs, filters):
    h0 = lrelu(conv2d(inputs, 3, filters, 1))
    h0 = conv2d(h0, 3, filters, 1)
    h0 = lrelu(tf.add(h0, inputs))
    return h0
# 定义判别器,reuse 表示是否可以重复使用具有相同名称的前一层权重
```

```python
def discriminator(image, reuse = None):
    with tf.variable_scope('discriminator', reuse = reuse):
        h0 = image
        f = 32
        #使用10个残差块
        for i in range(5):
            if i < 3:
                h0 = lrelu(conv2d(h0, 4, f, 2))
            else:
                h0 = lrelu(conv2d(h0, 3, f, 2))
            h0 = d_block(h0, f)
            h0 = d_block(h0, f)
            f = f * 2
        h0 = lrelu(conv2d(h0, 3, f, 2))
        #用于卷积神经网络全连接层前的预处理
        h0 = tf.contrib.layers.flatten(h0)
        #全连接层
        Y_ = tf.layers.dense(h0, units = LABEL)
        h0 = tf.layers.dense(h0, units = 1)
        #判别输出和分类输出
        return h0, Y_
```

2）生成器

生成器使用16个残差块。

```python
#定义残差块
def g_block(inputs):
    #定义激活函数 ReLU
    h0 = tf.nn.relu(batch_norm(conv2d(inputs, 3, 64, 1, use_bias = False)))
    h0 = batch_norm(conv2d(h0, 3, 64, 1, use_bias = False))
    h0 = tf.add(h0, inputs)
    return h0
#定义生成器
def generator(z, label):
    with tf.variable_scope('generator', reuse = None):
        d = 16
        #将噪声和标签拼接
        z = tf.concat([z, label], axis = 1)
        #全连接层
        h0 = tf.layers.dense(z, units = d * d * 64)
        #将h0变换为参数shape的形式
        h0 = tf.reshape(h0, shape = [-1, d, d, 64])
        #BN层
        h0 = tf.nn.relu(batch_norm(h0))
        shortcut = h0
        #使用16个残差块
        for i in range(16):
```

```
            h0 = g_block(h0)
        h0 = tf.nn.relu(batch_norm(h0))
        h0 = tf.add(h0, shortcut)
        # 亚像素卷积层
        for i in range(3):
            h0 = conv2d(h0, 3, 256, 1, use_bias = False)
            h0 = tf.depth_to_space(h0, 2)
            h0 = tf.nn.relu(batch_norm(h0))
        # 输出
        h0 = tf.layers.conv2d(h0, kernel_size = 9, filters = 3, strides = 1, padding = 'same', 
activation = tf.nn.tanh, name = 'g', use_bias = True)
        return h0
```

3. 优化损失函数

确定模型架构后进行编译，这是多类别的分类问题，因此使用交叉熵作为损失函数。由于所有标签都带有相似的权重，通常使用精确度作为性能指标。

```
# 损失函数
g = generator(noise, noise_y)
d_real, y_real = discriminator(X)
d_fake, y_fake = discriminator(g, reuse = True)
loss_d_real = tf.reduce_mean(sigmoid_cross_entropy_with_logits(d_real, tf.ones_like(d_
real)))
loss_d_fake = tf.reduce_mean(sigmoid_cross_entropy_with_logits(d_fake, tf.zeros_like(d_
fake)))
loss_g_fake = tf.reduce_mean(sigmoid_cross_entropy_with_logits(d_fake, tf.ones_like(d_
fake)))
loss_c_real = tf.reduce_mean(sigmoid_cross_entropy_with_logits(y_real, Y))
loss_c_fake = tf.reduce_mean(sigmoid_cross_entropy_with_logits(y_fake, noise_y))
# 判别器损失
loss_d = loss_d_real + loss_d_fake + BETA * loss_c_real
# 生成器损失
loss_g = loss_g_fake + BETA * loss_c_fake
# Adam 是常用的梯度下降方法，使用它来优化模型参数
update_ops = tf.get_collection(tf.GraphKeys.UPDATE_OPS)
# 定义优化器
with tf.control_dependencies(update_ops):
    optimizer_d = tf.train.AdamOptimizer(learning_rate = learning_rate, beta1 = 0.5)
.minimize(loss_d, var_list = vars_d)
    optimizer_g = tf.train.AdamOptimizer(learning_rate = learning_rate, beta1 = 0.5)
.minimize(loss_g, var_list = vars_g)
```

19.3.4 模型训练及保存

1. 模型训练

原始数据中各类标签分布不均匀，所以需要完整的迭代数据。

```python
# 初始化模型的参数
sess = tf.Session()
sess.run(tf.global_variables_initializer())
z_samples = np.random.uniform(-1.0, 1.0, [batch_size, z_dim]).astype(np.float32)
y_samples = get_random_tags()
for i in range(batch_size):
    y_samples[i, :28] = 0
    y_samples[i, i // 8 % 13] = 1 # hair color
    y_samples[i, i // 8 % 5 + 13] = 1 # hair style
    y_samples[i, i // 8 % 10 + 18] = 1 # eye color
samples = []
loss = {'d': [], 'g': []}
offset = 0
# 开始训练模型,迭代 3000 次
for i in tqdm(range(60000)):
    if offset + batch_size > X_all.shape[0]:
        offset = 0
    if offset == 0:
        data_index = np.arange(X_all.shape[0])
        np.random.shuffle(data_index)
        X_all = X_all[data_index, :, :, :]
        Y_all = Y_all[data_index, :]
    X_batch = X_all[offset: offset + batch_size, :, :, :]
    Y_batch = Y_all[offset: offset + batch_size, :]
    X_batch_perturb = X_batch + 0.5 * X_batch.std() * np.random.random(X_batch.shape)
    offset += batch_size
    n = np.random.uniform(-1.0, 1.0, [batch_size, z_dim]).astype(np.float32)
    ny = get_random_tags()
    _, d_ls = sess.run([optimizer_d, loss_d], feed_dict = {X: X_batch, X_perturb: X_batch_perturb, Y: Y_batch, noise: n, noise_y: ny, is_training: True})
    n = np.random.uniform(-1.0, 1.0, [batch_size, z_dim]).astype(np.float32)
    ny = get_random_tags()
    _, g_ls = sess.run([optimizer_g, loss_g], feed_dict = {noise: n, noise_y: ny, is_training: True})
    loss['d'].append(d_ls)
    loss['g'].append(g_ls)
    _, lr = sess.run([add_global, learning_rate])
    # 每迭代 500 次保存训练样本
    if i % 500 == 0:
        print(i, d_ls, g_ls, lr)
        gen_imgs = sess.run(g, feed_dict = {noise: z_samples, noise_y: y_samples, is_training: False})
        gen_imgs = (gen_imgs + 1) / 2
        imgs = [img[:, :, :] for img in gen_imgs]
        gen_imgs = montage(imgs)
        plt.axis('off')
        plt.imshow(gen_imgs)
```

```
            imsave(os.path.join(OUTPUT_DIR, 'sample_%d.jpg' % i), gen_imgs)
            plt.show()
            samples.append(gen_imgs)
#生成损失函数曲线图
plt.plot(loss['d'], label = 'Discriminator')
plt.plot(loss['g'], label = 'Generator')
plt.legend(loc = 'upper right')
plt.savefig('Loss.png')
plt.show()
```

训练样本如图19-7所示。

图19-7　训练样本

2. 模型保存

将训练好的模型参数进行保存，以便于下次继续用于训练或测试。

```
saver = tf.train.Saver()
saver.save(sess, './acgan', global_step=60000)
```

19.4 系统测试

本部分包括模型导入及调用、生成指定标签。

19.4.1 模型导入及调用

模型导入及调用的相关代码如下：

```
saver = tf.train.import_meta_graph('./acgan-60000.meta')
saver.restore(sess, tf.train.latest_checkpoint('./'))
```

19.4.2 生成指定标签

生成指定标签相关代码如下：

```
all_tags = ['blonde hair', 'brown hair', 'black hair', 'blue hair', 'pink hair', 'purple hair',
'green hair', 'red hair', 'silver hair', 'white hair', 'orange hair', 'aqua hair', 'grey hair', 'long
hair', 'short hair', 'twintails', 'drill hair', 'ponytail', 'blue eyes', 'red eyes', 'brown eyes',
'green eyes', 'purple eyes', 'yellow eyes', 'pink eyes', 'aqua eyes', 'black eyes', 'orange eyes',
'blush', 'smile', 'open mouth', 'hat', 'ribbon', 'glasses']
for i, tags in enumerate([['blonde hair', 'twintails', 'blush', 'smile', 'ribbon', 'red eyes'],
['red hair', 'long hair', 'blush', 'smile', 'open mouth', 'blue eyes']]):
    z_samples = np.random.uniform(-1.0, 1.0, [batch_size, z_dim]).astype(np.float32)
    y_samples = np.zeros([1, LABEL])
    for tag in tags:
        y_samples[0, all_tags.index(tag)] = 1
    y_samples = np.repeat(y_samples, batch_size, 0)
    gen_imgs = sess.run(g, feed_dict={noise: z_samples, noise_y: y_samples, is_training: False})
    gen_imgs = (gen_imgs + 1) / 2
    imgs = [img[:, :, :] for img in gen_imgs]
    gen_imgs = montage(imgs)
    gen_imgs = np.clip(gen_imgs, 0, 1)
    imsave('标签%d.jpg' % (i + 1), gen_imgs)
```

金发、双马尾、红眼睛的头像如图19-8所示；红发、长发、蓝眼睛的头像如图19-9所示。

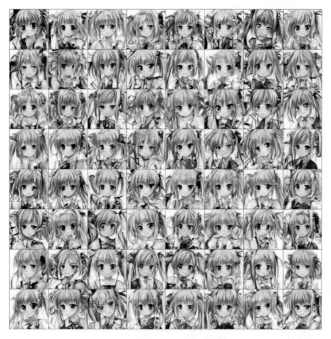

图 19-8　金发、双马尾、红眼睛的头像

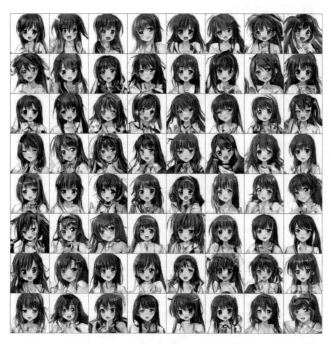

图 19-9　红发、长发、蓝眼睛的头像

项目 20 手势语言识别

PROJECT 20

本项目基于卷积神经网络，通过 Python 的翻转功能沿垂直轴翻转每个图像，实现手势语言识别的功能。

20.1 总体设计

本部分包括系统整体结构和系统流程。

20.1.1 系统整体结构

系统整体结构如图 20-1 所示。

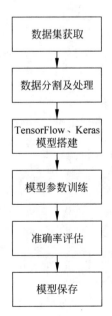

图 20-1 系统整体结构

20.1.2 系统流程

系统流程如图 20-2 所示。

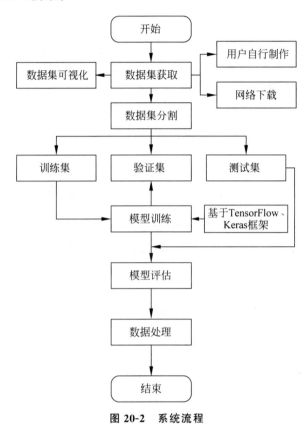

图 20-2　系统流程

20.2　运行环境

本部分包括 Python 环境、TensorFlow 环境和 OpenCV-Python 环境。

20.2.1　Python 环境

需要 Python 3.6 及以上配置,在 Windows 环境下推荐下载 Anaconda 完成 Python 所需的配置,下载地址为:https://www.anaconda.com/,也可以下载虚拟机在 Linux 环境下运行代码。

20.2.2 TensorFlow 环境

打开 Anaconda Prompt,输入清华仓库镜像,输入命令:

```
conda config -- add channels https://mirrors.tuna.tsinghua.edu.cn/anaconda/pkgs/free/
conda config - set show_channel_urls yes
```

创建 Python 3.5 的环境,名称为 TensorFlow,此时 Python 版本和后面 TensorFlow 的版本有匹配问题,此步选择 Python 3.x,输入命令:

```
conda create -n tensorflow python = 3.5
```

有需要确认的地方,都输入 y。
在 Anaconda Prompt 中激活 TensorFlow 环境,输入命令:

```
activate tensorflow
```

安装 CPU 版本的 TensorFlow,输入命令:

```
pip install -upgrade -- ignore-installed tensorflow
```

安装完毕。

20.2.3 OpenCV-Python 环境

安装 OpenCV 环境,输入命令:

```
pip install opencv_python-3.4.5-cp37-cp37m-win_amd64.whl
```

安装完毕。

20.3 模块实现

本项目包括 3 个模块:设置直方图、载入手势图像、模型训练及保存。下面分别给出各模块的功能介绍及相关代码。

20.3.1 设置直方图

输入命令 python set_hand_hist.py 运行 Python 文件,出现弹窗 Set hand histogram。设置直方图具有 50 个正方形(5×10)。将手放在其中,确保遮盖所有正方形。按下 C 键,出现 Thresh 窗口,具有皮肤颜色图像部分的白色补丁会出现在 Thresh 窗口中。获得良好的直方图后,按下 S 键保存图像,然后关闭所有窗口,如图 20-3 所示。

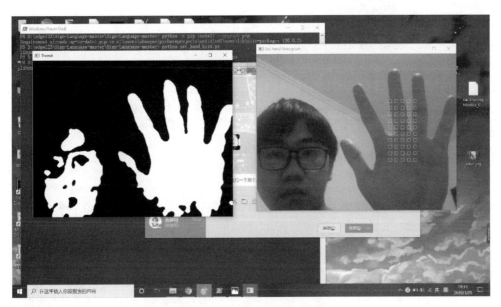

图 20-3 设置手部的直方图

set_hand_hist.py 相关代码如下：

```python
import cv2
import numpy as np
import pickle
def build_squares(img):                                          ＃确定扫描区域
    x, y, w, h = 420, 140, 10, 10
    d = 10
    imgCrop = None
    crop = None
    for i in range(10):
        for j in range(5):
            if np.any(imgCrop == None):
                imgCrop = img[y:y+h, x:x+w]
            else:
                imgCrop = np.hstack((imgCrop, img[y:y+h, x:x+w])) ＃print(imgCrop.shape)
            cv2.rectangle(img, (x,y), (x+w, y+h), (0,255,0), 1)
            ＃确定扫描区域,绿色,线条粗度为1
            x += w + d
        if np.any(crop == None):
            crop = imgCrop
        else:
            crop = np.vstack((crop, imgCrop))
        imgCrop = None
        x = 420
        y += h + d
```

```python
        return crop
def get_hand_hist():
    cam = cv2.VideoCapture(1)
    if cam.read()[0] == False:
        cam = cv2.VideoCapture(0)                           # 打开摄像头
    x, y, w, h = 300, 100, 300, 300
    flagPressedC, flagPressedS = False, False               # 设定按键 C、S
    imgCrop = None
    while True:
        img = cam.read()[1]                                 # 读取图片
        img = cv2.flip(img, 1)                              # 水平翻转
        img = cv2.resize(img, (640, 480))                   # 改变图片尺寸
        hsv = cv2.cvtColor(img, cv2.COLOR_BGR2HSV)          # 转换色彩空间
        keypress = cv2.waitKey(1)                           # 以 1ms 的频率不断刷新图像
        if keypress == ord('c'):
            hsvCrop = cv2.cvtColor(imgCrop, cv2.COLOR_BGR2HSV)
            flagPressedC = True                             # 按 C 键
            hist = cv2.calcHist([hsvCrop], [0, 1], None, [180, 256], [0, 180, 0, 256])
                                                            # 计算图像灰度直方图
            cv2.normalize(hist, hist, 0, 255, cv2.NORM_MINMAX)  # 进行归一化处理
        elif keypress == ord('s'):
            flagPressedS = True                             # 按 S 键
            break                                           # 停止程序
        if flagPressedC:
            dst = cv2.calcBackProject([hsv],[0,1],hist,[0, 180, 0, 256], 1)
                                                            # 将直方图反向投影
            dst1 = dst.copy()
            disc = cv2.getStructuringElement(cv2.MORPH_ELLIPSE,(10,10))
            # 创建椭圆形结构化元素
            cv2.filter2D(dst,-1,disc,dst)
            # 得到深度不变,按 disc 卷积核边缘复制的输出图像
            blur = cv2.GaussianBlur(dst, (11,11), 0)        # 进行高斯滤波
            blur = cv2.medianBlur(blur, 15)                 # 进行中值滤波
            ret,thresh = cv2.threshold(blur,0,255,cv2.THRESH_BINARY + cv2.THRESH_OTSU)
            # 用两种阈值类型将 blur 图像二值化
            thresh = cv2.merge((thresh,thresh,thresh))      # 将 BGR 通道合并
            #cv2.imshow("res", res)
            cv2.imshow("Thresh", thresh)                    # 用 Thresh 命名
        if not flagPressedS:
            imgCrop = build_squares(img)
        #cv2.rectangle(img, (x,y), (x + w, y + h), (0,255,0), 2)
        cv2.imshow("Set hand histogram", img)
    cam.release()
    cv2.destroyAllWindows()                                 # 删除窗口
    with open("hist", "wb") as f:
        pickle.dump(hist, f)                                # 自动关闭文件
get_hand_hist()
```

20.3.2 载入手势图片

加载 26 个字母和 10 个数字的每个手势以及包含 2400 张图像的数据集,通过 Python 实现,相关代码如下:

```
import cv2
from glob import glob
import numpy as np
import random
from sklearn.utils import shuffle
import pickle
import os
#添加新手势后,运行此程序以改变参数值,若没有添加则无须再次运行
def pickle_images_labels():
    images_labels = []
    images = glob("gestures/*/*.jpg")
    images.sort()
    for image in images:
        print(image)
        label = image[image.find(os.sep) + 1: image.rfind(os.sep)]
        img = cv2.imread(image, 0)
        images_labels.append((np.array(img, dtype = np.uint8), int(label)))
    return images_labels
images_labels = pickle_images_labels()
images_labels = shuffle(shuffle(shuffle(shuffle(images_labels))))
#将图像和标签顺序打乱
images, labels = zip(*images_labels)                           #重新组合图像和标签
print("Length of images_labels", len(images_labels))
#设置训练集图像
train_images = images[:int(5/6 * len(images))]                 #取前 5/6 的图像作为训练集
print("Length of train_images", len(train_images))
with open("train_images", "wb") as f:
    pickle.dump(train_images, f)
del train_images
#设置训练集标签
train_labels = labels[:int(5/6 * len(labels))]                 #取前 5/6 的标签作为训练集
print("Length of train_labels", len(train_labels))
with open("train_labels", "wb") as f:
    pickle.dump(train_labels, f)
del train_labels
#设置测试集图像
test_images = images[int(5/6 * len(images)):int(11/12 * len(images))]
#取中间 1/12 的图像作为测试集
print("Length of test_images", len(test_images))
with open("test_images", "wb") as f:
    pickle.dump(test_images, f)
```

```
del test_images
# 设置测试集标签
test_labels = labels[int(5/6 * len(labels)):int(11/12 * len(images))]
# 取中间 1/12 的标签作为测试集
print("Length of test_labels", len(test_labels))
with open("test_labels", "wb") as f:
    pickle.dump(test_labels, f)
del test_labels
# 设置验证集图像
val_images = images[int(11/12 * len(images)):]              # 取后 1/12 的图像作为验证集
print("Length of test_images", len(val_images))
with open("val_images", "wb") as f:
    pickle.dump(val_images, f)
del val_images
# 设置验证集标签
val_labels = labels[int(11/12 * len(labels)):]              # 取后 1/12 的标签作为验证集
print("Length of val_labels", len(val_labels))
with open("val_labels", "wb") as f:
    pickle.dump(val_labels, f)
del val_labels
```

读取代码成功，如图 20-4 所示。

图 20-4　读取代码成功

20.3.3 模型训练及保存

使用训练集训练模型,使模型能够识别不同手势。

```python
#模型生成
import numpy as np
import pickle
import cv2, os
from glob import glob
from keras import optimizers
from keras.models import Sequential
from keras.layers import Dense
from keras.layers import Dropout
from keras.layers import Flatten
from keras.layers.convolutional import Conv2D
from keras.layers.convolutional import MaxPooling2D
from keras.utils import np_utils
from keras.callbacks import ModelCheckpoint
from keras import backend as K
K.set_image_dim_ordering('tf')
os.environ['TF_CPP_MIN_LOG_LEVEL'] = '3'
def get_image_size():
    img = cv2.imread('gestures/1/100.jpg', 0)
    return img.shape
def get_num_of_classes():
    return len(glob('gestures/*'))
image_x, image_y = get_image_size()
def cnn_model():
    num_of_classes = get_num_of_classes()
    model = Sequential()
    model.add(Conv2D(16, (2,2), input_shape = (image_x, image_y, 1), activation = 'relu'))
                                                                               #卷积层
    model.add(MaxPooling2D(pool_size = (2, 2), strides = (2, 2), padding = 'same'))
                                                                               #最大池化层
    model.add(Conv2D(32, (3,3), activation = 'relu'))
    model.add(MaxPooling2D(pool_size = (3,3), strides = (3,3), padding = 'same'))
    model.add(Conv2D(64, (5,5), activation = 'relu'))
    model.add(MaxPooling2D(pool_size = (5,5), strides = (5,5), padding = 'same'))
    model.add(Flatten())                                                       #展平层
    model.add(Dense(128, activation = 'relu'))                                 #全连接层
    model.add(Dropout(0.2))                                                    #丢弃
    model.add(Dense(num_of_classes, activation = 'softmax'))
    sgd = optimizers.SGD(lr = 1e-2)
    model.compile(loss = 'categorical_crossentropy', optimizer = sgd, metrics = ['accuracy'])
    filepath = "cnn_model_keras2.h5"
    checkpoint1 = ModelCheckpoint(filepath, monitor = 'val_acc', verbose = 1, save_best_only
```

```python
        = True, mode = 'max')
        callbacks_list = [checkpoint1]
        from keras.utils import plot_model
        plot_model(model, to_file = 'model.png', show_shapes = True)
        return model, callbacks_list
def train():
    with open("train_images", "rb") as f:
        train_images = np.array(pickle.load(f))
    with open("train_labels", "rb") as f:                            #二进制格式打开,只读
        train_labels = np.array(pickle.load(f), dtype = np.int32)
    with open("val_images", "rb") as f:
        val_images = np.array(pickle.load(f))
    with open("val_labels", "rb") as f:
        val_labels = np.array(pickle.load(f), dtype = np.int32)
    train_images = np.reshape(train_images, (train_images.shape[0], image_x, image_y, 1))
    val_images = np.reshape(val_images, (val_images.shape[0], image_x, image_y, 1))
    train_labels = np_utils.to_categorical(train_labels)             #转换成虚拟变量
    val_labels = np_utils.to_categorical(val_labels)
    print(val_labels.shape)
    model, callbacks_list = cnn_model()
    model.summary()
    model.fit(train_images, train_labels, validation_data = (val_images, val_labels), epochs = 20, batch_size = 500, callbacks = callbacks_list)
    scores = model.evaluate(val_images, val_labels, verbose = 0)     #评估模型
    print("CNN Error: %.2f%%" % (100-scores[1] * 100))
    #model.save('cnn_model_keras2.h5')
train()
K.clear_session();
```

20.4 系统测试

本部分包括测试准确率及测试效果。

20.4.1 测试准确率

测试准确率和召回率达到99%甚至更高,意味着这个预测模型训练比较成功,如图20-5所示。

20.4.2 测试效果

将数据代入模型进行测试,分类的标签与原始数据进行显示和对比,如图20-6所示,可以得到验证:模型可以实现手势数字的识别。

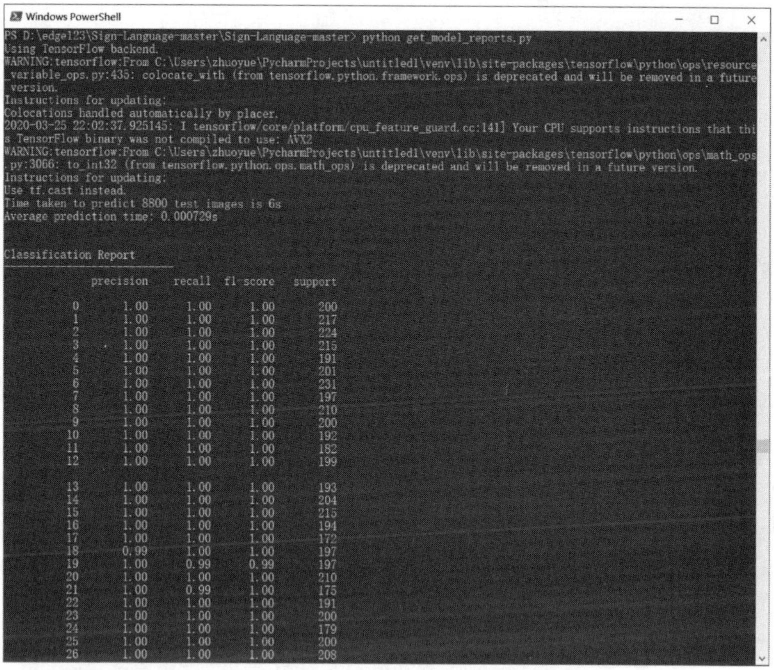

项目20 手势语言识别 369

图 20-5 测试准确率

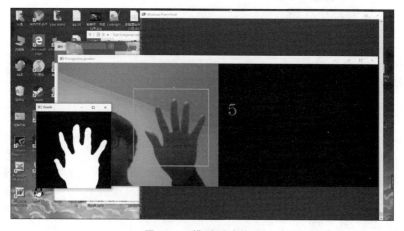

图 20-6 模型测试效果

图书资源支持

感谢您一直以来对清华大学出版社图书的支持和爱护。为了配合本书的使用，本书提供配套的资源，有需求的读者请扫描下方的"书圈"微信公众号二维码，在图书专区下载，也可以拨打电话或发送电子邮件咨询。

如果您在使用本书的过程中遇到了什么问题，或者有相关图书出版计划，也请您发邮件告诉我们，以便我们更好地为您服务。

我们的联系方式：

教学资源·教学样书·新书信息

地　　址：北京市海淀区双清路学研大厦 A 座 714

邮　　编：100084

电　　话：010-83470236　010-83470237

人工智能科学与技术
人工智能|电子通信|自动控制

资源下载：http://www.tup.com.cn

客服邮箱：tupjsj@vip.163.com

QQ：2301891038（请写明您的单位和姓名）

资料下载·样书申请

书圈

用微信扫一扫右边的二维码，即可关注清华大学出版社公众号。